BEHAVIOR OF MARINE
ANIMALS • VOLUME 5
Current Perspectives in Research

SHOREBIRDS

Breeding Behavior and Populations

BEHAVIOR OF MARINE ANIMALS

Current Perspectives in Research

Series Editors: Howard E. Winn and Bori L. Olla

Volume 1 Invertebrates
Edited by Howard E. Winn and Bori L. Olla

Volume 2 Vertebrates
Edited by Howard E. Winn and Bori L. Olla

Volume 3 Cetaceans
Edited by Howard E. Winn and Bori L. Olla

Volume 4 Marine Birds
Edited by Joanna Burger, Bori L. Olla, and Howard E. Winn

Volume 5 Shorebirds: Breeding Behavior and Populations
Edited by Joanna Burger and Bori L. Olla

Volume 6 Shorebirds: Migration and Foraging Behavior
Edited by Joanna Burger and Bori L. Olla

BEHAVIOR OF MARINE ANIMALS • VOLUME 5

Current Perspectives in Research

SHOREBIRDS

Breeding Behavior and Populations

Edited by

Joanna Burger

Rutgers University
Piscataway, New Jersey

and

Bori L. Olla

Oregon State University
Newport, Oregon
and Northwest and Alaska Fisheries Center
Seattle, Washington

PLENUM PRESS • NEW YORK AND LONDON

Library of Congress Cataloging in Publication Data

Winn, Howard Elliott, 1926–
Behavior of marine animals; current perspectives in research.

Vol. 4 edited by J. Burger, B. L. Olla, and H. E. Winn; v. 5– by J. Burger and B. L. Olla.
Includes bibliographical references and indexes.
CONTENTS: v. 1. Invertebrates.—v. 2. Vertebrates.—(etc.)—v. 5. Shorebirds, breeding behavior and populations.
1. Marine fauna—Behavior—Collected works. I. Olla, Bori L., joint author. II. Burger, Joanna, joint author. III. Title.
QL121.W5 591.5′2636 79-167675
ISBN 978-1-4684-4693-7 ISBN 978-1-4684-4691-3 (eBook)
DOI 10.1007/978-1-4684-4691-3

Softcover reprint of the hardcover 1st edition 1984

A Division of Plenum Publishing Corporation
233 Spring Street, New York, N.Y. 10013

CONTRIBUTORS

Joanna Burger Department of Biology and Bureau of Biological Research, Rutgers University, Piscataway, New Jersey 08854

P. R. Evans Department of Zoology, University of Durham, Durham DH1 3LE, England

Michael Gochfeld Department of Environmental and Community Medicine, UMDNJ–Rutgers Medical School, Piscataway, New Jersey 08854

Marshall A. Howe U.S. Fish and Wildlife Service, Patuxent Wildlife Research Center, Laurel, Maryland 20708

Joseph R. Jehl, Jr. Hubbs–Sea World Research Institute, Mission Bay, San Diego, California 92109

David B. Lank Department of Biology, University of North Dakota, Grand Forks, North Dakota 58202

Sarah Lenington Institute of Animal Behavior, Rutgers University, Newark, New Jersey 07102

Edward H. Miller Vertebrate Zoology Division, British Columbia Provincial Museum, and Biology Department, University of Victoria, Victoria, British Columbia V8V 1X4, Canada

Lewis W. Oring Department of Biology, University of North Dakota, Grand Forks, North Dakota 58202

M. W. Pienkowski Department of Zoology, University of Durham, Durham DH1 3LE, England

Stanley E. Senner Hawk Mountain Sanctuary Association, Kempton, Pennsylvania 19529

Jeffrey R. Walters Allee Laboratory of Animal Behavior, The University of Chicago, Chicago, Illinois 60637. *Present address:* Zoology Department, North Carolina State University, Raleigh, North Carolina 27695

INTRODUCTION

Among birds, shorebirds provide some of the more unique opportunities to examine basic problems in behavior, ecology, and evolution. This is in large measure due to the diversity, both behaviorally and ecologically, of a group closely related taxonomically and distributed throughout the world. The overall aim of these two volumes is to provide a representative selection of current research being conducted on shorebird behavior and ecology.

Traditionally, marine birds have included those species that breed in large colonies on offshore islands along coasts (see Volume 4 of this series). Although shorebirds have generally not been considered within this group, the fact that almost 40% of the species breed along coasts and more than 60% often or always spend the nonbreeding season in coastal habitats more than justifies their inclusion as marine birds (at least those species that totally or partially depend upon the marine environment). Their inclusion markedly increases species diversity in marine birds since shorebirds add about 217 species to the 280 that are traditionally thought of as marine.

Over the last twenty years there has been a veritable explosion in studies on shorebirds. Prior to this, there were few studies of breeding behavior and only cursory examinations of their migratory patterns. The increased interest in shorebirds was not the least caused by the increased awareness of the diversity and plasticity of their social systems. More pragmatically, the increased accessibility to researchers of Arctic breeding grounds as well as increase in the availability of fixed winged planes and helicopters for censusing populations and monitoring movements during the nonbreeding season also did much to foster the research effort.

Advances in shorebirds research have been of vital interest to those studying other animal groups, especially in the area of foraging behavior, with many of the paradigms and theories developed for optimal foraging theory being readily applicable to, and testable in shorebird species. Unlike the traditional marine species, their prey (primarily invertebrates) can be readily sampled both qualitatively and quantitatively. Further, since they forage primarily in estuaries it is possible to follow individuals in the nonbreeding season. The variability imposed by tidal regimes only make the system more valuable for field testing foraging theory.

Future studies on shorebirds will continue to center on examining breeding behavior and ecology, migration, and foraging behavior. However, more emphasis will be placed on examining the entire life cycle of particular species, comparing the success of populations that differ in breeding sites, breeding and nonbreeding habitats, and migration routes. Finally, the mechanisms that control behavioral and ecological plasticity in shorebirds should provide some of the more interesting avenues for future research.

J. Burger
B. Olla

PREFACE

Most chapters in these two volumes on the behavior of shorebirds include a balance between literature review, original data and synthesis. The research approaches of the authors range from in-depth study of a single species to a comparative one involving groups of species that are taxonomically or ecologically related.

The subject has been divided into two volumes, one dealing with breeding behavior and populations, and the second with migration and foraging behavior. The first two chapters of Volume 5 serve as an introduction to the two volumes and address the issues of shorebird classification and whether shorebirds are marine animals. Most of the remaining chapters in Volume 5 concern aspects of breeding behavior including breeding site fidelity, polyandrous mating systems, communication, parental behavior, and antipredator behavior. Two other aspects discussed in Volume 5 are population dynamics and conservation of shorebirds.

Volume 6 covers behavior aspects of shorebirds during the nonbreeding season and includes abiotic factors affecting migration, migratory behavior in the Western Palearctic and New World, foraging and activity patterns, food supplies, and spacing patterns of nonbreeding migrants. Taken together these two volumes provide paradigms for the further study of organisms that move freely between marine and terrestrial ecosystems, as well as summarizing current research with a particular group of birds. Some other important areas of shorebirds behavior (general mating systems, sexual dimorphism) have not been included because other reviews of these topics exist.

In recent times man's encroachment has been particularly strong on the beaches and shorelines of marine habitats used extensively by shorebirds. A volume elucidating aspects of their breeding, migration, and wintering behavior and ecology is particularly useful to managers and conservationists involved with coastal planning and protection of marine avifaunas. Their accessibility for study in breeding and nonbreeding habitats permits the identification of those factors which play a role in controlling population levels, a prerequisite for the application of rational management plans. For ornithologists, ecologists and behaviorists the two volumes provide a review of the concepts concerning shorebird behavior and the elucidation of the basic mechanisms affecting behavior.

We are especially grateful to Michael Gochfeld, Betty Green, Jill Grover, Brook Lauro, Cindy Paszkowski and Gary Braun for their help in indexing, and to the authors of Chapter 1 in Volume 5 for providing a shorebird classification that could be used as a basis for the other chapters in the two volumes.

J. Burger
B. Olla

CONTENTS

Chapter 1

THE CLASSIFICATION OF THE SHOREBIRDS OF THE WORLD

Michael Gochfeld

Department of Environmental and Community Medicine
UMDNJ–Rutgers Medical School
Piscataway, New Jersey 08854

Joanna Burger

Department of Biology and Bureau of Biological Research
Rutgers University
Piscataway, New Jersey 08854

and

Joseph R. Jehl, Jr.

Hubbs–Sea World Research Institute
Mission Bay, San Diego, California 92109

I. INTRODUCTION

The order Charadriiformes contains in addition to the gulls, terns, skuas, skimmers, and auks, a heterogeneous assemblage of families known as shorebirds in the New World and waders in the Old World. These terms, "shorebirds" and "waders," are slightly misleading in that many of the species, particularly in the tropics and Southern Hemisphere, infrequently wade and do not frequent shorelines. In fact, several taxa are characteristically desert or steppe birds (e.g., Thinocoridae, Cursoriinae).

The subdivision of this order into suborders has varied through the years (compare Peters, 1934; Bock, 1958; Jehl, 1968; Strauch, 1978). It is now widely agreed that the "shorebirds" comprise two separate lineages (Strauch, 1978). The Scolopaci includes the sandpipers, jacanas, and painted snipe. The second lineage, the Charadrii (1), includes the seedsnipe, sheathbills, plovers, oystercatchers, recurvirostrids, and their allies. Despite the fact that the shorebirds do not represent a monophyletic group, these two lines are sufficiently similar in morphology, behavior, and ecology that they can be treated together.

Several classifications of the major shorebird taxa have been advanced recently (see for example Peters, 1934; Bock, 1958; Jehl, 1968; Morony *et al.*, 1975; Strauch, 1978; Cracraft, 1981; Johnsgard, 1981; A.O.U., 1983), but few treat the entire group, and none takes all recent information into account. This chapter provides a list of shorebird taxa to achieve some uniformity of nomenclature for the contributions in this volume. We have not attempted to review systematic studies of the Charadriiformes, but instead present a species list, with an indication of appropriate higher categories. The classification used here at the tribal level and above was proposed by Jehl, and served as the basis for the species list. Our list largely agrees with nomenclature adopted by the American Ornithologists' Union (A.O.U., 1983) for the 6th edition of its Checklist.

It is always a problem to provide a uniform list of common or colloquial names, particularly for species with Holarctic, or in the case of some shorebirds, virtually cosmopolitan distribution. In general we have given a slight preference to names used in the Americas, and have further tampered with some names by inserting modifiers such as "Eurasian." For species with distribution confined to one continent, we have tended to adopt those names used locally in major avifaunal publications. We have used the following as sources for common names: A.O.U. (1983), Meyer de Schauensee (1966), Peterson *et al.* (1954), Mackworth-Praed and Grant (1962), King and Dickinson (1975), Slater (1972), and Johnsgard (1981). In the majority of cases, both common and scientific names are the same as those used in Clements (1981) who based his checklist on Morony *et al.* (1975).

The ranges given are abbreviated, and refer to the main areas of occurrence, omitting extralimital or occasional records. The word "to" generally designates migration from the first region(s) to the second. The following abbreviations are used: n, e, s, w, c, for northern, eastern, southern, western, central.

(1) For this suborder Cracraft (1981) has suggested the term "Charadriomorpha," which is, however, not parallel in structure to "Scolopaci," and we prefer the term "Charadrii," which has also been used for the infraorder comprising the oystercatchers and plovers.

II. LIST OF THE TAXA OF SHOREBIRDS

ORDER CHARADRIIFORMES

SUBORDER CHARADRII

INFRAORDER DROMADAE

DROMADIDAE (Crab Plover)

Crab Plover	*Dromas ardeola*	Indian Ocean shores

INFRAORDER THINOCORI

CHIONIDIDAE (Sheathbills)

Snowy Sheathbill	*Chionis alba*	Antarctica, Patagonia
Lesser (Black-faced) Sheathbill	*C. minor*	Antarctic islands

PLUVIANELLIDAE (Magellanic Plover)

Magellanic Plover	*Pluvianellus socialis*	s Patagonia

PEDIONOMIDAE (Plains Wanderer or Collared Hemipode)

Plains Wanderer (2)	*Pedionomus torquatus*	se & e Australia

THINOCORIDAE (Seedsnipes)

Rufous-bellied Seedsnipe	*Attagis gayi*	Andes
White-bellied Seedsnipe	*A. malouinus*	s Andes
Gray-breasted Seedsnipe	*Thinocorus orbignyianus*	s Andes
Least Seedsnipe	*T. rumicivorus*	w & s South America

INFRAORDER BURHINI

BURHINIDAE (Thick-knees and Stone Curlews)

Eurasian Stone Curlew	*Burhinus oedicnemus*	Eastern Hemisphere
Senegal Thick-knee	*B. senegalensis*	w & e Africa
Water Dikkop	*B. vermiculatus*	e & s Africa
Spotted Thick-knee (Cape Dikkop)	*B. capensis*	w & s Africa, mid-east

(2) Until recently classified in the Gruiformes, but Olson and Steadman (1981) show its relationship to the Charadriiformes. Olson and Feduccia (1980) have proposed inclusion of the Phoenicopteridae in the Charadriiformes next to the Recurvirostridae.

Double-striped Thick-knee	*B. bistriatus*	Neotropics
Peruvian Thick-knee	*B. superciliaris*	Ecuador to Chile
Australian Stone Curlew	*B. magnirostris*	s Australia
Great Stone Curlew	*Esacus recurvirostris*	India, Sri Lanka
Beach Stone Curlew	*E. magnirostris*	Malaysia, Australasia

INFRAORDER HAEMATOPODI
SUPERFAMILY HAEMATOPODOIDEA
HAEMATOPODIDAE (3) (Oystercatchers)

Eurasian (Pied) Oystercatcher	*Haematopus ostralegus*	Eastern Hemisphere
American Oystercatcher	*H. palliatus*	Western Hemisphere
American Black Oystercatcher	*H. bachmani*	w North America
African Black Oystercatcher	*H. moquini*	s Africa, Canaries
Variable Oystercatcher	*H. unicolor*	Australia, New Zealand
Magellanic Oystercatcher	*H. leucopodus*	s South America
Blackish Oystercatcher	*H. ater*	Peru, Chile, Argentina
Sooty Oystercatcher	*H. fuliginosus*	Australia

IBIDORHYNCHIDAE (Ibis-bill)

Ibis-bill	*Ibidorhynchus struthersii*	c Asia

RECURVIROSTRIDAE (4) (Stilts and Avocets)

Black-winged (Common) Stilt	*Himantopus himantopus*	s Europe, Asia, Africa
South American Stilt	*H. melanurus*	c & s South America

(3) Peters (1934) treated *palliatus, bachmani, moquini,* and *unicolor* as races of *ostralegus*. Johnsgard (1981) follows Peters (1934), except for treating *H. unicolor* as distinct. Morony *et al.* (1975) treated all as separate species. Hockey (1982) discusses status of *Haematopus moquini* and treats the extinct Canary Island form as a separate species *H. meadewaldoi*.

(4) Morony *et al.* (1975) listed eight species in the genus *Himantopus* following Mayr and Short (1970). Peters (1934) recognized only one species, *H. himantopus*. These are well-marked but allopatric forms, and are listed here separately, recognizing their questionable specific status. A.O.U. (1983) treats *Himantopus* as one species.

Black-necked Stilt	*H. mexicanus*	North America to Peru
Ceylon Stilt	*H. ceylonensis*	Sri Lanka
Pied Stilt	*H. leucocephalus*	Australia, New Zealand, New Guinea, Philippines, etc.
Hawaiian Stilt	*H. knudseni*	Hawaii
Black Stilt	*H. novaezeelandiae*	New Zealand
South African Stilt	*H. meridionalis*	South Africa
Banded Stilt	*Cladorhynchus leucocephalus*	sw & se Australia
Eurasian (Pied) Avocet	*Recurvirostra avosetta*	sw & e Europe, c Asia, e & s Africa
American Avocet	*R. americana*	North America
Red-necked Avocet	*R. novaehollandiae*	Australia, New Zealand
Andean Avocet	*R. andina*	c Andes
SUPERFAMILY CHARADRIOIDEA		
GLAREOLIDAE (Coursers and Pratincoles)		
Subfamily Cursoriinae (Coursers)		
Egyptian Plover	*Pluvianus aegyptius*	n Africa
Two-banded Courser	*Rhinoptilus africanus*	e & s Africa
Heuglin's Courser	*R. cinctus*	c & e Africa
Violet-tipped Courser	*R. chalcopterus*	Subsaharan Africa
Jerdon's Courser	*R. bitorquatus*	India
Cream-colored Courser	*Cursorius cursor*	n Africa to India
Indian Courser	*C. coromandelicus*	India, Sri Lanka
Temminck's Courser	*C. temminckii*	Subsaharan Africa
Subfamily Glareolinae (Pratincoles)		
Australian Pratincole	*Stiltia isabella*	Australia, New Guinea
Collared Pratincole	*Glareola pratincola*	Eastern Hemisphere
Eastern Pratincole	*G. maldivarium*	Asia, Australia
Black-winged Pratincole	*G. nordmanni*	Eastern Hemisphere
Madagascar Pratincole	*G. ocularis*	e Africa, Madagascar
White-collared Pratincole	*G. nuchalis*	e Africa & n South Africa

Gray Pratincole	*G. cinerea*	w Africa
Milky Pratincole	*G. lactea*	India to se Asia
CHARADRIIDAE (Lapwings and Plovers)		
Subfamily Vanellinae		
Tribe Hoploxypterini (Pied Lapwing)		
Pied Lapwing	*Hoploxypterus cayanus*	n & c South America
Tribe Vanellini (Lapwings and Plovers)		
Northern (Eurasian) Lapwing	*Vanellus vanellus*	Palearctic to India
Long-toed Lapwing	*V. crassirostris*	Sudan to Mozambique
Blacksmith Plover	*V. armatus*	e & s Africa
Spur-winged Plover	*V. spinosus*	w to e Africa
River Lapwing	*V. duvaucelii*	India to se Asia
Black-headed Plover	*V. tectus*	w, c, & e Africa
Yellow-wattled Lapwing	*V. malabaricus*	India, Sri Lanka, Nepal
White-crowned Plover	*V. albiceps*	Subsaharan Africa
Senegal Plover	*V. lugubris*	Subsaharan Africa
Black-winged Plover	*V. melanopterus*	e & s Africa
Crowned Plover	*V. coronatus*	e & s Africa
Wattled Plover	*V. senegallus*	Subsaharan Africa
Spot-breasted Plover	*V. melanocephalus*	Ethiopia
Brown-chested Plover	*V. superciliosus*	c & e Africa
Sociable Plover	*V. gregarius*	Asia & Africa
White-tailed Plover	*V. leucurus*	sw Asia, e Africa, India
Southern Lapwing	*V. chilensis*	South America
Andean Lapwing	*V. resplendens*	n & c Andes
Gray-headed Lapwing	*V. cinereus*	Japan to India, se Asia
Red-wattled Lapwing	*V. indicus*	w to se Asia, Indonesia
Javanese Wattled Lapwing	*V. macropterus*	Indonesia
Banded Plover	*V. tricolor*	c & s Australia

Masked Plover	*V. miles*	Australia, New Guinea, s New Zealand
Subfamily Charadriinae (Plovers and Dotterels) (5)		
Black-bellied (Grey) Plover	*Pluvialis squatarola*	Cosmopolitan
Greater (Eurasian) Golden Plover (6)	*P. apricaria*	w Palearctic
American Golden Plover (6)	*P. dominica*	Western Hemisphere
Pacific Golden Plover (6)	*P. fulva*	Alaska, Pacific islands, Australasia
Ringed Plover	*Charadrius hiaticula*	Palearctic to Africa, India
Semipalmated Plover	*C. semipalmatus*	Western Hemisphere
Long-billed Plover	*C. placidus*	e Asia & Japan to se Asia
Little Ringed Plover	*C. dubius*	Eastern Hemisphere (not Australia)
Thick-billed (Wilson's) Plover	*C. wilsonia*	s North America, Neotropics
Killdeer	*C. vociferus*	North America, Peru
Piping Plover	*C. melodus*	c & e North America
Black-banded Sandplover	*C. thoracicus*	Madagascar
Kittlitz's Plover	*C. pecuarius*	Subsaharan Africa, Madagascar
St. Helena Plover	*C. sanctaehelenae*	St. Helena (Gulf of Guinea)
Three-banded Plover	*C. tricollaris*	Subsaharan Africa, Madagascar

(5) The terms "Plover," "Sandplover," and "Dotterel" are widely used, sometimes interchangeably in reference to members of *Charadrius*. We use "Plover" here unless an alternative designation shows strong prominence in regional literature.

(6) We follow Kinsky and Yaldwyn (1981) and Connors (1983) in separating *P. fulva* as a distinct species, the Pacific Golden Plover, from *P. dominica*, the American Golden Plover. The alternative treatment adopted by the A.O.U. (1983) treats two species of golden plovers, the Greater, *P. apricaria*, and the Lesser, *P. dominica* (including *fulva*).

Snowy (Kentish) Plover (7)	*C. alexandrinus*	North America, Eurasia, n Africa
White-fronted Sandplover	*C. marginatus*	Subsaharan Africa, Madagascar
Peruvian Sandplover (7)	*C. occidentalis*	Coasts of Peru & Chile
Red-capped Dotterel (7)	*C. ruficapillus*	Australia
Malaysian Sandplover	*C. peronii*	se Asia, Indonesia, Philippines
Chestnut-banded Sandplover	*C. venustus*	e & s Africa
Collared Plover	*C. collaris*	Neotropical
Double-banded Dotterel	*C. bicinctus*	New Zealand to e Australia
Two-banded Plover	*C. falklandicus*	Patagonia to Brazil, Falklands
Puna Plover (8)	*C. alticola*	c Andes
Mongolian (Lesser Sand) Plover	*C. mongolus*	Eastern Hemisphere
Greater Sandplover	*C. leschenaultii*	c & e Asia to Africa, Australia
Red-breasted Dotterel	*C. obscurus*	n New Zealand
Caspian Plover	*C. asiaticus*	w & e Asia to Africa
Oriental Plover	*C. veredus*	Mongolia to Australia
Rufous-chested Dotterel	*C. modestus*	s Patagonia
Mountain Plover	*C. montanus*	c & w North America
Black-fronted Dotterel	*C. melanops*	Australia, New Zealand
Hooded Dotterel (9)	*C. cucullatus*	s Australia
Eurasian Dotterel	*C. morinellus*	Palearctic, se & c Asia to ne Africa

(7) Peters (1934) included *occidentalis, ruficapillus,* and *marginatus* in *C. alexandrinus.* Mayr and Short (1970) treated *ruficapillus* and *occidentalis* as distinct species. Johnsgard (1981) calls this the Sandplover including *occidentalis,* but not *ruficapillus* or *marginatus.*

(8) *C. alticola* was treated as a full species by Peters (1934) but was lumped with *C. falklandicus* by Bock (1958).

(9) Listed by Morony *et al.* (1975) as *Charadrius rubricollis,* and called Australian Plover (Clements, 1981). Hooded Dotterel is the usual name (Slater, 1972).

Shore Plover	*C. novaeseelandiae*	New Zealand (Chatham Islands)
Red-kneed Dotterel	*Erythrogonys cinctus*	Australia
Wrybill	*Anarhynchus frontalis*	New Zealand
Diademed Sandpiper-plover (10)	*Phegornis mitchellii*	c & s Andes
Australian Dotterel (11)	*Peltohyas australis*	c Australia
Tawny-throated (Slender-billed) Dotterel	*Oreopholus ruficollis*	Peru, Andes, Patagonia
SUBORDER SCOLOPACI		
SUPERFAMILY JACANOIDEA		
JACANIDAE (Jacanas)		
Lesser Jacana	*Microparra capensis*	c, e, & s Africa
African Jacana	*Actophilornis africana*	Subsaharan Africa
Madagascar Jacana	*A. albinucha*	Madagascar
Comb-crested Jacana (Lotus Bird)	*Irediparra gallinacea*	n & e Australia, New Guinea, East Indies
Pheasant-tailed Jacana	*Hydrophasianus chirurgus*	India, se Asia, Indonesia, Philippines
Bronze-winged Jacana	*Metopidius indicus*	India, Indonesia, se Asia
Northern Jacana	*Jacana spinosa*	Central America
Wattled Jacana	*J. jacana*	South America
ROSTRATULIDAE (Painted Snipes)		
Greater Painted Snipe	*Rostratula benghalensis*	Africa, Asia, Australia
South American Painted Snipe (12)	*R. semicollaris*	s South America
SUPERFAMILY SCOLOPACOIDEA		
SCOLOPACIDAE (Sandpipers, Stints, Curlew, Godwits, Snipes)		
Subfamily Scolopacinae		

(10) Relationship to the plovers confirmed by Zusi and Jehl (1970).
(11) The Australian or Inland Dotterel was included among the Coursers by Peters (1934), but Bock (1958) transferred it to the Charadriidae.
(12) Most often published under the genus *Nycticryphes*.

Tribe Tringini (Tattlers and Allies)

Spotted Redshank	*Tringa erythropus*	Palearctic, Africa, Orient
Common Redshank	*T. totanus*	Eastern Hemisphere
Marsh Sandpiper	*T. stagnatilis*	Eastern Hemisphere
Common Greenshank	*T. nebularia*	Eastern Hemisphere
Spotted (Nordmann's) Greenshank	*T. guttifer*	ne Asia to s & se Asia
Greater Yellowlegs	*T. melanoleuca*	Western Hemisphere
Lesser Yellowlegs	*T. flavipes*	Western Hemisphere
Green Sandpiper (13)	*T. ocrophus*	Palearctic, Africa, Orient
Solitary Sandpiper	*T. solitaria*	Western Hemisphere
Wood Sandpiper	*T. glareola*	Eastern Hemisphere
Willet	*Catoptrophorus semipalmatus*	North America to n South America
Terek Sandpiper	*Xenus cinereus*	Eastern Hemisphere
Common Sandpiper	*Actitis hypoleucos*	Eastern Hemisphere
Spotted Sandpiper	*A. macularia*	Western Hemisphere
Gray-tailed Tattler (14)	*Heteroscelus brevipes*	Siberia to s Asia, Australia, Melanesia
Wandering Tattler	*H. incanus*	Alaska & ? Siberia to Peru & sw Pacific

Tribe Phalaropodini (Phalaropes)

Wilson's Phalarope	*Phalaropus tricolor*	Western Hemisphere
Red-necked (Northern) Phalarope	*P. lobatus*	Cosmopolitan
Red (Grey) Phalarope	*P. fulicarias*	Cosmopolitan

Tribe Prosoboniini (Polynesian Sandpipers)

Tuamotu Sandpiper	*Prosobonia cancellata*	Polynesia (Tuamotus)
White-winged Sandpiper (15)	*P. leucoptera*	Society Islands (Tahiti & Eimeo)

Tribe Numeniini (Curlews)

(13) Although the original spelling by Linnaeus as *ocrophus* was an apparently unintentional misspelling of a name already in use (see Witherby *et al.*, 1940, p. 310), and although modern usage shows a strong preference for *ochropus*, the correct nomenclature is *ocrophus*.

(14) The Gray-tailed Tattler is also known as the Polynesian Tattler, Gray-rumped Sandpiper, and Siberian Tattler (Johnsgard, 1981). None of these names are quite satisfactory. These species are sometimes placed in *Tringa*.

(15) Extinct, known from only one specimen.

Upland Sandpiper	*Bartramia longicauda*	Western Hemisphere
Little Curlew	*Numenius minutus*	ne Asia to Australia
Eskimo Curlew	*N. borealis*	Western Hemisphere
Whimbrel	*N. phaeopus*	Cosmopolitan
Bristle-thighed Curlew	*N. tahitiensis*	Alaska to Pacific islands
Slender-billed Curlew	*N. tenuirostris*	nc Asia to Arabia, Africa
Eurasian Curlew	*N. arquata*	Palearctic, Africa, Orient
Far Eastern Curlew	*N. madagascariensis*	e Asia to Australia
Long-billed Curlew	*N. americanus*	w North America to Guatemala
Tribe Limosini (Godwits)		
Black-tailed Godwit	*Limosa limosa*	Eastern Hemisphere
Hudsonian Godwit	*L. haemastica*	Western Hemisphere
Bar-tailed Godwit	*L. lapponica*	Eastern Hemisphere
Marbled Godwit	*L. fedoa*	North America to Panama
Tribe Arenariini (Turnstones)		
Ruddy Turnstone	*Arenaria interpres*	Cosmopolitan
Black Turnstone	*A. melanocephala*	w North America
Tribe Calidridini (Sandpipers and Stints)		
Surfbird	*Aphriza virgata*	Alaska to South America
Red Knot	*Calidris canutus*	Cosmopolitan
Great Knot	*C. tenuirostris*	ne Asia, India, Australasia
Sanderling	*C. alba*	Cosmopolitan
Semipalmated Sandpiper	*C. pusilla*	Western Hemisphere
Western Sandpiper	*C. mauri*	Alaska & e Siberia to South America
Rufous-necked Stint	*C. ruficollis*	ne Palearctic to Africa, Australia
Little Stint	*C. minuta*	n Palearctic to Africa, Asia
Temminck's Stint	*C. temminckii*	Palearctic to Africa, Asia
Long-toed Stint	*C. subminuta*	e Palearctic to Australia
Least Sandpiper	*C. minutilla*	Western Hemisphere

White-rumped Sandpiper	*C. fuscicollis*	Western Hemisphere
Baird's Sandpiper	*C. bairdii*	Western Hemisphere
Pectoral Sandpiper (16)	*C. melanotus*	Cosmopolitan
Sharp-tailed Sandpiper	*C. acuminata*	e Siberia to Australasia & Polynesia
Purple Sandpiper	*C. maritima*	Holarctic
Rock Sandpiper	*C. ptilocnemis*	e Siberia, Alaska, Aleutians to n California
Dunlin	*C. alpina*	Cosmopolitan
Curlew Sandpiper	*C. ferruginea*	e Palearctic to Africa, Asia, Australia
Stilt Sandpiper	*C. himantopus*	Western Hemisphere
Spoon-billed Sandpiper	*Eurynorhynchus pygmeus*	e Asia to se Asia, India
Broad-billed Sandpiper	*Limicola falcinellus*	Eastern Hemisphere
Buff-breasted Sandpiper	*Tryngites subruficollis*	Western Hemisphere
Ruff	*Philomachus pugnax*	Eastern Hemisphere
Tribe Limnodromini (Dowitchers)		
Short-billed Dowitcher	*Limnodromus griseus*	Western Hemisphere
Long-billed Dowitcher	*L. scolopaceus*	Siberia & w Nearctic to Guatemala
Asiatic Dowitcher	*L. semipalmatus*	e Asia to India & se Asia
Tribe Gallinagonini (Snipes)		
Subantarctic Snipe (17)	*Coenocorypha aucklandica*	New Zealand (subantarctic islands)
Solitary Snipe	*Gallinago solitaria*	e Palearctic, Orient

(16) Parker (1982) has recently described Cox's Sandpiper *C. paramelanotus* based on two specimens and several photographs of birds apparently wintering in South Australia. Originally thought to represent a hybrid of *Calidris ferruginea* × *C. melanotus*, its status remains to be verified. Another sandpiper "*Tringa*" (= *Calidris*) *cooperi* is doubtless a hybrid but of undetermined parentage (Deignan, 1961).

(17) Includes *C. pusilla*, the Chatham Island Snipe.

Latham's (Japanese) Snipe (18)	*G. hardwickii*	Japan to Australia
Wood Snipe	*G. nemoricola*	Himalayas to India, se Asia
Pin-tailed Snipe	*G. stenura*	e Palearctic to Australia
Swinhoe's Snipe (19)	*G. megala*	c Asia to Australia
African Snipe	*G. nigripennis*	e & s Africa, Madagascar
Madagascar Snipe	*G. macrodactyla*	Madagascar, Mauritius
Great Snipe	*G. media*	n Palearctic to Africa
Common Snipe	*G. gallinago*	Cosmopolitan (not Australia)
Puna Snipe (20)	*G. andina*	c Andes
Noble Snipe	*G. nobilis*	n Andes
Giant Snipe	*G. undulata*	ne & c South America
Cordilleran Snipe (21)	*G. stricklandii*	s Andes
Andean Snipe (21)	*G. jamesoni*	n Andes
Imperial Snipe (22)	*G. imperialis*	Andes (Colombia, Peru)
Jacksnipe	*Lymnocryptes minimus*	Palearctic, Africa, India
Tribe Scolopacini (Woodcocks)		
Eurasian Woodcock	*Scolopax rusticola*	Palearctic to s Asia
Amami Woodcock	*S. mira*	Ryukyu Islands, Japan
Javanese (Dusky) Woodcock	*S. saturata*	Indonesia, New Guinea
Celebes Woodcock	*S. celebensis*	Celebes

(18) Also known as Australian Snipe.

(19) Also known as Chinese or Pin-tailed Snipe in Australia and Forest Snipe (Johnsgard, 1981).

(20) Puna Snipe was treated as a subspecies of *paraguaiae* (Peters, 1934), although that "species" has been lumped with *gallinago*. Meyer de Schauensee (1966) retains *andina* as a separate species (see discussion in Johnsgard, 1981).

(21) These two forms are sometimes treated as conspecific (e.g., Johnsgard, 1981), or placed in a separate genus *Chubbia*.

(22) Known only from two old Bogotá specimens [hence called Bogotá Snipe by Meyer de Schauensee (1966)] until recently rediscovered in Peru. Formerly placed in *Chubbia*.

Obi Woodcock	*S. rochussenii*	Moluccas (Obi Island)
American Woodcock	*S. minor*	e North America

ACKNOWLEDGMENTS

A preliminary draft of this list was circulated among the authors contributing to this volume. We thank them for their comments and suggestions. We especially thank Dr. K. C. Parkes for his careful and critical reading of the manuscript and B. L. Monroe, Jr., for clarifying nomenclatural issues.

REFERENCES

A.O.U., 1983, Checklist of North American Birds, 6th ed., American Ornithologist's Union, Lawrence, Kansas.

Bock, W., 1958, A generic review of the plovers (Charadriinae: Aves), *Bull. Mus. Comp. Zool. Harv. Univ.* **118**:27–97.

Clements, J. F., 1981, *Birds of the World: A Checklist,* 3rd ed., Two Continents Publishing Group, New York.

Connors, P. G., 1983, Taxonomy, distribution, and evolution of the Golden Plovers *Pluvialis dominica* and *Pluvialis fulva, Auk,* **100**:607–620.

Cracraft, J., 1981, Toward a phylogenetic classification of the recent birds of the world (class Aves), *Auk* **98**:681–714.

Deignan, H. G., 1961, Types specimens of birds in the United States National Museum, *U.S. Natl. Mus. Bull.* **221**.

Hockey, P. A. R., 1982, *Haematopus meadewaldoi*: A good species, *Bull. Br. Ornithol. Club* **102**:77–83.

Jehl, J. R., Jr., 1968, Relationships in the Charadrii (shorebirds): A taxonomic study based on color patterns of the downy young, *San Diego Soc. Nat. Hist. Mem.* **3**:1–54.

Johnsgard, P. A., 1981, *The Plovers, Sandpipers, and Snipes of the World,* University of Nebraska Press, Lincoln.

King, B., and Dickinson, E. C., 1975, *A Field Guide to the Birds of South-east Asia,* Collins, Glasgow.

Kinsky, F. C., and Yaldwyn, J. C., 1981, The bird fauna of Niue Island, southwest Pacific, with special notes on the White-tailed Tropicbird and Golden Plover, *Natl. Mus. N.Z. Misc. Ser. No. 2.*

Mackworth-Praed, C. W., and Grant, C. H. B., 1962, *Birds of the Southern Third of Africa,* Vol. 1, Longmans, Green, New York.

Mayr, E., and Short, L. L., 1970, Species taxa of North American birds, *Publ. Nuttall Ornithol. Club.* **9**:1–127.

Meyer de Schauensee, R., 1966, *The Species of Birds of South America,* Livingston, Narberth, Pa.

Morony, J. J., Jr., Bock, W. J., and Farrand, J., Jr., 1975, *Reference List of the Birds of the World,* American Museum of Natural History, New York.

Olson, S. L., and Feduccia, A., 1980, Relationships and evolution of flamingos (Aves: Phoenicopteridae), *Smithson. Contrib. Zool.* **316**.

Olson, S. L., and Steadman, D. W., 1981, The relationships of the Pedionomidae (Aves: Charadriiformes), *Smithson. Contrib. Zool.* **337**:1–25.

Parker, S. A., 1982, A new sandpiper of the genus *Calidris, South Aust. Nat.* **56**:63.

Peters, J. L., 1934, *Check-list of Birds of the World,* Vol. 2, Harvard University Press, Cambridge, Mass.

Peterson, R. T., Mountfort, G., and Hollom, P. A. D., 1954, *A Field Guide to the Birds of Britain and Europe,* Houghton, Mifflin, Boston.

Slater, P., 1972, *A Field Guide to Australian Birds: Non-passerines,* Oliver & Boyd, Edinburgh.

Strauch, J. G., Jr., 1978, The phylogeny of the Charadriiformes (Aves): A new estimate using the method of character compatability analysis, *Trans. Zool. Soc. London* **34**:263–345.

Witherby, H. F., Jourdain, F. C. R., Ticehurst, N. F., and Tucker, B. W., 1940, *The Handbook of British Birds,* Vol. 5, Witherby, London.

Zusi, R. L., and Jehl, J. R., Jr., 1970, The systematic position of *Aechomorhynchus, Prosobonia,* and *Phegornis* (Charadriiformes: Charadrii), *Auk* **87**:760–780.

Chapter 2

SHOREBIRDS AS MARINE ANIMALS

Joanna Burger

Department of Biology and Bureau of Biological Research
Rutgers University
Piscataway, New Jersey 08854

I. INTRODUCTION

Marine biologists usually consider marine animals to include those organisms that live in the seas, bays, and estuaries. Traditionally, they have studied marine mammals (such as seals, dolphins, porpoises, whales), marine fishes, and a myriad of invertebrates whose entire life cycles are normally restricted to salt water habitats. Even within the traditionally defined marine fauna, some organisms such as sea lions spend a considerable portion of their breeding season on land, although they are usually restricted to offshore or coastal islands free from terrestrial predators.

Birds that spend their nonbreeding lives at sea have also been considered marine animals. Many of these species, like marine mammals, breed on offshore islands, and continue to feed from the sea during their breeding season. From a historical tradition, ornithologists have considered marine birds to include all the species belonging to families that contain predominantly marine species. Thus, all members of the following families have been considered marine birds: Spheniscidae (penguins), Diomedeidae (albatrosses), Procellariidae (fulmars, petrels, shearwaters), Hydrobatidae (storm petrels), Pelecaniodidae (diving petrels), Phaethontidae (tropicbirds), Pelecanidae (pelicans), Sulidae (gannets, boobies), Phalacrocoracidae (cormorants), Anhingidae (anhingas), Fregatidae (frigatebirds), Chionididae (sheathbills), Stercorariidae (skuas), Laridae (gulls, terns), Rynchopidae (skimmers), and Alcidae (alcids).

In fact, many of these species breed entirely on islands along the coasts or in estuaries, and some breed far inland on lakes and marshes

(Franklin's Gull *Larus pipixcan*, Ring-billed Gull *L. delawarensis*). These species depend upon the sea at other times of the year when they are not breeding, and many remain at sea or along the coasts during the years before they are mature and able to breed. Other avian species, traditionally considered marine because most members of their families are marine, may nest in freshwater habitats (e.g., some species of *Anhinga*).

Shorebirds, or waders, in the order Charadriiformes, have not traditionally been considered marine birds, even though their name implies some relationship to coastal shores. Although some species of shorebirds breed along the shores in sand dunes, or on sandy or rocky beaches (Figs. 1, 2), most species breed in interior regions along lakes and rivers, in marshes, on the tundra, or even in dry habitats (Fig. 3). Ornithologists have often studied shorebirds during the breeding season, ignoring them during the nonbreeding season when they may migrate long distances away. Even the study of migration of shorebirds often has concentrated on their migration pattern and on the factors initiating migration, and not on the functional aspects of their existence as estuarine or marine birds. Until recently, the study of shorebirds in the nonbreeding season was limited to recording numbers on a particular mudflat, bay, or estuary. Shorebirds, however, spend two-thirds to three-quarters of the year on migration routes and wintering grounds (see MacLean and Holmes, 1971; Elliot *et al.*, 1976; Kelly and Cogswell, 1979), suggesting that their existence during this period plays an important role in shaping their population structure, social system, and evolution.

More recently, biologists have begun to examine the ecology of wintering shorebirds, concentrating on habitat selection, survival strategies, and foraging behavior. Foraging behavior studies on migrant shorebirds have necessitated examinations of prey abundance and availability (see studies by Evans, 1976; Goss-Custard, 1979, 1980; Pienkowski, 1980, 1981). Because the prey organisms of shorebirds are generally marine invertebrates, shorebirds are finally being studied as parts of estuarine and marine ecosystems (Fig. 4). Their abundance and distribution among habitats are directly influenced by similar factors as are their marine prey (temperature, tides, winds; see Burger, Vol. 6 of this series), and by prey abundance and availability. Indeed, Wolff (1969) went so far as to examine the distribution of shorebirds along salinity gradients, and found some clear relationships. When considered in the context of shorebirds foraging on marine organisms in tidal marshes and bays, they surely are marine animals. That many species of shorebirds spend about 70% of their yearly cycle in tidal environments entitles them to consideration as vital parts of the marine ecosystem.

Fig. 1. Piping Plover nest along sandy beaches. They nest in the sand, often gathering shells for their nests (Long Island, N.Y.).

Fig. 2. Eurasian Oystercatcher (A) and Northern Lapwing (B) both nest along coasts in sandy beaches or on grass adjacent to the beach (Ravenglass, England).

Fig. 3. Gray-breasted Seedsnipe (A) and Tawny-throated Dotterel (B) both nest on dry, inland areas. [Photos taken by M. Gochfeld in Argentina.]

Fig. 4. Red Phalarope (A) are the most pelagic species as they migrate and overwinter at sea (Hudson Bay, Canada). Migrant shorebirds (B) frequently stop over at coastal estuaries and marshes (Jamaica Bay, N.Y.).

Nonetheless, the scope of the use of marine habitats by shorebirds has not been reviewed in detail, although Pitelka (1979) examined Pacific coast shorebirds. He summarized shorebird species of the world, and concluded that 80% of the New World shorebirds are maritime, but he did not separate breeding and nonbreeding habitats, nor did he give a species account of all shorebirds of the world. In order to understand the relationship of shorebirds to marine habitats, it is important to know their breeding and nonbreeding habitats, and I examine all shorebird species in this chapter. Further, I discuss the importance of shorebirds in marine avifaunas, and the importance of marine habitats to shorebirds.

II. MARINE AVIFAUNAS

Assessing the abundance of marine avifaunas entails censusing birds in bays and estuaries along coasts, and in the open oceans. Early data on seabird numbers come from the casual notes or diaries of seafarer's or others traveling in ships (for review see Brown, 1980). Gradually, ornithologists began taking advantage of scheduled cruise ships, ferries, and military ships to gather data on the distribution of marine birds (see Murphy, 1936). Most data on the distribution of marine birds at sea were gathered from ships employed for other purposes such as fishing (Sinclair, 1978; Dare, 1981), studying seals or other marine mammals (Cline *et al.*, 1969), or collecting oceanographic or fisheries information (Bartonek and Gibson, 1972). In such cases, scientists observing birds usually have no control over where the ship goes, its speed, or the length of stops at any location. Such logistical problems make it difficult to evaluate avian use of the oceans either spatially or temporally. In only a few cases has the primary objective been the study of marine birds and mammals (Jehl, 1973a; Harrison, 1979).

Study of the abundance and distribution of marine birds has also been attempted by censusing them from a fixed point on shore (e.g., Marchant, 1976; Henty, 1976), or from a fishing vessel in a bay or estuary (Bourne, 1970). These census methods give information on inshore populations of seabirds and on local movements. Some investigators have attempted to census seabirds both inshore and offshore (see Jehl, 1973a), but this approach is rare.

From the above discussion it is clear that the acquisition of information on marine birds is difficult logistically. The differences in the nature of marine ecosystems from location to location make it imperative that ornithologists relate seabird numbers to the physical, chemical, and biological characteristics of the oceans and such data must be obtained from systematically censused, large areas of the ocean. In addition to these logistical problems, the past biases of ornithologists in excluding some avian species found at sea make it difficult to evaluate the data on "marine birds." Most authors do not report on the numbers of all species found at sea, nor do they state if they have reported on all species found at sea. There are exceptions: Bartonek and Gibson (1972) report on numbers of species traditionally considered marine as well as on Arctic Loon (*Gavia stellataa*), ducks, and several shorebirds. Such a broad list of species suggests that their censuses included all species. Although quantitative, the data are given by categories (for example, 101–1000, 1001 +), making it difficult to evaluate the importance of various components of the marine avifauna. Four of the thirty-two species reported were shorebirds, and Red Phalarope (*Phalaropus fulicarius*) were more abundant than over half of the other species. Brown (1980) similarly reports sightings of Red Phalarope in the Bay of Fundy. Harrison (1979) also noted that phalaropes are part of the bird–whale feeding association he observed.

Nonetheless, most censuses of seabirds, either from land or from the open ocean, do not give quantitative data on shorebirds. Thus, it is difficult to assess the importance of shorebirds in the marine avifauna, either of inshore or of offshore faunas. However, I found several studies of coastal areas, bays, or estuaries where shorebird numbers were given (see Table I). In general, gulls and terns comprise 20–94% of the inshore marine avifaunas; ducks, geese, and swan comprise 0-17%; and shorebirds account for 6–66% of the inshore avifaunas. As might be expected, shorebirds comprise a small percentage of the total when marine birds are censused during the breeding season (6% at the Wash in England), but a much larger proportion in the winter (58% in Weston, England). These data are imperfect in the sense that the object of these studies (except for Burger *et al.*, 1982) was not to census *all* birds in marine environments, but to census "marine birds." Two studies are available where shorebirds were censused along with the other "marine birds" (see Table I). Jehl *et al.* (1973) censused Golfo San Jose on the Argentine coast and found that shorebirds accounted for 19% of the birds. Kevin Powers (personal communication) has attempted to estimate the number of all birds in the shelf waters off the northeastern United States in the four seasons of the year. This is an admirable undertaking as he is not merely censusing birds and adding up totals, but attempting to accurately estimate population numbers. Phalaropes were the only shorebirds considered in his calculations (Fig. 4A). Nonetheless, phalaropes comprised 21% of the birds in the spring, but only 1% in the fall migration period. His data clearly indicate

Table I. The Composition of Coastal Avifaunas, and the Contribution of Shorebirds

Location	Season	Seabirds (%) (excluding gulls and terns)	Gulls and terns (%)	Ducks, geese, swans (%)	Shorebirds (%)	Total number censused	Source
Shelf waters[a] off NE United States	Spring	41	38	0	21	3,840,000	K. Powers (1982, personal communication)
	Fall	48	59	0	1	3,866,000	
Gulfo San Jose (Argentina Coast)[b]	4–9 August	40	36	3	19	13,572	Jehl *et al.* (1973)
Wales (estuaries)[c]	Monthly counts	0	20	14	66	201,890	Prater (1981)
Seven, England[c]	Monthly counts	0	39	7	54	141,158	Prater (1981)
Solway Firth, England[c]	Monthly counts	1	63	8	28	158,015	Prater (1981)
Wash, England[c]	Breeding season	0	94	0	6	23,957	Prater (1981)
Langstone Harbour, England[c]	Winter	0	35	17	48	48,830	Prater (1981)
Weston, England[d]	December–February 1980	0	12	6	58	60,005	P. Lack (personal communication)
Mollendo, Peru[e]	1971–1975 (annual)	5	76	5	14	845,399	Hughes (1979)
Barnegat Bay, New Jersey[f]	April–July 1977	0	14	14	12	14,710	Burger *et al.* (1982)
Cape Town, South Africa[g]	1-year counts	0	49	0	47	227	McLachlan *et al.* (1980)

[a] 35–44°N; seabirds = fulmars, shearwaters, petrels, auks.
[b] Other 9% are grebes, flamingos, herons, and egrets.
[c] Selected species, used highest monthly counts.
[d] Other 23% are half Starlings (*Sturnella vulgaris*).
[e] Counted Charadriiformes only.
[f] Other 70% are herons, egrets, ibises, and passerines (salt marsh counts).
[g] Based on number of birds/km for 12 months.

the importance of seasonal considerations. Powers, however, did not include loons, ducks, or other shorebirds in his calculations.

From Table I it is clear that shorebirds are important components of the inshore marine avifauna, and can be important in the offshore avifauna at least at some times of the year. The available censuses have not been conducted to answer one key question—how important are various bird groups (or species) to marine avifaunas? Shorebirds are not the only birds that use marine ecosystems that are frequently excluded from censuses of marine birds; loons, grebes, and ducks are also omitted. Because all of these species are living on the oceans for a considerable period of the year, their role must also be considered when examining marine ecosystems. Certainly when management and conservation decisions are made regarding bays, estuaries, inshore and offshore areas, the impact on these species groups (shorebirds, loons, grebes, ducks) should be considered.

III. SHOREBIRDS IN MARINE ECOSYSTEMS

In the above section I considered the role of shorebirds as components of marine avifaunas. Far more data are available on the number of shorebirds using marine ecosystems, particularly inshore bays and estuaries. That is, just as some ornithologists have censused species traditionally considered marine (ignoring shorebirds), others have censused shorebirds without recording the numbers of gulls, terns, and other marine species (see papers in Pitelka, 1979; Prater, 1981). Counts made of shorebirds on bays or estuaries were usually made of migrants during the winter. Table II lists counts in a few representative areas. It is not exhaustive, and is meant to convey the range of numbers found in estuaries, and the methods of recording such information.

More recently, biologists have attempted to census large geographical areas both to determine population numbers, and to determine which areas are critical stopover or staging areas for migrant shorebirds. Spaans (1978) estimated numbers of shorebirds using the Surinam coast in winter (Table III). Morrison and Harrington (1979) are conducting an international shorebird survey to monitor numbers of species from Hudson Bay to Central and South America (see Table II). Their survey scheme uses volunteers to census shorebirds every 2 weeks during fall migration at as many locations as possible. This type of census clearly delineates important staging areas, and gives estimates of numbers using particular marine and estuarine ecosystems, but does not provide estimates of total population sizes.

Table II. Selected Shorebird Censuses Indicating Number of Shorebirds Censused

Census area	Census time	Species counted	Number censused	Source
Large census areas				
British estuaries	Winter	All shorebirds	2,000,000	Prater (1981)
Jarves and Hudson bays[a]	Autumn migration	Semipalmated Sandpiper, Red Knot, Hudsonian Godwit	27,857	Morrison and Harrington (1979)
Gulf of St. Lawrence to Bermuda[a]	Autumn migration	Semipalmated Sandpiper, Red Knot, Hudsonian Godwit	404,637	Morrison and Harrington (1979)
Copper River Delta, Alaska	Spring	All shorebirds	20,000,000	Isleib (1979)
NW Africa	Winter	Knot	135,000	Dick *et al*. (1976)
Southern Africa	Winter	Knot	10,000	Dick *et al*. (1976)
Western Europe	Winter	Knot	609,000	Dick *et al*. (1976)
Small census areas				
Wash, England	Peak winter counts	10 species[b]	172,520	Goss-Custard *et al*. (1977a)
Eastern Cape, South Africa	12 months	6 species[c]	8.80/km	McLachlan *et al*. (1980)
Southwestern Cape, Africa	Austral summer	29 species	119,008	Summers *et al*. (1977)
Nelson Lagoon	1 July–1 Dec. fall	All shorebirds	326,000	Gill and Jorgenson (1979)
Part of Copper River Delta, Alaska	Summer	All shorebirds	250,000/mile^2	Senner (1979)
Plymouth, Mass.	Early August daily counts	All shorebirds	>5,000	Schneider (1978)

[a] Maritime and International Shorebird Summary.
[b] In order of abundance: *Calidris canutus*, *C. alpina*, *Haematopus ostralegus*, *Limosa lapponica*, *Numenius arquata*, *Tringa totanus*, *Pluvialis squatarola*, *Calidris alba*, *Arenaria interpres*, *Charadrius hiaticula*.
[c] In order of abundance: *Haematopus moquini*, *Charadrius marginatus*, *Calidris alba*, *Pluvialis squatarola*, *Numenius phaeopus*, *Tringa nebularia*.

Table III. Maximum Winter Counts for Surinam Coast[a,b]

Over 100,000	2500–10,000
Lesser Yellowlegs	Wimbrel
Short-billed Dowitcher	Solitary Sandpiper
Semipalmated Sandpiper	Below 2500
50,000–100,000	Red Knot
Greater Yellowlegs	Upland Sandpiper
Least Sandpiper	Sanderling
Willet	Western Sandpiper
10,000–50,000	
Semipalmated Plover	
Spotted Sandpiper	
Ruddy Turnstone	
White-rumped Sandpiper	
Stilt Sandpiper	

[a] After Spaans (1978).
[b] Scientific names given in Table IV.

In England, biologists are attempting to survey shorebird numbers during the winter at bays and estuaries where shorebirds winter (see Prater, 1981). Prater's (1981) book provides estimates of all major estuaries in Britain and Ireland, but he notes that even the highest counts underestimate the numbers of birds using any estuary. By censusing all coastline areas throughout the winter, he was able to estimate that 2 million shorebirds overwinter in British estuaries (see Table II). By comparison, Isleib (1979) estimated that 20 million shorebirds migrate through the Copper River Delta of Alaska during spring migration.

Another approach is to estimate the number of individuals of one species over a large geographical area. Dick *et al.* (1976) examined wintering populations of Knots in Europe and Africa (see Table II). As is clear from Table II, large numbers of shorebirds frequent coastal areas in the winter. The relative importance of inland compared to coastal sites for migrant and wintering populations of shorebirds is seldom examined, although only 8% of the shorebirds reported on the maritime and international shorebird survey in the New World were reported from the interior regions. This figure may be low because few shorebird biologists regularly census interior regions and birds are not as concentrated as they are along the coastal shores.

In summary, large numbers of shorebirds are regularly recorded in coastal areas, particularly during the migration periods. Important staging areas, such as the Copper River Delta in Alaska, have millions of shorebirds moving through during spring migration. Such staging areas are providing essential feeding habitat for shorebirds to lay down fat stores for the next stage of migration, and for necessary reproductive activities (see Senner, 1979; Isleib, 1979).

IV. HABITAT SELECTION IN SHOREBIRDS

There are 216 species of shorebirds in 14 families (two suborders: Charadrii and Scolopaci). Shorebirds exhibit a wide range of life histories including species that are nonmigratory, some that are nomadic, and many that are migratory. I examined habitat selection in shorebirds by reviewing the relevant literature and from personal communication with knowledgeable investigators, dividing habitat selection into breeding, migrating, and wintering habitats. I was able to find information on the breeding habitats of 215 species, and on the wintering habitats of 203 species. The data on habitat selection are presented in Table IV. I intentionally have used the phrases of the original authors to describe habitats, rather than use uniform terms for similar habitats, because the original terms more accurately describe the habitats. Further, Table IV indicates the species that are nomadic or nonmigratory ("NM" in the table). There is a great deal of variation in the extent a species exhibits a particular pattern. For example, in some species all individuals migrate, whereas in others some migrate and others do not. I discuss each family below. Scientific names for all shorebirds mentioned in the text are given in Table IV. For the following discussion, refer to Table IV.

A. Taxonomic Summary

1. Dromadidae

This family contains only the Crab Plover, which is primarily coastal, living along the Indian Ocean during the entire year.

2. Chionididae

Both species of sheathbills are coastal and nest in sub-Antarctica and Patagonia, often in penguin (*Eudyptes* and *Aptenodytes*) colonies (Burger, 1979). Although most individuals are nonmigratory (Burger, 1979), some Snowy Sheathbills migrate from Antarctic regions to rocky beaches near seal and cormorant colonies (J. R. Jehl, Jr., personal communication).

Table IV. Habitats of Shorebirds of the World

Species	Habitats			Sources
	Nesting	Migrating	Wintering	
Dromadidae				
Crab Plover (*Dromas ardeola*)	Sea beaches (in burrows)	Sandy beaches, mudflats	Coastal sand banks, mangrove swamps	Rand (1936), Ripley (1961), McLachlan and Liversidge (1978), J. R. Jehl, Jr. (personal communication)
Chionididae				
Snowy Sheathbill (*Chionis alba*)	Coasts of Antarctic Peninsula, Falkland Islands	Coast	Sandy and rocky beaches	Johnson and Goodall (1965), Humphrey *et al.* (1970), Burger (1979), J. R. Jehl, Jr. (personal communication)
Lesser Sheathbill (*Chionis minor*)	Southern Indian Ocean, in crevices under lava boulders	NM	Beaches	Burger (1979)
Pluvianellidae				
Magellanic Plover (*Pluvianellus socialis*)	Lakes in steppe grassland, brackish lakes (not coastal)	Inland lakes, coastal areas (some sedentary)	Coastal bays and river mouths, debris-covered sandy beaches (tidal)	Johnson and Goodall (1965), Jehl (1975)
Pedionomidae				
Plain Wanderer (*Pedionomus torquatus*)	Grassland and open plains in arid regions	NM	Grassland and open plains in arid regions	Slater (1971)

Thinocoridae				
Rufous-bellied Seedsnipe (*Attagis gayi*)	High Andes near snow line, nests near water	Local movement only	Floors of mountain valleys	Johnson and Goodall (1965)
White-bellied Seedsnipe (*Attagis malouinus*)	Mountain alpine zone, desolate moorland	Mountains	Lowlands, rarely coastal flats	Johnson and Goodall (1965), Humphrey *et al.* (1970)
Gray-breasted Seedsnipe (*Thinocorus orbignyianus*)	High elevations, semidesert, marshy areas	Grasslands	Pampas grasslands	Johnson and Goodall (1965), Maclean (1969)
Least Seedsnipe (*Thinocorus rumicivorus*)	Coastal hills, open country in lowlands, llanos, few feed in *Salicornia* tide flats	Grasslands	Upland grassland, gravel coastal plain, exposed tideflats	Maclean (1969), Humphrey *et al.* (1970), Jehl *et al.* (1973), Hughes (1979), Myers and Myers (1979), Myers (1980a)
Burhinidae				
Eurasian Stone Curlew (*Burhinus oedicnemus*)	Uncultivated fields, rocky plateaus, near marshes, open semidesert	NM	Uncultivated fields near rivers, and sandbanks, thorn scrub	Etchecopar and Hue (1967), Ali and Ripley (1969), King and Dickinson (1975)
Senegal Thick-knee (*Burhinus senegalensis*)	Sandbanks of rivers		Edges of wet areas, cultivated gardens	Etchecopar and Hue (1967)
Water Dikkop (*Burhinus vermiculatus*)	Perennial lakes or rivers, sandy or rocky inlets	NM	Near lakes or rivers, or rocky inlets near water	Mackworth-Praed and Grant (1962), McLachlan and Liversidge (1978)
Spotted Thick-knee (*Burhinus capensis*)	Dry open fields, scrub country	NM	Dry open fields	McLachlan and Liversidge (1978)
Double-striped Thick-knee (*Burhinus bistriatus*)	Tropical dry forest, grassy plains	NM	Tropical dry forest, grassy plains	Slud (1964)

(*Continued*)

Table IV. (*Continued*)

Species	Habitats: Nesting	Habitats: Migrating	Habitats: Wintering	Sources
Peruvian Thick-knee (*Burhinus superciliaris*)	Coastal desert (some feed on tidal mudflats)		Coastal deserts, coastal mudflats	Hughes (1979)
Australian Stone Curlew (*Burhinus magnirostris*)	Reefs, beaches, coastal mudflats, grassy woodland		Reefs, beaches, coastal mudflats	Slater (1971), Frith (1976)
Great Stone Curlew (*Esacus recurvirostris*)	Near rivers, seacoasts	NM	Near rivers, reefs, seacoasts, estuaries	Smythies (1953), King and Dickinson (1975), B. King (personal communication)
Beach Stone Curlew (*Esacus magnirostris*)	Sand and shingle beaches, near rivers	NM	Sandy, tidal beaches, near rivers	Ripley (1944), Hoogerwerf (1969), King and Dickinson (1975)
Haematopodidae				
Eurasian Oystercatcher (*Haematopus ostralegus*)	Rocky mussel beds, sandy beaches on coast, coastal islands, sand dunes, few in inland brackish marshes	Tidal mudflats	Salt marshes, tidal mudflats, sandy and rocky beaches	Lind (1965), Bianki (1967), Harris (1967), Wolff (1969), Dare and Mercer (1973), Evans (1981), J. Burger (unpublished data), B. King (personal communication)
American Oystercatcher (*Haematopus palliatus*)	Sandy beaches, mudflats, salt marshes	Tidal mudflats	Tidal mudflats, sandy beaches	Myers and Myers (1979), Myers (1980a), J. Burger (unpublished data)
American Black Oystercatcher (*Haematopus bachmani*)	Rocky intertidal, rocky shore	Rocky intertidal	Rocky intertidal, tidal mudflats	Recher (1966), Hartwick (1974), Morrell *et al.* (1979), Hartwick and Blaylock (1979)

African Black Oystercatcher (*Haematopus moquini*)	Sandy and rocky beaches, limited to seashore	NM	Rocky and sandy tidal beaches	Pringle and Cooper (1977), McLachlan and Liversidge (1978)
Variable Oystercatcher (*Haematopus unicolor*)	Rocky and sandy tidal beaches	Some local movement	Rocky and sandy tidal beaches	Falla *et al.* (1966), Stidolph (1973), Jones (1979)
Magellanic Oystercatcher (*Haematopus leucopodus*)	Freshwater lakes and lagoons, sandy mudflats	Local movement to coast	Beaches and adjacent upland	Humphrey *et al.* (1970), Jehl (1975)
Blackish Oystercatcher (*Haematopus ater*)	Rocky and shingle beaches		Coastal sandy and rocky beaches	Humphrey *et al.* (1970) Jehl (1975), Hughes (1979)
Sooty Oystercatcher (*Haematopus fuliginosus*)	Rocky or sandy beaches along coast or on offshore islands	NM	Rocky shores, beaches, mudflats, estuaries	Frith (1976), Johnsgard (1981)
Ibidorhynchidae				
Ibis-bill (*Ibidorhynchus struthersii*)	Mountainous river valleys, 1600–4400 m above sea level	NM	Stony riverbeds	Dement'ev *et al.* (1969), King and Dickinson (1975), Johnsgard (1981)
Recurvirostridae				
Black-winged Stilt (*Himantopus himantopus*)	Freshwater marshes, estuaries, sand dunes (may feed in tidal areas)	NM	Tidal sloughs, few in freshwater ponds	Johnson and Goodall (1965), Frith (1976), Summers *et al.* (1977)
South American Stilt (*Himantopus melanurus*)	Shallow lakes and mouths of rivers, swamps, marshes (feeds in tidal sloughs)	NM	Inland lagoons, broad intertidal mudflats, tidal sloughs	Johnson and Goodall (1965), Myers and Myers (1979), J. Burger (unpublished data)
Black-necked Stilt (*Himantopus mexicanus*)	Meadows and wet grasslands, freshwater ponds, coastal ponds	Tidal mudflats, inland marshes	Tidal mudflats, salt marshes	Bent (1927), Recher (1966), Hamilton (1975), Smith and Stiles (1979), J. R. Jehl, Jr. (personal communication)

(Continued)

Table IV. (*Continued*)

Species	Habitats			Sources
	Nesting	Migrating	Wintering	
Ceylon Stilt (*Himantopus ceylonensis*)	Lakes, dry margins of salt pans	Nomadic	Tidal and fresh marshes, lakes, pans, fields	Ali and Ripley (1969)
Pied Stilt (*Himantopus leucocephalus*)	Seashore, boulder banks; saline lagoons, paddocks, river beds	Moves toward coast	Most coastal, few inland	Falla *et al.* (1966)
Hawaiian Stilt (*Himantopus knudseni*)	Lagoons			Munro (1964)
Black Stilt (*Himantopus novaezeelandiae*)	Inland, dry shingle beds, adjacent to swamps	NM	Inland, dry shingle beds, adjacent to swamps, tidal mudflats	Sibson and McKenzie (1943), Falla *et al.* (1966)
South African Stilt (*Himantopus meridionalis*)	Marshy ground	NM	Marshy ground	McLachlan and Liversidge (1978)
Banded Stilt (*Cladorhynchus leucocephalus*)	Sand banks, inland salt flats	Salt lakes	Salt lakes, brackish estuaries (tidal)	Slater (1971)
Eurasian Avocet (*Recurvirostra avosetta*)	Marshes and reed beds, estuaries and seacoast islands		Brackish areas, sandy and mixed intertidal	Makkink (1936), Wolff (1969), Summers *et al.* (1977), de Bie and Zijlstra (1979)
American Avocet (*Recurvirostra americana*)	Along alkaline or saline lakes, dry mudflats	Sandy beaches of lakes, salt marshes	Tidal mudflats, salt marshes	Bent (1927), Recher (1966), Hamilton (1975), Bradstreet *et al.* (1977), Gerstenberg (1979)

Red-necked Avocet (*Recurvirostra novaehollandiae*)	Shallow, saline lakes, tussocks near shallow water	Lakes and marshes	Coastal marshes, estuaries, mudflats, tidal inlets	Falla *et al*. (1966), Slater (1971), Frith (1976)
Andean Avocet (*Recurvirostra andina*)	Puna zone of Andes over 3600 m above sea level, near salt lakes in barren hills	NM	Puna zone over 3600 m above sea level	Johnson and Goodall (1965), Johnsgard (1981)
Glareolidae				
Egyptian Plover (*Pluvianus aegyptius*)	Along rivers in sand bars, sandy islands			Howell (1979)
Two-banded Courser (*Rhinoptilus africanus*)	Dry pans in salt veldt	NM	Dry pans in salt velt	McLachlan and Liversidge (1978)
Heuglin's Courser (*Rhinoptilus cinctus*)	Dry, thorny scrub, Mopane, Miombo woodland	NM	Bushland in low-rainfall areas	Mackworth-Praed and Grant (1962), Britton (1980), Williams and Arlott (1980)
Violet-tipped Courser (*Rhinoptilus chalcopterus*)	Bushes in grassland	Local movement only	Bushes and grassland	McLachlan and Liversidge (1978)
Jerdon's Courser (*Rhinoptilus bitorquatus*)	Hilly country, rocky and undulating ground, thin forest, high scrub	NM	Hilly country, high scrub forest	Greenway (1958), Woodcock (1980)
Cream-colored Courser (*Cursorius cursor*)	Semidesert, coastal plateau	NM	Semidesert	Etchecopar and Hue (1967), Bannerman and Bannerman (1968), Dement'ev *et al*. (1969)
Indian Courser (*Cursorius coromandelicus*)	Dry country, ploughed fields	NM	Fields and dry country	Ali (1969)

(*Continued*)

Table IV. *(Continued)*

Species	Habitats			Sources
	Nesting	Migrating	Wintering	
Temminck's Courser (*Cursorius temminckii*)	Short grass veld, open country, on recently burned areas			Benson *et al*. (1971), McLachlan and Liversidge (1978)
Australian Pratincole (*Stiltia isabella*)	Blacksoil plains, inland scrub, near damp places, temperate and subtropical grasslands	Local movement only	Gibber plains	Maclean (1973), Frith (1976)
Collared Pratincole (*Glareola pratincola*)	Sandbanks near water, sandflats near coastal (tidal) areas		Riverbanks, river sandbanks, flats in coastal areas	McLachlan and Liversidge (1978), J. Burger (unpublished data)
Eastern Pratincole (*Glareola maldivarium*)	Marshy areas, rice stubble	NM, but some migratory	Dried ricefields, grass meadows; fallow fields	Hoogerwerf (1969), King and Dickinson (1975), B. King (personal communication)
Black-winged Pratincole (*Glareola nordmanni*)	Steppes, lake floodplain, meadows	Saline vegetation	Grassland	Dement'ev *et al*. (1969), McLachlan and Liversidge (1978)
Madagascar Pratincole (*Glareola ocularis*)	Savannah and marshes	Tidal mudflats	Savannah	Rand (1936)
White-collared Pratincole (*Glareola nuchalis*)	Near rivers and lakes, on rocks	NM	Near rivers and lakes	McLachlan and Liversidge (1978)
Gray Pratincole (*Glareola cinerea*)	Smaller streams with sandbanks		Sandbanks of rivers, estuaries, tidal lagoons	Bannerman (1958), Serle *et al*. (1977)
Milky Pratincole (*Glareola lactea*)	Near rivers, marshes, river sandbanks	NM	Near rivers, marshes; shingle banks of rivers	Smythies (1953), Ali (1969), King and Dickinson (1975), B. King (personal communication)

Charadriidae				
Pied Lapwing (*Hoploxypterus cayanus*)	Along stream and ponds in open country and savannah		Along streams and ponds	Cherrie (1916), Johnsgard (1981)
Northern Lapwing (*Vanellus vanellus*)	Inland, marshes and peat bogs, coastal islands, sand dunes	Inland	Inland, some tidal estuaries, brackish mudflats	Harris (1967), Wolff (1969), Evans (1981), Dyrcz *et al.* (1981), J. Burger (unpublished data)
Long-toed Lapwing (*Vanellus crassirostris*)	Short grass plains, marshes, swamps			McLachlan and Liversidge (1978), Walters (1979)
Blacksmith Plover (*Vanellus armatus*)	Grass near rivers or streams	NM	Rocky intertidal	Summers *et al.* (1977), McLachlan and Liversidge (1978)
Spur-winged Plover (*Vanellus spinosus*)	On islets or mudflats in rivers	NM	Sand and shingle banks of rivers	Ripley (1961), Mackworth-Praed and Grant (1962)
River Lapwing (*Vanellus duvaucelii*)	River banks	NM	River banks	King and Dickinson (1975)
Black-headed Plover (*Vanellus tectus*)	Arid thorn bush, often near rivers	NM	Arid and thorn bush	North (1937), Williams (1964)
Yellow-wattled Lapwing (*Vanellus malabaricus*)	Barren wasteland, fallow fields	NM	Marshes	Ripley (1961), Jayakar and Spurway (1968), Ali (1969), Johnsgard (1981)
White-crowned Plover (*Vanellus albiceps*)	Mud or sandbanks along rivers	NM	Mud or sandbanks along rivers	McLachlan and Liversidge (1978)
Senegal Plover (*Vanellus lugubris*)	Coastal dry grassland, recently burned fields			McLachlan and Liversidge (1978), Johnsgard (1981)

(Continued)

Table IV. *(Continued)*

Species	Habitats: Nesting	Habitats: Migrating	Habitats: Wintering	Sources
Black-winged Plover (*Vanellus melanopterus*)	Wet grassland; burned areas (not tidal)	Wet grassland, coastal tideflats	Coastal flats, wet grassland	McLachlan and Liversidge (1978)
Crowned Plover (*Vanellus coronatus*)	Open veld, short grass, burned areas	NM	Dry upland, freshwater marshes	Summers *et al.* (1977), McLachlan and Liversidge (1978), Ade (1979)
Wattled Plover (*Vanellus senegallus*)	Near rivers, vleis, marshes, or other water, semidesert	Local movement only	Near rivers, marshes	Mackworth-Praed and Grant (1962), Little (1967), McLachlan and Liversidge (1978)
Spot-breasted Plover (*Vanellus melanocephalus*)	Marshy grasslands and moorlands with giant heath	NM	Marshy grasslands and moorlands with giant heath and tussock grass	Urban (1978)
Brown-chested Plover (*Vanellus superciliosus*)	Orchard bush, savannah grassland	Savannah	Southern savannah	Johnsgard (1981)
Sociable Plover (*Vanellus gregarius*)	Steppe and salt pans, inland brackish lakes	Cultivated ground, desert	Semicultivated areas	Ripley (1961), Etchecopar and Hue (1967), Khrokov (1978)
White-tailed Plover (*Vanellus leucurus*)	Dry fields, often near marshes	Marshes	Marshes	Ripley (1961), Etchecopar and Hue (1967), Johnsgard (1981)
Southern Lapwing (*Vanellus chilensis*)	Inland marshes, upland grassland, pampas (may feed in salt marshes)	NM	Grassland, freshwater marshes, salt marshes, tidal mudflats	Humphrey *et al.* (1970), Myers and Myers (1979)
Andean Lapwing (*Vanellus resplendens*)	Puna in high Andes, slopes			Johnsgard (1981)

Gray-headed Lapwing (*Vanellus cinereus*)	Marshy areas, rice stubble, river flats		Marshy areas	Ripley (1961), King and Dickinson (1975)
Red-wattled Lapwing (*Vanellus indicus*)	Rice stubble, rivers, marshes, open forest	NM	Marshes, near rivers, rice stubble	King and Dickinson (1975)
Javanese Wattled Lapwing (*Vanellus macropterus*)	Agricultural lands near marshes			Johnsgard (1981)
Banded Plover (*Vanellus tricolor*)	Inland and coastal plains, dry interior	Nomadic	Dry interior	Slater (1971), Frith (1976)
Masked Plover (*Vanellus miles*)	Grassland and mudflats, grazing land, lakes and swamps	NM	Grassland, tidal mudflats	Thomas (1969), Slater (1971), Johnsgard (1981)
Greater Golden Plover (*Pluvialis apricaria*)	Arctic tundra	Mostly inland, also tidal marshes	Tidal estuaries, inland grasslands	Evans (1981)
Lesser Golden Plover (*Pluvialis dominica*) Including: American Golden Plover (*P. dominica*), and Pacific Golden Plover (*P. fulva*)	Tundra plains and grasslands, often near coasts	Tundra, inland lakes, tidal marshes	Salt marshes, mudflats, tidal shores, pampas marshes, fallow fields, Pacific islands	Drury (1961), Sauer (1962), Weir and Cooke (1976), Baker (1977), Myers and Myers (1979), Connors *et al.* (1979), Robertson and Dennison (1979), Myers and Pitelka (1980), J. R. Jehl, Jr. (personal communication)
Black-bellied Plover (*Pluvialis squatarola*)	Inland arctic plain, Arctic tundra	Inland lakes, tidal marshes and mudflats	Salt marshes, tidal beaches	Weir and Cooke (1976), Myers and Myers (1979), Myers and Pitelka (1980), Schneider and Harrington (1981), Duffy *et al.* (1981)

(Continued)

Table IV. *(Continued)*

Species	Habitats			Sources
	Nesting	Migrating	Wintering	
Ringed Plover (*Charadrius hiaticula*)	Sandy beaches with pebbles, some inland in Arctic stony fields	Tidal marshes	Tidal marshes and mudflats, sandy beaches	Bianki (1967), Pringle and Cooper (1977), Meltofte (1979), Evans (1981)
Semipalmated Plover (*Charadrius semipalmatus*)	Tundra marshes and grassland	Inland lakes, salt marshes, tidal beaches, mudflats	Coastal marshes, tidal mudflats and beaches	Weir and Cooke (1976), Spaans (1978), Burger *et al*. (1977, 1982), Baker (1979), Myers and Pitelka (1980), Duffy *et al*. (1981)
Long-billed Plover (*Charadrius placidus*)	Sandy river shorelines, some on coastal sandy beaches		Rivers, tidal mudflats	Dement'ev *et al*. (1969), King and Dickinson (1975), Johnsgard (1981)
Little Ringed Plover (*Charadrius dubius*)	Freshwater marshes, river banks, few on coastal gravel beaches		Coastal salt marshes, freshwater marshes	Bent (1929), Etchecopar and Hue (1967)
Thick-billed Plover (*Charadrius wilsonia*)	Sandy reefs and ocean beaches, sand dunes near salt water, or sandy areas of brackish water	Sandy beaches	Tidal mudflats, sandy beaches	Bent (1929), Strauch and Abele (1979), Johnsgard (1981)
Killdeer (*Charadrius vociferus*)	Grassy fields, cultivated land, coastal lagoons and beaches (interdune area)	Inland ponds, salt marshes	Salt marshes, intertidal areas, sandy beaches	Urner and Storer (1949), Brooks (1967), Myers *et al*. (1979), Page *et al*. (1979), Lenington (1980), J. Burger (unpublished data)

Piping Plover (*Charadrius melodus*)	Sandy beaches (both salt and freshwater) outer tidal sand spits	Tidal beaches	Tidal beaches, sandy beaches	Wilcox (1959), Godfrey (1966), Burger *et al.* (1977), Bradstreet *et al.* (1977)
Black-banded Sandplover (*Charadrius thoracicus*)	Muddy flats at river mouths, along coasts	Coastal areas, coastal marshes	Tidal flats, tidal *Salicornia* flats	Rand (1936), Johnsgard (1981)
Kittlitz's Plover (*Charadrius pecuarius*)	Sandy and rocky beaches, sandbanks, (few dry velt)	NM	Sandy and rocky beaches	Hall (1959), Pringle and Cooper (1977)
St. Helena Plover (*Charadrius sanctaehelenae*)	Open grassland or plains		Grasslands, ploughed fields	Pitman (1965), Johnsgard (1981)
Three-banded Plover (*Charadrius tricollaris*)	Inland lakes, pans, and rivers, along seashore	NM	Rocky intertidal zone	Martin (1972), Summers *et al.* (1977), McLachlan and Liversidge (1978), Tyler (1978)
Snowy Plover (*Charadrius alexandrinus*)	Coastal sandy beaches, sandy beaches along inland saltflats	Sandy beaches	Sandy beaches, intertidal mudflats; fallow fields	Etchecopar and Hue (1967), Wolff (1969), Hobbs (1972), Nakazawa (1979), Gerstenberg (1979), Myers *et al.* (1979), B. King (personal communication)
White-fronted Sandplover (*Charadrius marginatus*)	Sandy and shingle beaches	NM	Sandy beaches, rocky intertidal	Winterbottom (1963), Summers *et al.* (1977), Pringle and Cooper (1977), McLachlan and Liversidge (1978)

(*Continued*)

Table IV. *(Continued)*

Species	Habitats: Nesting	Migrating	Wintering	Sources
Peruvian Sandplover (*Charadrius occidentalis*)	Sandy beaches	NM	Sandy beaches	Koepcke (1970)
Red-capped Dotterel (*Charadrius ruficapillus*)	Sandy coastal areas, *Juncus* flats (some on salt lakes)	NM	Coastal areas, some on salt lakes	Hobbs (1972), Frith (1976)
Malaysian Sandplover (*Charadrius peronii*)	Coral and sand beaches	NM	Coral and sand beaches, mudflats	Hoogerwerf (1969), King and Dickinson (1975)
Chestnut-banded Sandplover (*Charadrius venustus*)	Coastal and inland salt pans, vleis	NM	Sandy beaches, coastal marshes, freshwater wetlands	Summers *et al.* (1977), McLachlan and Liversidge (1978)
Collared Plover (*Charadrius collaris*)	Sandy coasts, river edges	NM	Sandy coasts, tidal mudflats, river edges	Strauch and Abele (1979), Myers and Myers (1979)
Double-banded Dotterel (*Charadrius bicinctus*)	Coastal bays, ocean beaches, riverbeds by ploughed fields	Tidal mud and sand flats	Tidal mud and sand flats	Falla *et al.* (1966), Frith (1976), Robertson and Dennison (1979)
Two-banded Plover (*Charadrius falklandicus*)	Inland streams, lakes, and canals (often feeds in tidal mudflats)	Canals, tidal marshes	Sandy beaches, mudflats, salt marshes, coastal freshwater marshes	Humphrey *et al.* (1970), Myers and Myers (1979), Myers *et al.* (1979)
Puna Plover (*Charadrius alticola*)	Salty and fresh lagoons at high altitudes	NM	Salty and freshwater lagoons	Johnson and Goodall (1965), Hoy (1967)
Mongolian Plover (*Charadrius mongolus*)	Inland mud and flats	Lakes and seacoasts	Tidal mudflats and sandflats	Slater (1971), Nielsen (1971), King and Dickinson (1975)
Greater Sandplover (*Charadrius leschenaultii*)	Inland flats of clay or solonetz	Tidal mud and sand banks, sand and coral beaches	Coastal areas, sand and coral beaches, tidal mud and sand banks	Dement'ev *et al.* (1969), Hoogerwerf (1969), McLachlan and Liversidge (1978), Summers and Waltner (1979), Johnsgard (1981)

Red-breasted Dotterel (*Charadrius obscurus*)	Among coastal dunes, pebble beaches, amid sandy spits	NM	Ocean beaches, sandy flats in harbors, estuaries	Falla *et al.* (1966)
Caspian Plover (*Charadrius asiaticus*)	Inland on bare ground, semideserts, steppes, some on tidal mudflats	Open grassland, grassy fields	Open grassland, grassy fields, tidal mudflats	Slater (1971), McLachlan and Liversidge (1978)
Oriental Plover (*Charadrius veredus*)	Inland flats		Dry game pastures, river marshes, coastal marshes; grass on airstrips	Hoogerwerf (1969), Nielsen (1971), King and Dickinson (1975), B. King (personal communication)
Rufous-chested Dotterel (*Charadrius modestus*)	Dry grasslands on mountaintops, peat bogs, some near coasts in Tierra del fuego	Coastal or mountain regions	Upland shortgrass areas, rarely in salt marshes and rocky coasts	Johnson and Goodall (1965), Humphrey *et al.* (1970), Myers and Myers (1979), J. R. Jehl, Jr. (personal communication)
Mountain Plover (*Charadrius montanus*)	Shortgrass prairie		Semidesert, may be far from water	Graul (1973), Johnsgard (1981)
Black-fronted Dotterel (*Charadrius melanops*)	River shingles, lagoon edges, few in estuaries		River shingles, rarely along tidal shores	Falla *et al.* (1966), Frith (1976)
Hooded Dotterel (*Charadrius cucullatus*)	Windswept sandy tidal beaches, some in salt pans	Sandy beaches	Tidal, sandy beaches	Frith (1976), Johnsgard (1981)
Eurasian Dotterel (*Charadrius morinellus*)	Northern tundra, heaths, barrens not far from tree line	Tundra, steppes	Mountain plateaus, along seashores	Dement'ev *et al.* (1969), Johnsgard (1981)

(Continued)

Table IV. *(Continued)*

Species	Habitats			Sources
	Nesting	Migrating	Wintering	
Shore Plover (*Charadrius novaeseelandiae*)	Rock crevices or burrows in open grassland	Rocky shore, *Salicornia* tidal marshes	Tidal mudflats, rocky shores, fields, brackish ponds	Falla *et al.* (1966), Barlow *et al.* (1972), Phillips (1977), J. R. Jehl, Jr. (personal communication)
Red-kneed Dotterel (*Erythrogonys cinctus*)	Edges of swamps and lagoons, few are coastal		Inland swamps	Slater (1971), Frith (1976)
Wrybill (*Anarhynchus frontalis*)	Inland shingle on river beds	Coastal areas	Coastal areas, often at rivermouths	Falla *et al.* (1966)
Diademed Sandpiper Plover (*Phegornis mitchellii*)	Mountain streams and lakes, on sand spits	Local movement only	Mountain streams and lakes	Johnson and Goodall (1965)
Australian Dotterel (*Peltohyas australis*)	Arid places, open places, clay pans, gravel fields, gibber plains	Gibber plains, saltbush in desert (nomadic)	Flat open country	Maclean (1973, 1976)
Tawny-throated Dotterel (*Oreopholus ruficollis*)	Pampas grassland, sand dunes, coastal to 4000 m above sea level	Semiarid hill slopes and coastal zone	Upland grassland, coastal zone	Johnson and Goodall (1965), Humphrey *et al.* (1970), Myers and Myers (1979)
Jacanidae				
Lesser Jacana (*Microparra capensis*)	Quiet lagoons, lakes, and streams	NM	Near lakes, lagoons, streams	McLachlan and Liversidge (1978)
African Jacana (*Actophilornis africana*)	Freshwater marshes and lakes	Freshwater lagoons, weed-fringed lakes	Freshwater lagoons, weed-fringed lakes	Steyn (1973), McLachlan and Liversidge (1978)
Madagascar Jacana (*Actophilornis albinucha*)	Marshes of lakes and rivers		Marshes of lakes and rivers	Rand (1936)

Comb-crested Jacana (*Irediparra gallinacea*)	Floating vegetation on ponds		Freshwater ponds	Slater (1971)
Pheasant-tailed Jacana (*Hydrophasianus chirurgus*)	Swamps below 1500 m above sea level	Swamps (some NM)	Freshwater swamps at lower elevations	Ripley (1961), Hoogerwerf (1969), Ali (1969)
Bronze-winged Jacana (*Metopidius indicus*)	Freshwater marshes and swamps	NM	Marshes, swamps	Hoogerwerf (1969)
Northern Jacana (*Jacana spinosa*)	Ponds in premontane wet forest	Tidal mudflats	Tidal mudflats	Jenni and Betts (1978), Smith and Stiles (1979)
Wattled Jacana (*Jacana jacana*)	Freshwater marshes, littoral marshes		Freshwater marshes, littoral marshes	Myers and Myers (1979)
Rostratulidae				
Greater Painted Snipe (*Rostratula benghalensis*)	Marshy habitat, swamp edges with reeds	Sometimes nomadic	Reed marshes, inland mudflats	Muller (1975), Summers *et al.* (1977), McLachlan and Liversidge (1978), B. King (personal communication)
South American Painted Snipe (*Rostratula semicollaris*)	Freshwater marshes, some in estuaries, pampas marshes		Low-lying marsh habitats, estuaries	Myers and Myers (1979), Johnsgard (1981), J. Burger (unpublished communication)
Scolopacidae				
Spotted Redshank (*Tringa erythropus*)	Marshes, swamps, areas near Arctic timberline	Salt lakes, marshes, estuaries	Tidal mudflats, beaches, salt pans, river mudflats, marshes	Etchecopar and Hue (1967), Wolff (1969), Johnsgard (1981), B. King (personal communication)
Common Redshank (*Tringa totanus*)	Coastal salt marshes, river marshes, peat bogs, inland salt marshes	Coastal pastures, tidal mudflats	Tidal mudflats, coastal pastures and estuaries, rocky intertidal, inland mudflats	Bent (1927), Wolff (1969), Goss-Custard (1979), Dyrcz *et al.* (1981), Baker (1981), Evans (1981), B. King (personal communication)

(*Continued*)

Table IV. (*Continued*)

Species	Habitats			Sources
	Nesting	Migrating	Wintering	
Marsh Sandpiper (*Tringa stagnatilis*)	Freshwater marshes, steppes	Muddy steppe lakes	Freshwater marshes, some in tidal estuaries, salt pans	Summers *et al.* (1977), Johnsgard (1981), B. King (personal communication)
Common Greenshank (*Tringa nebularia*)	In marshes with nearby trees	Freshwater areas, marshes, saltwater habitats	Sandy beaches, tidal mudflats, estuaries, rocky intertidal, also inland lakes	Etchecopar and Hue (1967), Dement'ev *et al.* (1969), Wolff (1969), Summers *et al.* (1977), Pringle and Cooper (1977), Tree (1979)
Spotted Greenshank (*Tringa guttifer*)	Littoral swampy, larch forests and tidal lagoons (tree nests)		Coastal mudflats, salt pans	King and Dickinson (1975), Nechaev (1980), B. King (personal communication)
Greater Yellowlegs (*Tringa melanoleuca*)	Muskeg in bogs, pools, lakes	Inland lakes and ponds, tidal flats	Sandy beaches, mudflats, salt marshes, freshwater coastal marshes	Godfrey (1966), Recher (1966), Brooks (1967), Weir and Cooke (1976), Burger *et al.* (1977), Myers and Myers (1979)
Lesser Yellowlegs (*Tringa flavipes*)	Muskeg	Inland lakes and ponds, tidal flats, salt marshes	Coastal marshes, tidal mudflats, flooded fields (may be inland in South America)	Brooks (1967), Weir and Cooke (1976), Burger *et al.* (1977), Spaans (1978), Myers and Myers (1979), Baker (1979), J. R. Jehl, Jr. (personal communication)

Green Sandpiper (*Tringa ocrophus*)	Boreal forests, alder swamps, krummholtz zone	Sewage farms, marshes, river banks, salt marshes	Along rivers and streams, some in tidal flats, marshes	Dement'ev *et al.* (1969), McLachlan and Liversidge (1978), Johnsgard (1981), B. King (personal communication)
Solitary Sandpiper (*Tringa solitaria*)	Woodland pools, muskeg pools	Inland lakes and ponds, few on tidal mudflats, salt marshes	Brackish and freshwater ponds, rarely tidal areas	Urner and Storer (1949), Godfrey (1966), Brooks (1967), Oring (1973), Weir and Cooke (1976), Spaans (1978)
Wood Sandpiper (*Tringa glareola*)	Bogs, swampy heathland, marshes in coniferous forests, near coasts	Coastal backwaters	Coastal salt marshes, grassy marshes, streams, pools, rice fields	Bannerman (1961), Dement'ev *et al.* (1969), Ali (1969), Hoogerwerf (1969), Summers *et al.* (1977), McLachlan and Liversidge (1978)
Willet (*Catoptrophorus semipalmatus*)	Inland grassland and alkali flats, salt marshes (tidal), coral-sandy beaches	Along large lakes, tidal flats, estuaries	Tidal mudflats and sandflats, rocky intertidal	Vogt (1938), Recher (1966), Phelps (1975), Bradstreet *et al.* (1977), Burger and Shisler (1978), Spaans (1978), Page *et al.* (1979), Burger and Gochfeld (1980)
Terek Sandpiper (*Xenus cinereus*)	Marshy openings in coniferous forests, also scrub tundra, along seacoasts	Muddy beaches near mangroves	Estuaries and mudflats, lagoons, muddy beaches near mangroves	Hosking and Ferguson-Lees (1959), King and Dickinson (1975), Frith (1976), Summers *et al.* (1977), Johnsgard (1981)

(*Continued*)

Table IV. *(Continued)*

Species	Habitats: Nesting	Migrating	Wintering	Sources
Common Sandpiper (*Actitis hypoleucos*)	Seashores near riverbanks and streams	Sandy beaches, near pools in steppe marshes	Estuaries, sandy beaches, rocky intertidal salt marshes; inland salt flats	Bianki (1967), Wolff (1969), Pringle and Cooper (1977), Summers *et al*. (1977), B. King (personal communication)
Spotted Sandpiper (*Actitis macularia*)	Grasslands, uplands, upland on barrier beach islands, prairie lakes	Inland ponds, salt marshes, tidal flats	Coastal lakes and rivers, rocky intertidal, tidal mudflats, mangrove swamps	Urner and Storer (1949), Brooks (1967), Gochfeld (1971), Hays (1972), Spaans (1978), Myers *et al*. (1979), Maxson and Oring (1980)
Grey-tailed Tattler (*Heteroscelus brevipes*)	Alpine zone, forest tundra	Tidal mudflats	Coral atolls, sand beaches, estuaries, mudflats	Baker (1951), Falla *et al*. (1966), Frith (1976), Johnsgard (1981)
Wandering Tattler (*Heteroscelus incanus*)	Gravel bars, along mountain streams (also seen feeding pelagically)	Ocean islands and open oceans, tidal mud and sand flats	Mud and sand flats, rocky intertidal	Murie (1924), Recher (1966), Falla *et al*. (1966), Bartonek and Gibson (1972), Gerstenberg (1979)
Wilson's Phalarope (*Phalaropus tricolor*)	Shallow sloughs in prairie grassland	Alkaline lakes and salt ponds, inland lakes and ponds, tidal mudflats	High-altitude alkaline lakes, a few on tidal salt marshes and coastal ponds, pampas marshes	Hohn (1967), Brooks (1967), Burger and Howe (1975), Weir and Cooke (1976), Myers and Myers (1979), Gerstenberg (1979), J. R. Jehl, Jr. (personal communication)

Red-necked Phalarope (*Phalaropus lobatus*)	Arctic coastal plain, tundra grassland, forest tundra, coastal islands, sandy seashores	Alkaline lakes, inland lakes and ponds, estuaries, open oceans	Mostly open oceans, few on salt marshes, lagoons	Larson (1960), Recher (1966), Bianki (1967), Bartonek and Gibson (1972), Hilden and Vuolanto (1972), Weir and Cooke (1976), Baker (1977), Summers *et al.* (1977), Page *et al.* (1979), Myers and Pitelka (1980), J. R. Jehl, Jr. (personal communication)
Red Phalarope (*Phalaropus fulicarius*)	Tundra grass marshes, lakes, grassy fields, moss-sedge tundra, coastal islands (generally not far from sea, some seen pelagically)	Open oceans, few on lakes, littoral zones, sand beaches	Open ocean	Bent (1927), Larson (1960), Bartonek and Gibson (1972), Kistchinski (1975), Weir and Cooke (1976), Gerstenberg (1979), Meltofte (1979), Connors *et al.* (1979), Brown (1980), Ridley (1980), J. R. Jehl, Jr. (personal communication)
Tuamotu Sandpiper (*Prosobonia cancellata*)	Open shingle or gravel atolls, lagoons	NM		Greenway (1958)
White-winged Sandpiper (*Prosobonia leucoptera*)	Small brooks, river banks, fast-flowing streams	NM	Riverbanks, along fast-flowing streams	Holyoak (1974)

(Continued)

Table IV. *(Continued)*

Species	Habitats			Sources
	Nesting	Migrating	Wintering	
Upland Sandpiper (*Bartramia longicauda*)	Grasslands and prairies	Uplands and grassy places	Pampas, few on sandy beaches (tidal)	Guiguet (1962), Bradstreet *et al.* (1977), Spaans (1978), J. R. Jehl, Jr. (personal communication)
Little Curlew (*Numenius minutus*)	Burned clearings in larch forest, a timberline species, or along valleys of small rivers		Tidal mudflats, grasslands, grass on airstrips	King and Dickinson (1975), Johnsgard (1981), Labutin *et al.* (1982), King (personal communication)
Eskimo Curlew (*Numenius borealis*)	Marshy tundra	Along Mississippi River in spring, tidal areas in fall	Pampas (grasslands)	Johnson and Goodall (1965), Jehl and Smith (1970)
Whimbrel (*Numenius phaeopus*)	Marshy tundra	Inland lakes, salt marshes, tidal flats	Sandy beaches, rocky intertidal, tidal mudflats	Urner and Storer (1949), Recher (1966), Jehl and Smith (1970), Weir and Cooke (1976), Pringle and Cooper (1977), Myers *et al.* (1979)
Bristle-thighed Curlew (*Numenius tahitiensis*)	Flat tundra ridges	Transoceanic	Intertidal, algae-covered mudflats, sandflats of lagoons	Bakus (1967), Johnson (1979), Johnsgard (1981)
Slender-billed Curlew (*Numenius tenuirostris*)	Bogs or swamps on steppe, forest edge	Interior mudflats	Freshwater mudflats, tidal mudflats	Etchecopar and Hue (1967), Johnsgard (1981)
Eurasian Curlew (*Numenius arquata*)	Marshy moors, dry heathland, coastal pastures, estuaries	Lowland, coastal pastures, estuaries	Coastal pastures and estuaries, sandy beaches, mudflats, freshwater marshes	Wolff (1969), Evans (1981), Johnsgard (1981)

Eastern Curlew (*Numenius madagascariensis*)	Mossy marshes, reed grass thicket	Fields, sandy shores, meadows, estuaries	Tidal sandflats and mudflats, mud under mangroves, tidal estuaries	Dement'ev *et al.* (1969), Robertson and Dennison (1979), Johnsgard (1981)
Long-billed Curlew (*Numenius americanus*)	Short-grass prairies	Inland lakes, estuaries, tidal mudflats	Tidal sandflats and mudflats, salt marshes	Recher (1966), Graul (1971), Page and Whitacre (1975), Stenzel *et al.* (1976), J. R. Jehl, Jr. (personal communication)
Black-tailed Godwit (*Limosa limosa*)	Coastal marshes, river marshes and peat bogs, on heather	Freshwater-tidal marshes	Freshwater-tidal marshes, rocky intertidal, mudflats	Bent (1927), Wolff (1969), King and Dickinson (1975), Summers *et al.* (1977), Dyrcz *et al.* (1981)
Hudsonian Godwit (*Limosa haemastica*)	Grass tundra, and sedge meadows next to tidal or fluvial shores	Inland lakes, tidal areas and salt marshes, sandy beaches	Mudflats, salt marshes, freshwater and salt marshes	Urner and Storer (1949), Hagar (1966), Jehl and Smith (1970), Weir and Cooke (1976), Baker (1977), Myers and Myers (1979), J. R. Jehl, Jr. (personal communication)
Bar-tailed Godwit (*Limosa lapponica*)	Inland, Arctic plain	Tidal mudflats, sandflats	Sandy beaches, tidal mudflats	Wolff (1969), Gill and Jorgenson (1979), Myers and Pitelka (1980), Evans (1981)
Marbled Godwit (*Limosa fedoa*)	Inland grassland, prairie	Freshwater ponds, coastal areas, salt marshes	Tidal mudflats, sandy beach, salt marsh	Urner and Storer (1949), Godfrey (1966), Recher and Recher (1969), Gerstenberg (1979), Myers *et al.* (1979), Wishart and Sealy (1980)

(*Continued*)

Table IV. (*Continued*)

Species	Habitats			Sources
	Nesting	Migrating	Wintering	
Ruddy Turnstone (*Arenaria interpres*)	Tundra (drier areas), often nests near coast or on coastal islands (in marram grass)	Coastal tundra, inland lakes, salt marshes, littoral zones	Rocky intertidal, salt marshes, sandy and mud beaches, intertidal	Urner and Storer (1949), Larson (1960), Bianki (1967), Nettleship (1972), Bartonek and Gibson (1972), Weir and Cooke (1976), Beven and England (1977), Myers and Myers (1979), Connors *et al.* (1979), Meltofte (1979), Evans (1981), Baker (1981)
Black Turnstone (*Arenaria melanocephala*)	Rocky shores	Estuaries, tidal areas of rocky beaches	Rocky intertidal	Godfrey (1966), Page and Whitacre (1975), Page *et al.* (1979), Myers *et al.* (1979), J. R. Jehl, Jr. (personal communication)
Surfbird (*Aphriza virgata*)	Alpine tundra, rockslides	Tidal mudflats	Rocky intertidal	Recher (1966), Jehl (1968), Myers *et al.* (1979), Johnsgard (1981)
Red Knot (*Calidris canutus*)	Arctic tundra (dry areas), grassy fields, some coastal	Inland lakes (on sandy beaches), tidal areas	Tidal mudflats, sandy beaches	Weir and Cooke (1976), Meltofte (1979), Duffy *et al.* (1981), Evans (1981), J. R. Jehl, Jr. (personal communication)

Great Knot (*Calidris tenuirostris*)	Mountain tundra	Tidal flats	Tidal mudflats, seacoasts	Dement'ev *et al*. (1969), King and Dickinson (1975), Frith (1976)
Sanderling (*Calidris alba*)	Rocky or grassy inland tundra	Tidal mudflats	Tidal beaches, rocky intertidal, salt marshes	Meltofte (1979), Myers *et al*. (1979), Connors *et al*. (1981), Schneider and Harrington (1981), Evans (1981)
Semipalmated Sandpiper (*Calidris pusilla*)	Arctic tundra and littoral, Arctic plain	Inland lakes, coastal areas, tidal mudflats, salt marshes	Sandy beaches, tidal mudflats, salt marshes	Urner and Storer (1949), Weir and Cooke (1976), Baker (1977, 1979), Morrison and Harrington (1979), Connors *et al*. (1979), Schneider and Harrington (1981), Duffy *et al*. (1981)
Western Sandpiper (*Calidris mauri*)	Arctic tundra and littoral, Arctic marshes	Inland lakes, Arctic tundra and littoral, coastal mudflats	Sand beaches, salt marshes, tidal mudflats	Brooks (1967), Ashmole (1970), Holmes (1971), Weir and Cooke (1976), Myers *et al*. (1979), Connors *et al*. (1979), Myers and Pitelka (1980), Duffy *et al*. (1981)
Rufous-necked Sandpiper (*Calidris ruficollis*)	Dry areas of subarctic tundra	Tidal sand and mud flats	Tidal sand and mud flats, salt marshes, freshwater marshes, salt pans	Dement'ev *et al*. (1969), Thomas and Dartnall (1971), Portenko (1972), Frith (1976), B. King (personal communication)

(*Continued*)

Table IV. *(Continued)*

Species	Habitats			Sources
	Nesting	Migrating	Wintering	
Little Stint (*Calidris minuta*)	Mosses and sedges of subarctic tundra	Freshwater lakes, sandy intertidal beach	Sandy and rocky intertidal estuaries, salt pans	Bengtson and Svensson (1968), Fuchs (1975), Summers *et al.* (1977), B. King (personal communication)
Temminck's Stint (*Calidris temminckii*)	Boreal forest, forest tundra, coastal meadows and sandy beaches		Tidal mudbanks, mudflats, coastal swamps, freshwater swamps, riverflats	Etchecopar and Hue (1967), Hildén (1975, 1978), B. King (personal communication)
Long-toed Stint (*Calidris subminuta*)	Boggy areas of taiga		Tidal mudflats, paddy fields, dried-up swamps, marshes	Hoogerwerf (1969), Dement'ev *et al.* (1969), B. King (personal communication)
Least Sandpiper (*Calidris minutilla*)	Taiga, marshy plains, freshwater and brackish ponds, tidal ponds	Inland lakes and ponds, salt marshes, and tidal flats	Tidal mudflats, ponds near coast (freshwater and brackish)	Bent (1927), Urner and Storer (1949), Brooks (1967), Weir and Cooke (1976), Baker (1977, 1979), Myers *et al.* (1979)
White-rumped Sandpiper (*Calidris fuscicollis*)	Arctic coastal plain, wet tundra	Inland lakes, brackish ponds, tidal mudflats	Tidal mudflats, salt marshes, freshwater ponds and pools	Parmelee *et al.* (1968), Weir and Cooke (1976), Myers and Myers (1979), Myers and Pitelka (1980)

Baird's Sandpiper (*Calidris bairdii*)	Arctic tundra and littoral, Arctic plain	Arctic tundra, Arctic littoral, inland lakes, prairie, mudflats	Salt marshes, coastal freshwater marshes, tidal mudflats	Brooks (1967), Weir and Cooke (1976), Jehl (1979), Myers and Myers (1979), Connors *et al.* (1979), Myers and Pitelka (1980)
Pectoral Sandpiper (*Calidris melanotus*)	Arctic tundra fields, mainly along coast, on Arctic plain, on tide flats	Coastal tundra, inland lakes and ponds, salt marshes	Coastal freshwater marshes, rarely tidal beaches	Bent (1927), Brooks (1967), Weir and Cooke (1976), Connors *et al.* (1979), Myers and Myers (1979), Myers and Pitelka (1980), Burger *et al.* (1982)
Sharp-tailed Sandpiper (*Calidris acuminata*)	Wet tundra	Grassy marshes, flooded fields	Coastal upland tidal mudflats	Godfrey (1966), King and Dickinson (1975), Gerstenberg (1979), Johnsgard (1981)
Purple Sandpiper (*Calidris maritima*)	Wet grassy tundra, some nest just above high-tide mark, often feeds on rocky intertidal	Most rocky shores and jetties, few on inland lakes	Rocky beaches and jetties	Bent (1927), Urner and Storer (1949), Sutton and Parmelee (1955), Wolff (1969), Bengtson (1970), Weir and Cooke (1976), J. R. Jehl, Jr. (personal communication)
Rock Sandpiper (*Calidris ptilocnemis*)	Upland tundra in mosses and lichens, just behind beaches in sand and gravel	Rocky intertidal	Rocky intertidal	Hanna (1921), Gerstenberg (1979), Gill and Jorgenson (1979), Johnsgard (1981)

(*Continued*)

Table IV. *(Continued)*

Species	Habitats			Sources
	Nesting	Migrating	Wintering	
Dunlin (*Calidris alpina*)	Arctic tundra, wet grasslands, grassy tidelands along Arctic coast	Inland lakes, tundra, coastal areas, salt marshes	Tidal mudflats, coastal salt and freshwater marshes	Bent (1927), Urner and Storer (1949), Holmes (1966), Weir and Cooke (1976), Connors *et al.* (1979), Meltofte (1979), Baker (1979), Myers *et al.* (1979), Evans (1981)
Curlew Sandpiper (*Calidris ferruginea*)	Tundra	Freshwater ponds, tidal marshes and mudflats	Tidal marshes and mudflats, brackish lakes, inland mudflats	Bannerman and Bannerman (1968), Fuchs (1975), Elliott *et al.* (1976), Puttick (1979), B. King (personal communication)
Stilt Sandpiper (*Calidris himantopus*)	Sedge meadows of tundra, wet to dry grassy meadows	Inland lakes and ponds, brackish swamps	Tidal salt marshes, coastal freshwater marshes, brackish swamps	Brooks (1967), Jehl and Smith (1970), Jehl (1973b), Weir and Cooke (1976), Baker (1977), Spaans (1978), Myers and Myers (1979), Gerstenberg (1979)
Spoon-billed Sandpiper (*Eurynorhynchus pygmeus*)	Freshwater marshes and gravelly tundra	Tidal seacoast	Tidal mudflats	Bent (1927), Dement'ev *et al.* (1969), Portenko (1972), King and Dickinson (1975)

Broad-billed Sandpiper (*Limicola falcinellus*)	Marshes in coniferous forest, few at edge of tundra	Boggy lake shores, rice fields, seashore, salt marsh creeks	Estuaries, salt pans, mudflats, brackish lagoons	Nisbet (1961), Dement'ev *et al.* (1969), Ali and Ripley (1969), Summers *et al.* (1977), Johnsgard (1981), B. King (personal communication)
Buff-breasted Sandpiper (*Tryngites subruficollis*)	Arctic coastal plain and marshy tundra	Inland lakes and ponds, mudflats, salt marshes	Pampas marshes and grassland	Urner and Storer (1949), Weir and Cooke (1976), Myers (1980b), Myers and Pitelka (1980)
Ruff (*Philomachus pugnax*)	River marshes and peat bogs, maritime meadows, grassy lake shores	Coastal, maritime marshes (ludi in Russia)	Tidal and brackish marshes, sandy and rocky intertidal, inland mudflats	Bianki (1967), Wolff (1969), Shepard (1976), Summers *et al.* (1977), Prater (1979), Dyrcz *et al.* (1981), B. King (personal communication)
Short-billed Dowitcher (*Limnodromus griseus*)	Tundra marshes and fields, tundra grassland	Inland lakes, tidal mudflats, salt marshes	Tidal mudflats, sandy beaches, coastal marshes	Urner and Storer (1949), Baker (1977, 1979), Duffy *et al.* (1981), Schneider and Harrington (1981)
Long-billed Dowitcher (*Limnodromus scolopaceus*)	Arctic tundra and plains (near ponds)	Tundra, ponds, tidal mudflats, littoral coasts, salt marshes	Ponds, tidal mudflats	Urner and Storer (1949), Pitelka (1950), Burger *et al.* (1977), Connors *et al.* (1979), Myers and Pitelka (1980)

(Continued)

Table IV. *(Continued)*

Species	Habitats			Sources
	Nesting	Migrating	Wintering	
Asiatic Dowitcher (*Limnodromus semipalmatus*)	Marshes, damp meadows, grassy floodplains		Estuaries, tidal mudflats, rice paddies	Ripley (1961), Dement'ev *et al.* (1969), King and Dickinson (1975)
Subantarctic Snipe (*Coenocorypha aucklandica*)	Bush or shrub	NM	Bush and shrub	Falla *et al.* (1966)
Solitary Snipe (*Gallinago solitaria*)	Pebbly stream beds of glacier rivulets (alpine zone)	Forest peat bogs, mountain streams	Forest peat bogs, mountain rivers	Yamashina (1961), Dement'ev *et al.* (1969), King and Dickinson (1975)
Latham's Snipe (*Gallinago hardwickii*)	Grasslands, rice and vegetable farms, peat swamps	Coastal muddy fields, lake shores	Freshwater lagoons and swamps with heavy grass cover	Wolfe (1954), Yamashina (1961), Frith (1976)
Wood Snipe (*Gallinago nemoricola*)	Nests at timberline	Marshes and streams	Heavy cover (scrub or grass) along mountain streams, reed swamps, wooded marshes	Ripley (1961), Ali (1969), King and Dickinson (1975), Johnsgard (1981), B. King (personal communication)
Pin-tailed Snipe (*Gallinago stenura*)	Alpine tundra with stones and birch trees, grassy meadows	Marshy sites	Marshes and bogs	Hoogerwerf (1969), Dement'ev *et al.* (1969)
Swinhoe's Snipe (*Gallinago megala*)	Meadows bordered by forests, marshes, dry clearings in taiga	Marshy areas	Marshy areas, paddy fields	Dement'ev *et al.* (1969), King and Dickinson (1975)

African Snipe (*Gallinago nigripennis*)	Vleis or marshes with flooded short grass		Reed marshes	Summers *et al.* (1977), McLachlan and Liversidge (1978)
Madagascar Snipe (*Gallinago macrodactyla*)	Grass and sedge marshes, swamps			Rand (1936)
Great Snipe (*Gallinago media*)	Inland marshes and damp meadows, along lake and river shores	Marshes and swamps	Marshy areas	Etchecopar and Hue (1967), Lemnell (1978), Johnsgard (1981)
Common Snipe (*Gallinago gallinago*)	Peat bogs, muskeg, inland tundra and coastal plain, freshwater marshes, upland fields, very few in saltings and sand dunes	Freshwater marshes, inland pools, tidal areas and salt marshes	Estuaries, freshwater marshes, swamps, brackish areas, flooded fields	Urner and Storer (1949), Brooks (1967), Wolff (1969), Jehl and Smith (1970), Mason and Macdonald (1976), Gerstenberg (1979), Myers and Myers (1979), Myers and Pitelka (1980), Dyrcz *et al.* (1981), Burger *et al.* (1982)
Puna Snipe (*Gallinago andina*)	Along rivers in puna zone, marshes	NM	Marshes and swamps	Johnson and Goodall (1965), Johnsgard (1981)
Noble Snipe (*Gallinago nobilis*)	Grassy bogs and wet savannahs of paramo			Johnsgard (1981)
Giant Snipe (*Gallinago undulata*)	Marshy pastures and savannahs			Johnsgard (1981)
Cordilleran Snipe (*Gallinago stricklandii*)	Bogs and marshes in lowland grassland, in highland paramo marshes	NM	Bogs and marshes in lowland, grassland, paramo	Johnson and Goodall (1965), Humphrey *et al.* (1970), Johnsgard (1981)

(Continued)

Table IV. *(Continued)*

Species	Habitats: Nesting	Migrating	Wintering	Sources
Andean Snipe (*Gallinago jamesoni*)	Paramo zone, marshes	NM	Paramo zone, marshes	Chapman (1926), Phelps and Phelps (1958)
Imperial Snipe (*Gallinago imperialis*)	Andean timberline zone, moss of elfin forest	NM	Andean timberline zone, moss of elfin forest	Johnsgard (1981)
Jacksnipe (*Lymnocryptes minimus*)	Marshes, open mud swamps in coniferous forest, horsetail thickets	Grassy marshes along rivers	Marshes	Etchecopar and Hue (1967), Dement'ev *et al.* (1969), Johnsgard (1981)
Eurasian Woodcock (*Scolopax rusticola*)	Dense mixed forests of pine, fir, aspen, birch	Forests	Forests, but coastal when its cold	Bent (1927), Etchecopar and Hue (1967), King and Dickinson (1975)
Amami Woodcock (*Scolopax mira*)	Second-growth woods and orchards	NM	Second-growth woods and orchards	Yamashina (1961)
Javanese Woodcock (*Scolopax saturata*)	Montane forests, forest in alpine zone			Johnsgard (1981)
Celebes Woodcock (*Scolopax celebensis*)				
Obi Woodcock (*Scolopax rochussenii*)	High mature forests			Johnsgard (1981)
American Woodcock (*Scolopax minor*)	Soft substrates in young forests with scattered openings, Alder swamp thickets	Marshy areas	Soft substrates in young forests, coastal marshes	Bent (1927), Meanley (1975), Johnsgard (1981)

3. Pluvianellidae

This family contains only one species, which breeds inland near lakes, and migrates to coastal bays, rivers, and sandy beaches. Some individuals are sedentary. The Magellanic Plover breeds in southern South America (Jehl, 1975).

4. Pedionomidae

The Plains Wanderer, the only species in this family, is nonmigratory and resides on grasslands and open plains in Australia (Slater, 1971).

5. Thinocoridae

There are four species of seedsnipes and all breed in South America. Three species nest inland and only one (Least Seedsnipe) sometimes nests in coastal regions. The Least Seedsnipes often winter in coastal areas.

6. Burhinidae

There are nine species of thick-knees and stone curlews and they are generally nonmigratory. They are nearly cosmopolitan, except none nest in North America north of Mexico (see Gochfeld *et al.*, this volume). Most species nest in dry open fields, semideserts, or grassy fields, although some nest near rivers or beaches. They seem to prefer to nest near water when it is available.

7. Haematopodidae

Oystercatchers are cosmopolitan, and are the most marine shorebird family in that all eight species (depending upon your taxonomy) usually nest along the coasts in rocky or sandy habitats, and some even nest in salt marsh islands in bays (Burger, unpublished data) and on offshore islands (Harris, 1967; Hartwick, 1974), often in colonies of *Larus* gulls (Fig. 5). In New Jersey salt marshes, oystercatchers often nest on wrack with gulls (*L. atricilla*), terns (*Sterna hirundo*), and Black Skimmers (*Rynchops nigra*), and are often flooded out by late spring high tides. When nest loss due to flooding occurs early in the nesting cycle, oystercatchers will re-lay (Burger, unpublished data). For most species of oystercatchers, all sightings are in the littoral zone (Bent, 1929; Bianki, 1967; Myers and Myers, 1979; Myers, 1980a; Burger, unpublished data), but in England they have started to nest inland (Heppleston, 1972). Indeed, inland-breed-

Fig. 5. American Oystercatchers nest in *Spartina* salt marshes in the grass or on marsh wrack and feed along the sandy or on rocky intertidal areas (Atlantic coast of North America).

ing oystercatchers were more successful than coastal breeders in one study (Heppleston, 1972).

In England, considerable attention has been devoted to studying the feeding habits of Eurasian Oystercatchers because they feed extensively on cockles (*Cardium edule*) and mussels (*Mytilus edulis*) that are also used for food by man (Bent, 1929; Heppleston, 1972; Hulscher, 1976; Hartwick and Blaylock, 1979; Zwarts and Drent, 1981; Goss-Custard *et al.*, 1981). Some individuals do, however, feed in fields at high tide (Heppleston, 1971; Dare and Mercer, 1973).

8. Ibidorhynchidae

The Ibis-bill is a nonmigratory shorebird inhabiting mountainous riverbeds at elevations of over 1600 m in central Asia (Ali and Ripley, 1969; Dement'ev *et al.*, 1969).

9. Recurvirostridae

There are 13 species of stilts and avocets in the world, their distribution is cosmopolitan, and many are colonial (see Bent, 1927; Bryant, 1947, 1948; Bianki, 1967; Ali and Ripley, 1969; Dement'ev *et al.*, 1969; Hamilton, 1975). They generally nest in inland lakes and marshes although 5 of 13 species sometimes nest along the coast (Table IV). Over half of the species are nonmigratory, remaining on inland lakes and marshes, or on coastal marshes. Seven species sometimes or often occur along the coasts in the nonbreeding season although even then they tend to feed in freshwater marshes (refer to Table IV).

10. Glareolidae

There are 16 species of coursers and pratincoles, all limited in range to the Old World. Fourteen species always breed in inland areas in dry country (coursers) or along streams, marshes, or other wet habitats (refer to Table IV). Pratincoles are particularly likely to nest in small colonies along stream beds (Smythies, 1953; Dement'ev *et al.*, 1969) and even on coastal estuaries (e.g., Collared Pratincole, J. Burger and M. Gochfeld, unpublished data), where they hawk insects over the water (see Cheke, 1980). Most species are nonmigratory, remaining in inland habitats throughout the nonbreeding season. Of all the shorebird families, there is generally the least information available on the breeding behavior and habitat selection of coursers and pratincoles. Descriptions of habitats are limited, and little is published on each species. Apparently, pratincoles change colony sites frequently (Dement'ev *et al.*, 1969), presumably an adaptation to nesting in unstable habitats such as riverbeds, riverbanks, or sand pits, which can disappear from one year to the next (see McNicholl, 1975).

11. Charadriidae

There are 64 species of lapwings, plovers, and dotterels, and they are truly cosmopolitan, some species breeding on every continent (Gochfeld *et al.*, this volume). Over half (34 or 64) of the species always nest inland, and only 21% predominantly nest along the coasts. Inland species primarily nest in grasslands near rivers, marshes, or swamps. A few inland species (such as Yellow-wattled Lapwing nest in arid wastelands, and a number of species nest in burned-over areas (Senegal Plover, Black-winged Plover, Crowned Plover; refer to Table IV). Some species (such as Killdeer, Banded Plover, "Lesser Golden Plover"; see Table IV) nest both inland and along the coasts, but in selecting coastal habitats they often nest in similar areas (tundra marshes, dry upland), whereas other species nest in estuarine habitats such as salt marshes or sand dunes (Northern Lapwing). Northern Lapwing will also nest on offshore coastal islands with gulls (*Larus*, Harris, 1967). Still other species nest predominantly in marine habitats including sandy or muddy beaches, rocky beaches and dunes, or sand flats (Ringed Plover, Thick-billed Plover, Piping Plover, Kittlitz's Plover, Red-capped Dotterel, Snowy Plover, White-fronted Sandplover, Peruvian Sandplover, Malaysian Sandplover, Collared Plover, Double-banded Dotterel, Red-breasted Dotterel, Hooded Dotterel; see Table IV). There are significant differences among genera in the occurrence of strictly marine species. All of the species in the family Charadriidae that are predominantly marine in the breeding season are in the genus *Charadrius*, and not in either *Pluvialis* or *Vanellus* (refer to Table IV; 3 ts 2 contingency table, $\chi^2 = 14.4$, $df = 2$, $p < 0.001$). Indeed, some of these species nest in precarious situations, where nests are exposed to high tides. Bianki (1967) reported that 47% (of 40) of the nests of Ringed Plover were washed away by tides in Kandalaksha Bay, Russia. Snowy Plover (Bianki, 1967) and Piping Plover (J. Burger, unpublished data) also lose some nests to high tides.

Only 58% of the shorebirds in Charadriidae migrate (Tables IV, VI), and those that migrate predominantly go to coastal areas (Tables IV, V). There are 24 species in the genus *Vanellus*, and 72% (of the 18 species for which information is available) are nonmigratory, whereas none of *Pluvialis* are nonmigratory, and only 31% of the *Charadrius* species (of

Table V. Summary of Nesting and Wintering Habitats of Shorebirds

Family	Number of species	Breeding season			Nonbreeding season		
		Inland[a]	Both	Coastal[b]	Inland[a]	Both	Coastal[b]
Dromadidae	1			1			1
Chionididae	2			2			2
Pluvianellidae	1	1					1
Pedionomidae	1	1			1	2	
Thinocoridae	4	3	1		2		
Burhinidae	9	5	3	1	5	2	2
Haematopodidae	8		2	6			8
Ibidorhynchidae	1	1			1		
Recurvirostridae	13	8	5		3	3	7
Glareolidae	16[c]	14	2		12	2	
Charadriidae	64[c]	34	17	13	21	13	26
Jacanidae	8	7	1		6	1	1
Rostratulidae	2	1	1		1	1	
Scolopacidae	86[c]	56	23	6	17	12	50
Total Number	216	131	55	29	69	36	98
Percent		61	25	14	34	18	48

[a] Most individuals of that species nest inland.
[b] Most individuals of that species nest on coast.
[c] Information not known for all species.

30 for which information is known) are nonmigratory. These differences are significant (2 × 2 contingency table, $\chi^2 = 9.09$, $df = 1$, $p < 0.02$).

Similar differences exist among genera with respect to wintering habitats: 75% of *Vanellus* species are inland in the winter and 25% utilize both inland and coastal habitats; whereas for *Pluvialis* and *Charadrius*, 13% are inland in winter, 21% are both inland and coastal, and 66% are predominantly coastal (2 × 3 contingency table, $\chi^2 = 25.3$, $df = 2$, $p < 0.001$).

Thus, in summary, less than half of the Charadriidae species sometimes or always nest along the coasts, 42% are nonmigratory, and 66% sometimes or always winter along the coasts.

12. Jacanidae

Of the eight species of jacanas, seven always nest inland and only one species (Wattled Jacana) sometimes nests in coastal areas (Table IV). Jacanas feed from vegetation on the surface or edge of the water (Jenni and Betts, 1978). Most species are nonmigratory, nomadic, or migrate

Table VI. Summary of Use of Coastal Areas by Shorebirds

Family	Number of species	Percent sometimes or always nest on coast	Percent nonmigratory[a]	Percent sometimes or always winter on coast
Dromadidae	1	100	0	100
Chionididae	2	100	50[b]	100
Pluvianellidae	1	0	100	0
Thinocoridae	4	25	25[b]	50
Pedionomidae	1	0	100	0
Burhinidae	9	44	100[a]	44
Haematopodidae	8	100	57	100
Ibidorhynchidae	1	0	100	0
Recurvirostridae	13	42[a]	54	82[a]
Glareolidae	15	13	83[a]	14[a]
Charadriidae	64	47	42[a]	65[a]
Jacanidae	8	13	40[a]	25
Rostratulidae	2	50	100[a]	50
Scolopacidae	86	34[a]	11[a,c]	78[a]
Total	216	84	62[d]	135
Percent		39	34	66

[a] Of species for which information is known.
[b] Some individuals of a species are migratory, some not.
[c] Six of eight nonmigratory scolopacids are snipe or woodcock.
[d] Information available on 176 species.

only short distances, and their breeding ranges are limited to Central and South America, Asia, and Africa. In the nonbreeding seasons, jacanas primarily remain on freshwater marshes and lagoons, although Northern and Wattled Jacanas also use tidal mudflats or marshes (refer Table IV).

13. Rostratulidae

The two species in this family, restricted in range to southern South America, Africa, Asia, and Australia, predominantly inhabit freshwater and swamps although some South American Painted Snipe nest in estuaries (see Table IV). Both species are nonmigratory, although some individuals of the Greater Painted Snipe are nomadic (Table IV), and regularly visit coastal marshes (M. Gochfeld, personal communication).

14. Scolopacidae

Forty percent of the shorebirds in the world belong to this cosmopolitan family (refer to Gochfeld *et al.*, this volume, for ranges). The major groups in the family are godwits (*Limosa*), curlews (*Numenius*), phalaropes (*Phalaropus*), *Tringa* sandpipers, *Calidris* sandpipers, and the snipes (*Gallinago*), dowitchers (*Limnodromus*), and woodcocks (*Scolopax*). Over 65% of the species always nest inland, in habitats ranging from wet marshes, muskegs, and salt lakes to soft substrate in forests (some snipes and woodcocks; Table IV, Fig. 6). Only 13% of the species usually nest along the coast (Table VI). There are significant differences among the genera with respect to nesting habitats: more *Calidris* sandpipers nest along the coasts and more snipes nest inland than expected on the basis of the family characteristics (Table VII). Common Snipe will sometimes nest along the coast; Mason and Macdonald (1976) found that 6% of 563 nests were located in saltings and on sand dunes.

Only 11% of the Scolopacidae are nonmigratory (Table VI). But again, there are significant generic differences: more *Calidris* sandpipers migrate, and fewer snipes migrate than expected (Table VII). Scolopacids often migrate at considerable distances offshore, McNeil and Burton (1973) reported that Knot, White-rumped Sandpiper, Least Sandpiper, Semipalmated Sandpiper, and Short-billed Dowitchers migrate mainly by offshore routes to reach South American wintering grounds, although most go to their breeding grounds across the Caribbean Sea and through the interior United States. They banded 1912 shorebirds (of 21 species) in a 2-year period on Sable Island 160 km east of the Nova Scotia coast (McNeil and Burton, 1977). Similar offshore migratory routes are known for phalaropes (Brown, 1980; K. Powers, personal communication; Table I).

Habitat selection in the nonbreeding season is particularly important for scolopacids, as many of the sandpipers nesting in northern tundra areas spend only 2 to 3 months on the tundra (Fig. 7). Weather constraints (and thus food constraints) make it difficult to arrive until June (see Pitelka *et al.*, 1974; Pitelka, 1979). Connors *et al.* (1979) noted that in Alaska, most shorebirds (20 residents), except Red Phalaropes, used mostly tundra until late July when they switched to littoral habitats. Both Red Phalaropes and turnstones used littoral areas significantly in July. Several authors have noted that shorebirds spend 8 months or more away from the breeding grounds (MacLean and Holmes, 1971; Elliott *et al.*, 1976; Kelly and Cogswell, 1979). Many species spent more than two-thirds of the year at a wintering area in San Francisco Bay (Kelly and Cogswell, 1979). Kelly and Cogswell (1979) marked birds, and found that the mean interval of absence from San Francisco Bay was 137 days ($N = 9$, range 92–181) for Marbled Godwit and 91 days ($N = 14$, range 70–124) for Willet. Such data can be gathered because shorebirds show fidelity to

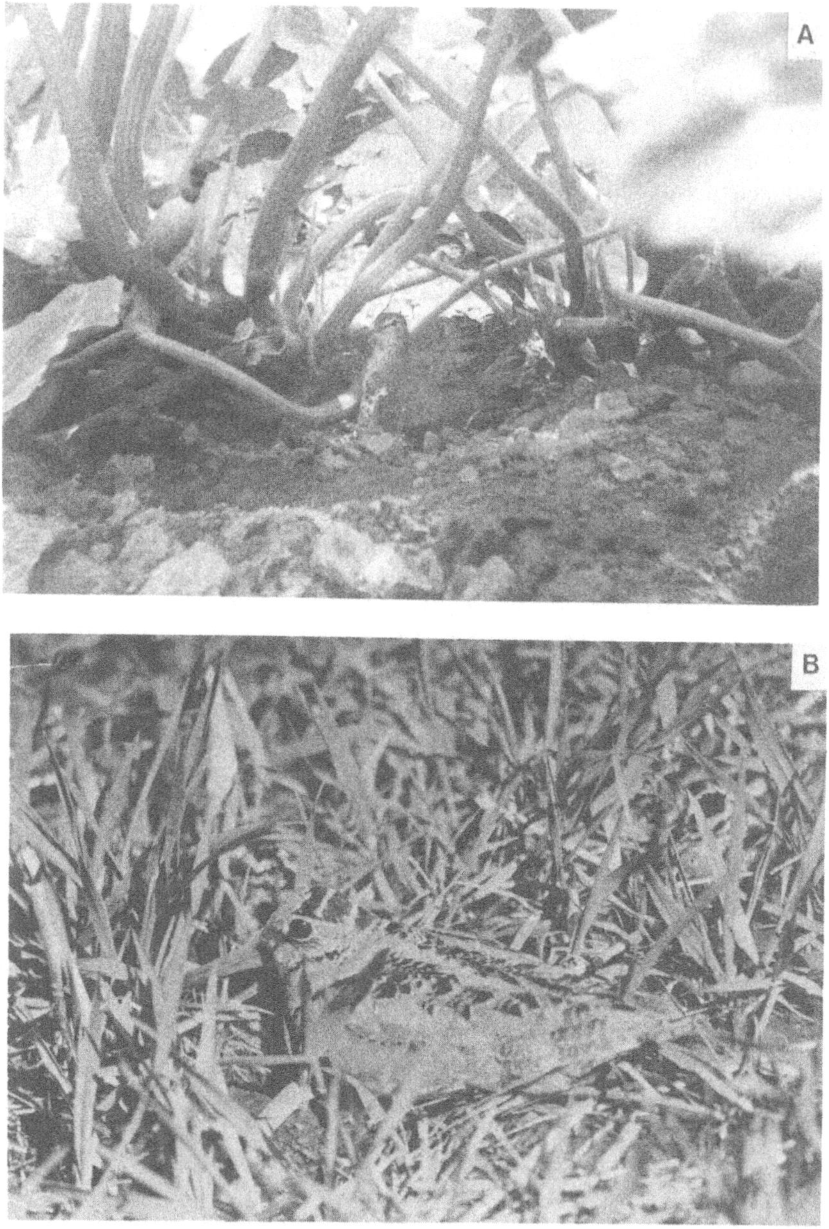

Fig. 6. Spotted Sandpiper (A) nest inland in a variety of habitats (Niskayana, N.Y.) as well as nesting in coastal areas. Similarly, American Woodcock (B) generally nest away from coasts in woods or fields (Buffalo, N.Y.).

Table VII. Characteristics of Major Groups of Scolopacidae

	Number of species	Percent always nest inland	Percent migratory	Percent coastal when nonbreeding[a]
Tringa sandpiper	10	70 (7)[b]	100 (10)	100 (10)
Phalaropes	3	33 (1)	100 (3)	100 (3)[c]
Curlews	8	88 (7)	100 (8)	88 (7)
Godwits	4	75 (3)	100 (4)	100 (4)
Calidris sandpipers	19	42 (8)	100 (19)	100 (19)
Dowitchers	3	100 (3)	100 (3)	100 (3)
Snipes	17	94 (16)	62 (8)[d]	7 (1)[e]
Woodcocks	6	100 (5)[f]	66 (2)[g]	66 (2)[h]
χ^2		24.9[i]	22.2	51.5
p		0.005	0.005	0.005

[a] Always or sometimes coastal.
[b] Number of species in parentheses.
[c] Most are pelagic.
[d] Information only available for 13 species.
[e] Information only available for 14 species.
[f] Information only available for 5 species.
[g] Information only available for 3 species.
[h] Information only available for 3 species.
[i] χ^2 values from 8 × 2 contingency tables, $df = 7$. This table includes only 70 of the 86 species; not included are genera with only one or two species.

specific wintering sites as well as breeding places (Kelly and Cogswell, 1979; Smith and Stiles, 1979; Pitelka, 1979).

Although 78% of scolopacids sometimes or always use coasts in the nonbreeding season (Table VI), there are significant differences among genera (Table VII): *Calidris* sandpipers use coastal habitats more, and snipes use coastal habitats less than expected on the basis of general family characteristics. In general, the three aspects (breeding habitat, migratory behavior, wintering habitat) are related. For example, *Calidris* sandpipers are generally Arctic nesters, many nest along the coast, and they migrate to predominantly tidal habitats; whereas snipes are inland nesters in marshes or woods in more temperate areas, many are non-migratory, and they generally remain in inland habitats.

The phalaropes deserve special mention as they are the most pelagic shorebirds during the nonbreeding season, and many nest near the coasts in tundra areas (refer to Table IV). As K. Powers (personal communication; Table I) has indicated, large numbers (he estimated over 700,000) occur in the shelf waters during spring migration. Brown (1980) has extensively studied the nonbreeding distribution of Red Phalaropes, particularly in response to plankton concentrations. He reported a direct re-

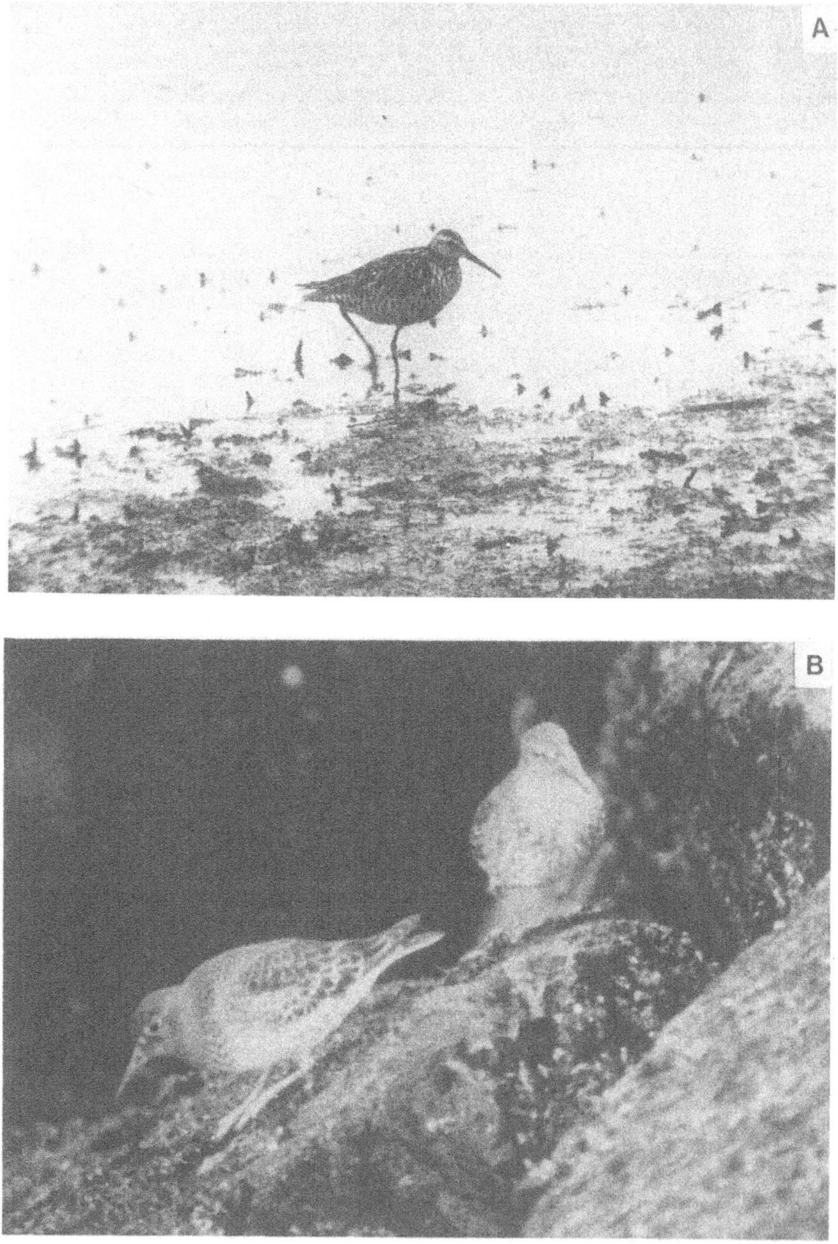

Fig. 7. Migrant Stilt Sandpiper (A) (Long Island, N.Y.) and Purple Sandpiper (B) (Long Island, N.Y.) forage in coastal areas. [Purple Sandpiper photo by M. Gochfeld.]

lationship between the number of Red Phalaropes and plankton dry weights at Brier Island, Nova Scotia; and he found a similar relationship between Northern Phalaropes and plankton off Deer Island, Nova Scotia (Brown, 1980). Brown reported that Red Phalaropes spend 75% of the year at sea, whereas some marine mammals spend only 50% of the year at sea. He clearly has shown that Red Phalaropes are associated with fronts, upwellings, and plankton numbers, thus they respond to localized marine habitats, and as such are clearly part of the marine ecosystem (Brown, 1980).

B. Comparisons and Discussion

Shorebirds are cosmopolitan, nesting on all continents, and in many habitats from dry desert areas (thick-knees and coursers) to wet marshes (jacanas). Similarly, they nest in temperate, tropical, and tundra biomes. In comparing the habitat selection of shorebirds, there are differences among families in breeding and wintering habitats, and in the degree they are migratory (refer to Tables V and VI). Some families are entirely coastal (Dromadidae, Chionididae, Haematopodidae), others are entirely inland nesters (Pluvianellidae, Pedionomidae, Ibidorhynchidae). The two largest families generally nest inland: Charadriidae (53%) and Scolopacidae (66%). Similar differences exist in winter habitat selection: all species of Dromadidae, Chionididae, Pluvianellidae, and Haematopodidae always or sometimes winter along the coast, while all species of Pedionomidae and Ibidorhynchidae never occur along the coasts (refer to Table VI). Shorebirds can be divided into two suborders that seem to have evolved as separate lines (J. R. Jehl, Jr., personal communication). There are differences in habitat selection between the two groups: there are significantly more coastal species during the breeding season in Charadriomorpha than Scolopaci; and there are significantly more coastal species of Scolopaci than Charadrii during the nonbreeding season (Table VIII). This difference occurs because more charadriomorphs are nonmigratory (41%), while 85% of scolopacids are migratory.

Shorebirds in general seem to prefer wet or moist habitats for nesting whether they nest inland or along the coast. Many species nest in marshes or along rivers and lakes or bodies of water regardless of their location with respect to coasts. Even seedsnipes that nest in dry country nest near water when possible. Further, many species nest near or in salt water, whether it be along the coasts, or in salt lakes high in the Andes (e.g., some avocets). Overall, 39% of all species of shorebirds sometimes or always nest along the coast, and 66% always or sometimes winter along

Table VIII. Summary of Habitats by Suborders

	Suborder	
	Charadriomorpha	Scolopaci
Number of families	11	3
Number of species	120	96
Breeding season		
Inland	67 (56%)	64 (68%)
Both	30 (25%)	25 (26%)
Coastal[a]	23 (19%)	6 (6%)
Nonbreeding season (winter)		
Inland	45 (39%)	24 (27%)
Both	23 (20%)	14 (16%)
Coastal[b]	47 (41%)	51 (57%)
Nonmigratory[c]	49 (41%)	12 (15%)

[a] There are significantly more strictly coastal species in Charadriomorpha than Scolopaci ($\chi^2 = 6.85$, $df = 1$, $p < 0.01$).
[b] There are significantly more strictly coastal species in Scolopaci than Charadriomorpha ($\chi^2 = 5.17$, $df = 1$, $p < 0.02$).
[c] Significantly more charadriomorphs are nonmigratory than are scolopacids ($\chi^2 = 26.10$, $df = 1$, $p < 0.001$).

the coast (Table VI). Thus, it is clear that an important component of shorebird species use the coasts for nesting and wintering. Coastal areas are more important as wintering areas than as nesting areas, but as shorebirds spend 60% or more of the year on wintering grounds, coastal areas are exceedingly important to their life histories.

ACKNOWLEDGMENTS

I thank M. Gochfeld, J. R. Jehl, Jr., and B. G. Murray, Jr. for valuable comments on the manuscript. Several people contributed information or references for the data in Table IV, and I would like to thank them: P. A. Buckley, M. Gochfeld, J. R. Jehl, Jr., B. King, M. Kleinbaum, and G. Tudor. I thank J. Sherwood for typing the manuscript, and D. J. Gochfeld for help with the literature cited.

REFERENCES

Ade, B., 1979, Some observations on the breeding of Crowned Plovers (1977), *Bokmakierie* 31:9–16.

Ali, S., 1969, *Birds of Kerala*, pp. 102–133, Oxford University Press, London.
Ali, S., and Ripley, S. D., 1969, *Handbook of the Birds of India and Pakistan*, Vol. 3, Oxford University Press, London.
Ashmole, M. J., 1970, Feeding of Western and Semipalmated Sandpipers in Peruvian winter quarters, *Auk* **87**:131–135.
Baker, J. M., 1981, Winter feeding rates of Redshank *Tringa totanus* and turnstone *Arenaria interpres* on a rocky shore, *Ibis* **123**:85–87.
Baker, M. C., 1977, Shorebird food habits in the eastern Canadian Arctic, *Condor* **79**:56–62.
Baker, M. C., 1979, Morphological correlates of habitat selection in a community of shorebirds (*Charadriiformes*), *Oikos* **33**:121–126.
Baker, R. H., 1951, The avifauna of Micronesia, its origin, evolution and distribution, *Univ. Kans. Publ.* **3**:1–359.
Bakus, G. J., 1967, Changes in the avifauna of Fanning Island, Central Pacific, between 1924 and 1963, *Condor* **69**:207–209.
Bannerman, D. A., 1958, *Larger Birds of West Africa*, Penguin, Baltimore.
Bannerman, D. A., 1961, *The Birds of the British Isles*, Vol. 10, Oliver & Boyd, Edinburgh.
Bannerman, D. A., and Bannerman, W. M., 1968, *History of the Birds of the Cape Verde Islands*, Oliver & Boyd, Edinburgh.
Barlow, M. L., Muller, P. M., and Sutton, R. R., 1972, Breeding data on the Spur-winged Plover in Southland, New Zealand, *Notornis* **19**:212–249.
Bartonek, J. C., and Gibson, D. D., 1972, Summer distribution of pelagic birds in Bristol Bay, Alaska, *Condor* **74**:416–422.
Bengtson, S., 1970, Breeding behaviour of the Purple Sandpiper *Calidris maritima* in west Spitsbergen, *Ornis Scand.* **1**:17–25.
Bengston, S. A., and Svensson, B., 1968, Feeding habits of *Calidris alpina* L. and *C. minuta* Leisl. (Aves) in relation to the distribution of marine shore invertebrates, *Oikos* **19**:152–157.
Benson, C. W., Brooke, R. K., Dowsett, R. J., and Irwin, M. P. S., 1971, *The Brids of Zambia*, Collins, Glasgow.
Bent, A. C., 1927, Life histories of North American shorebirds, Part 1, *U.S. Natl. Mus. Bull.* **142.**
Bent, A. C., 1929, Life histories of North American shorebirds, Part 2, *U.S. Natl. Mus. Bull.* **146.**
Beven, G., and England, M. D., 1977, Studies of less familiar birds, 181. Turnstone, *Br. Birds* **70**:23–32.
Bianki, V. V., 1967, Gulls, shorebirds and alcids of Kandalaksha Bay, *Proc. Kandalaksha State Res.* **6**:1–247.
Bourne, W. R. P., 1970, The seabirds of the Firth Clyde in September 1970, *Seabird Rep.* **1970**:26–32.
Bradstreet, M. S. W., Page, G. W., and Johnston, W. G., 1977, Shorebirds at Long Point, Lake Eries, 1966–1971: Seasonal occurrence, habitat preference and variation in abundance, *Can. Field Nat.* **91**:225–236.
Britton, P. L. (ed.), 1980, *Birds of East Africa, Their Habitats, Status and Distribution*, Nairobi East African Natural History Society, Nairobi.
Brooks, W. S., 1967, Food and feeding habits of autumn migrant shorebirds at a small midwestern pond, *Wilson Bull.* **79**:307–315.
Brown, R. G. B., 1980, Seabirds as marine animals, in: *Behavior of Marine Animals*, Vol. 4 (J. Burger, B. L. Olla, and H. E. Winn, eds.), pp. 1–40, Plenum Press, New York.
Bryant, C. E., 1947, Notes on Avocets breeding near Melbourne, *Emu* **46**:241–245.

Bryant, C. E., 1948, More observations on nesting avocets, *Emu* **48**:89–92.

Burger, A. E., 1979, Breeding biology, moult and survival of Lesser Sheathbill *Chionis minor* at Marion Island, *Ardea* **67**:1–14.

Burger, J., and Gochfeld, M., 1980, Opportunistic scavenging by shorebirds: Feeding behavior and aggression, *J. Field Ornithol.* **51**:373–375.

Burger, J., and Howe, M., 1975, Notes on winter feeding behavior and molt in Wilson's Phalaropes, *Auk* **92**:442–451.

Burger, J., and Shisler, J., 1978, Nest-site selection of Willets in a New Jersey salt marsh, *Wilson Bull.* **90**:599–607.

Burger, J., Howe, M. A., Hahn, D. C., and Chase, J., 1977, Effects of tide cycles on habitat selection and habitat partitioning by migrating shorebirds, *Auk* **94**:743–758.

Burger, J., Shisler, J. K., and Lesser, F. H., 1982, Avian utilization on six salt marshes in New Jersey, *Biol. Conserv.* **23**:187–212.

Chapman, F. M., 1926, The distribution of bird-life in Ecuador, *Bull. Am. Mus. Nat. Hist.* **55**.

Cheke, R. A., 1980, A small breeding colony of the Rock Pratincole *Glareola nuchalis liberiae* in Togo, *Bull. Br. Ornithol. Club* **100**:175–178.

Cherrie, G. K., 1916, A contribution to the ornithology of the Orinoco region, *Sci. Bull. Brooklyn Mus.* **2**:133–374.

Cline, D. R., Siniff, D. B., and Erickson, A. W., 1969, Summer birds of the pack ice in the Weddell Sea, Antarctica, *Auk* **86**:701–716.

Connors, P. G., Myers, J. P., and Pitelka, F. A., 1979, Seasonal habitat use by Arctic Alaskan shorebirds, in: *Studies in Avian Biology No. 2* (F. A. Pitelka, ed.), pp. 101–111, Cooper Ornithological Society, Allen Press, Lawrence, Kans.

Connors, P. G., Myers, J. P., Connors, C. S. W., and Pitelka, F. A., 1981, Interhabitat movements by Sanderlings in relation to foraging profitability and the tidal cycle, *Auk* **98**:49–64.

Dare, P. J., 1981, Notes on seabirds attending a commercial trawler fishing in shelf waters off Ireland in Summer, *Seabird Rep.* **1977–1981**:110–114.

Dare, P. J., and Mercer, A. J., 1973, Foods of the oystercatcher in Morecambe Bay, Lancashire, *Bird Study* **20**:173–184.

de Bie, S., and Zijlstra, M., 1979, Some remarks on the behaviour of the avocet (*Recurvirostra avosetta* L.) in relation to different breeding places, *Ardea* **67**:68–69.

Dement'ev, G. P., Gladkov, N. A., and Spangenberg, E. P., 1969, *The Birds of the Soviet Union,* Vol. 5, Israel Program for Scientific Translations, Jerusalem.

Dick, W. J. A., Pienkowski, M. W., Waltner, M., and Minton, C. D. T., 1976, Distribution and geographical origins of Knot *Calidris canutus* wintering in Europe and Africa, *Ardea* **64**:22–47.

Drury, W. H., Jr., 1961, The breeding biology of shorebirds on Bylot Island, Northwest Territories, Canada, *Auk* **78**:176–219.

Duffy, D. C., Atkins, N., and Schneider, D. C., 1981, Do shorebirds compete on their wintering grounds?, *Auk* **98**:215–229.

Dyrcz, A., Witkowski, J., and Okulewicz, J., 1981, Nesting of "timid" waders in the vicinity of "bold" ones as an antipredation adaptation, *Ibis* **123**:542–545.

Elliott, C. C. H., Waltner, M., Underhill, L. G., Pringle, J. S., and Dick, W. J. A., 1976, The migration system of the Curlew Sandpiper *Calidris ferruginea* in Africa, *Ostrich* **47**:191–213.

Etchecopar, R. D., and Hue, F., 1967, *The Birds of North Africa,* Oliver & Boyd, Edinburgh.

Evans, P. R., 1976, Energy balance and optimal foraging strategies in shorebirds: Some implications for their distribution and movement in the non-breeding season, *Ardea* **64:**117–139.

Evans, P. R., 1981, Migration and dispersal of shorebirds as a survival strategy, in: *Feeding and Survival Strategies of Estuarine Organisms* (N. V. Jones and W. J. Wolff, eds.), pp. 275–290, Plenum Press, New York.

Falla, R. A., Sibson, R. B., and Turbott, E. G., 1966, *A Field Guide to the Birds of New Zealand,* Houghton Mifflin, Boston.

Frith, H. J., 1976, *Reader's Digest Complete Book of Australian Birds,* Reader's Digest, Sydney.

Fuchs, E., 1975, Observations sur les resources alimentaires et à l'alimentation des becasseaux variable, minute et cocroli *Calidris alpina, minuta* et *ferruginea* en Mediterranée, au passage et pendant l'hivernage, *Alauda Rev. Int. Ornithol.* **43:**55–69.

Gerstenberg, R. H., 1979, Habitat utilization by wintering and migrating shorebirds on Humboldt Bay, California, in: *Studies in Avian Biology No. 2* (F. A. Pitelka, ed.), pp. 33–40, Cooper Ornithological Society, Allen Press, Lawrence, Kans.

Gill, R., Jr., and Jorgensen, P. D., 1979, A preliminary assessment of timing and migration of shorebirds along the northcentral Alaska peninsula, in: *Studies in Avian Biology No. 2* (F. A. Pitelka, ed.), pp. 113–123, Cooper Ornithological Society, Allen Press, Lawrence, Kans.

Gochfeld, M., 1971, Notes on a nocturnal roost of Spotted Sandpipers in Trinidad, West Indies, *Auk* **88:**167–168.

Godfrey, W. E., 1966, The birds of Canada, *National Museum of Canada, Bull. No.* **203:**1–428.

Goss-Custard, J. D., 1979, The energetics of foraging Redshank, *Tringa totanus,* in: *Studies in Avian Biology No. 2* (F. A. Pitelka, ed.), pp. 247–257, Cooper Ornithological Society, Allen Press, Lawrence, Kans.

Goss-Custard, J. D., 1980, Competition for food and interference among waders, *Ardea* **68:**31–52.

Goss-Custard, J. D., Jones, R. E., and Newbery, P. E., 1977, The ecology of the Wash. I. Distribution and diet of wading birds (Charadrii), *J. Appl. Ecol.* **14:**681–700.

Goss-Custard, J. D., Durell, S. E. A., McGrorty, S., Reading, C. J., and Clarke, R. T., 1981, Factors affecting the occupation of mussel (*Mytilus edulis*) beds by oystercatchers (*Haematopus ostralegus*) on the Exe Estuary, Devon, in: *Feeding and Survival Strategies of Estaurine Organisms* (N. V. Jones and W. J. Wolff, eds.), pp. 217–230, Plenum Press, New York.

Graul, W. D., 1971, Observations at a Long-billed Curlew nest, *Auk* **88:**182–184.

Graul, W. D., 1973, Adaptive aspects of the Mountain Plover social system, *Living bird* **12:**69–94.

Greenway, J. C., 1958, *Extinct and Vanishing Birds of the World,* American Committee for International Wild Life Protection, New York.

Guiguet, C. J., 1962, *The Birds of British Columbia,* (3) *The Shorebirds,* British Columbia Prov. Mus. Handbook No. 8, Victoria, British Columbia.

Hagar, J. A., 1966, Nesting of the Hudsonian Godwit at Churchill, Manitoba, *Living Bird* **6:**5–43.

Hall, K. R. L., 1959, Nest records and additional behavior for Kittlitz's Sandplover *Charadrius pecuarias* in the S. W. Cape Province, *Ostrich* **30:**33–38.

Hamilton, R. B., 1975, *Comparative Behavior of the American Avocet and the Black-necked Stilt* (*Recurvirostridae*), A.O.U. Monograph No. 17, Allen Press, Lawrence, Kans.

Hanna, G. O., 1921, The Pribilof Sandpiper, *Condor* **23:**50–57.

Harris, M. P., 1967, The biology of oystercatchers *Haematopus ostralegus* on Skokholm Island, S. Wales, *Ibis* **109**:180–193.
Harrison, G. S., 1979, The association of marine birds and feeding Gray Whales, *Condor* **81**:93–95.
Hartwick, E. B., 1974, Breeding ecology of the Black Oystercatcher (*Haematopus bachmani* Audubon), *Syesis* **7**:83–92.
Hartwick, E. B., and Blaylock, W., 1979, Winter ecology of a Black Oystercatcher population, in: *Studies in Avian Biology No. 2* (F. A. Pitelka, ed.), pp. 207–215, Cooper Ornithological Society, Allen Press, Lawrence, Kans.
Hays, H., 1972, Polyandry in the Spotted Sandpiper, *Living Bird* **11**:43–57.
Henty, C. J., 1976, Seabird movements off the Cherbourg Peninsula, April 1958, *Seabird Rep.* **1975–76**:52–54.
Heppleston, P. B., 1971, The feeding ecology of oystercatchers *Haematopus ostralegus* L. in winter in northern Scotland, *J. Anim. Ecol.* **40**:651–672.
Heppleston, P. B., 1972, The comparative breeding ecology of oystercatchers (*Haematopus ostralegus* L.) in inland and coastal habitats, *J. Anim. Ecol.* **41**:23–51.
Hildén, O., 1975, Breeding system of Temminck's Stint *Calidris temminckii, Ornis Fenn.* **52**:117–146.
Hildén, O., 1978, Population dynamics in Temminck's Stint *Calidris temminckii, Oikos* **30**:17–38.
Hildén, O., and Vuolanto, S., 1972, Breeding biology of the Red-necked Phalarope *Phalaropus lobatus* in Findland, *Ornis Fenn.* **49**:57–85.
Hobbs, J. N., 1972, Breeding of the Red-capped Dotterel at Fletcher's Lake, Dareton, NSW, *Emu* **72**:121–125.
Hohn, E. O., 1967, Observations on the breeding biology of Wilson's Phalarope (*Steganopus tricolor*) in central Alberta, *Auk* **84**:220–244.
Holmes, R. T., 1966, Breeding ecology and annual cycle adaptations of the Red-backed Sandpiper (*Calidris alpina*) in northern Alaska, *Condor* **68**:3–46.
Holmes, R. T., 1971, Density, habitat and the mating system of the Western Sandpiper (*Calidris mauri*), *Oecologia* (*Berlin*) **7**:191–208.
Holyoak, D. T., 1974, Lex oiseaux des iles de la societe, *Oiseau Rev. Fr. Ornithol.* **44**:1–181.
Hoogerwerf, A., 1969, On the ornithology of the rhino sanctuary Udjong Kulon in west Jave (Indonesia), *Nat. Hist. Bull. Siam. Soc.* **23**:1–135.
Hosking, E., and Ferguson-Lees, I. J., 1959, Photographic studies of some less familiar birds: Terek Sandpiper, *Br. Birds* **52**:85–90.
Howell, T. R., 1979, Breeding biology of the Egyptian Plover *Pluvianus aegyptius, Univ. Calif. Berkeley Publ. Zool.* **113**:1–76.
Hoy, G., 1967, The eggs and nesting ground of the Puna Plover, *Auk* **84**:130–131.
Hughes, R. A., 1979, Notes on Charadriiformes of the south coast of Peru, in: *Studies in Avian Biology No. 2* (F. A. Pitelka, ed.), pp. 49–53, Cooper Ornithological Society, Allen Press, Lawrence, Kans.
Hulscher, J. B., 1976, Localisation of cockles (*Cardium edule* L.) by the oystercatcher (*Haematopus ostralegus* L.) in darkness and daylight, *Ardea* **64**:292–310.
Humphrey, P. S., Bridge, D., Reynolds, P. W., and Peterson, R. T., 1970, *Birds of Isla Grandes* (*Tierra de Fuego*), University of Kansas Museum of Natural History, Lawrence, Kans.
Isleib, M. E., 1979, Migratory shorebird populations on the Copper River Delta and eastern Prince William Sound, Alaska, in: *Studies in Avian Biology No. 2* (F. A. Pitelak, ed.), pp. 125–130, Cooper Ornithological Society, Allen Press, Lawrence, Kans.

Jayakar, S. D., and Spurway, H., 1968, The Yellow-wattled Lapwing, *Vanellus malabaricus* (Boddaert), a tropical dry-season nester. III. Two further seasons' breeding, *J. Bombay Nat. Hist. Soc.* **65**:369–383.

Jehl, J. R., Jr., 1968, The systematic position of the Surfbird *Aphriza virgata*, *Condor* **70**:206–210.

Jehl, J. R., Jr., 1973a, The distribution of marine birds in Chilean waters in winter, *Auk* **90**:114–135.

Jehl, J. R., Jr., 1973b, Breeding biology and systematic relationships of the Stilt Sandpiper, *Wilson Bull.* **85**:115–147.

Jehl, J. R., Jr., 1975, *Pluvianellus socialis:* Biology, ecology, and relationships of an enigmatic Patagonian shorebird, *Trans. San Diego Soc. Nat. Hist.* **18**:31–73.

Jehl, J. R., Jr., 1979, The autumnal migration of Baird's Sandpiper, in: *Studies in Avian Biology No. 2* (F. A. Pitelka, ed.), pp. 44–68, Cooper Ornithological Society, Allen Press, Lawrence, Kans.

Jehl, J. R., Jr., and Smith, B. A., 1970, *Birds of the Churchill Region, Manitoba,* pp. 1–87, Special Publication No. 1, Manitoba Museum of Man and Nature, Winnipeg.

Jehl, J. R., Jr., Rumboll, M. A. E., and Winter, J. P., 1973, Winter bird populations of Golfo San Jose, Argentina, *Bull. Br. Ornithol. Club* **93**:56–63.

Jenni, D. A., and Betts, B. J., 1978, Sex differences in nest construction, incubation and parental behavior in the polyandrous American Jacana (*Jacana spinosa*), *Anim. Behav.* **26**:207–218.

Johnsgard, P. A., 1981, *The Plovers, Sandpipers and Snipes of the World,* University of Nebraska Press, Lincoln.

Johnson, A. W., and Goodall, J. D., 1965, *The Birds of Chile,* Platt Establecimientos Graficos S. A., Buenos Aires.

Johnson, O. W., 1979, Biology of shorebirds summering on Eniwetok Atoll, in: *Studies in Avian Biology No. 2* (F. A. Pitelka, ed.), pp. 193–205, Cooper Ornithological Society, Allen Press, Lawrence, Kans.

Jones, A., 1979, Notes on the behaviour of Variable Oystercatchers, *Notornis* **26**:47–52.

Kelly, P. R., and Cogswell, H. L., 1979, Movements and habitat use by wintering populations of Willets and Marbled Godwits, in: *Studies in Avian Biology No. 2* (F. A. Pitelka, ed.), pp. 69–82, Cooper Ornithological Society, Allen Press, Lawrence, Kans.

Khrokov, V. V., 1978, The Sociable Plover (Krechetka), *Priroda (Sofia)* **1978**:92–96.

King, B. F., and Dickinson, E. C., 1975, *Birds of South-east Asia,* Collins, Glasgow.

Kistchinski, A. A., 1975, Breeding biology and behaviour of the Gray Phalarope *Phalaropus fulicarius* in East Siberia, *Ibis* **117**:285–301.

Koepcke, M., 1970, *The Birds of the Department of Lima, Peru,* Livingston Publishers, Wynnewood, Pa.

Labutin, Y. V., Leonovitch, V. V., and Veprintsev, B. N., 1982, The Little Curlew *Numenius minutus* in Siberia, *Ibis* **124**:302–309.

Larson, S., 1960, On the influence of the Arctic fox *Alopex lagopus* on the distribution of Arctic birds, *Oikos* **11**:276–305.

Lemnell, P. A., 1978, Social behaviour of the Great Snipe *Capella media* at the arena display, *Ornis Scand.* **9**:146–163.

Lenington, S., 1980, Bi-parental care in Killdeer: An adaptive hypothesis, *Wilson Bull,* **92**:8–20.

Lind, H., 1965, Parental feeding in the oystercatcher, *Dan. Ornithol. Foren. Tidsskr.* **59**:1–31.

Little, J. de V., 1967, Some aspects of the behaviour of the Wattled Plover *Afribyx senegallus* (Linnaeus), *Ostrich* **38**:259–280.

Mackworth-Praed, C. W., and Grant, C. H. B., 1962, *Birds of the Southern Third of Africa,* Longmans, London.
McLachlan, G. R., and Liversidge, R., 1978, *Roberts Birds of South Africa,* John Voelcker Bird Book Fund, Cape Town.
McLachlan, G. R., Woldridge, T., Schramm, M., and Kuhn, M., 1980, Seasonal abundance, biomass and feeding of shorebirds on sandy beaches on the Eastern Cape, South Africa, *Ostrich* **51**:44–52.
Maclean, G. L., 1969, A study of seedsnipe in southern Africa, *Living Bird* **8**:33–80.
Maclean, G. L., 1973, A review of the biology of the Australian desert waders, *Stiltia* and *Peltohyas, Emu* **73**:61–70.
Maclean, G. L., 1976, A field study of the Australian dotterel, *Emu* **76**:207–215.
MacLean, S. F., Jr., and Holmes, R. T., 1971, Bill length, wintering areas, and taxonomy of North American Dunlins, *Calidris alpina, Auk* **88**:893–901.
McNeil, R., and Burton, J., 1973, Dispersal of some southbound migrating North American shorebirds away from the Magdalen Islands, Gulf of St. Lawrence and Sable Island, Nova Scotia, *Caribb. J. Sci.* **13**:257–278.
McNeil, R., and Burton, J., 1977, Southbound migration of shorebirds from the Gulf of St. Lawrence, *Wilson Bull.* **89**:167–171.
McNicholl, M. K., 1975, Larid site tenacity and group adherence in relation to habitat, *Auk* **92**:98–104.
Makkink, G. F., 1936, An attempt at an ethogram of the European Avocet, *Ardea* **25**:1–74.
Marchant, S., 1976, A seawatch on the southern coast of New South Wales, *Emu* **76**:9–18.
Martin, J., 1972, Nesting habits of our three resident sandplovers, *Bokmakierie* **24**:40–41.
Mason, C. F., and Macdonald, S. M., 1976, Aspects of the breeding biology of the snipe, *Bird Study* **23**:33–38.
Maxson, S. J., and Oring, L. W., 1980, Breeding season, time and energy budgets of the polyandrous Spotted Sandpiper, *Behaviour* **74**:200–263.
Meanley, B., 1975, *Birds and Marshes of the Chesapeake Bay Country,* Tidewater Publishers, Cambridge, Md.
Meltofte, H., 1979, The population of waders *Charadriidae* at Danmarks Haun, northeast Greenland, 1975, *Dan. Ornithol. Foren. Tidsskr.* **73**:69–94.
Morrell, S. H., Huber, H. R., Lewis, T. J., and Ainley, D. G., 1979, Feeding ecology of Black Oystercatchers on South Farallon Island, California, in: *Studies in Avian Biology No. 2* (F. A. Pitelka, ed.), pp. 185–186, Cooper Ornithological Society, Allen Press, Lawrence, Kans.
Morrison, R. I. G., and Harrington, B. A., 1979, Critical shorebird resources in James Bay and eastern North America, *Transactions of the 44th North American Wildlife and Natural Resources Conferences,* pp. 498–507, Wildlife Management Institute, Washington, D.C.
Muller, K. A., 1975, Threat displays of the Australian Painted Snipe, *Emu* **75**:28–30.
Munro, G. C., 1964, *Birds of Hawaii,* Tuttle, Rutland, Vt.
Murie, O. J., 1924, Nesting records of the Wandering Tattler and Surfbird in Alaska, *Auk* **41**:231–237.
Murphy, R. C., 1936, *Oceanic Seabirds of South America,* Vol. II, Macmillan Co., New York.
Myers, J. P., 1980a, The pampas shorebird community: Interactions between breeding and nonbreeding members, in: *Migrant Birds in the Neotropics: Ecology, Behavior, Distribution and Conservation* (A. Keast and E. S. Morton, eds.), pp. 37–49, Smithsonian Institution Press, Washington, D.C.

Myers, J. P., 1980b, Territoriality and flocking by Buff-breasted Sandpipers: Variations in non-breeding dispersion, *Condor* **82:**241–250.

Myers, J. P., and Myers, L. P., 1979, Shorebirds of coastal Buenos Aires Province, Argentina, *Ibis* **121:**186–200.

Myers, J. P., and Pitelka, F. A., 1980, Seasonal abundance and habitat use patterns of shorebirds at two sites in northern Alaska, *Wader Study Group Bull.* **29:**28–30.

Myers, J. P., Connors, P. G., and Pitelka, F. A., 1979, Territoriality in non-breeding shorebirds, in: *Studies in Avian Biology No. 2* (F. A. Pitelka, ed.), pp. 231–246, Cooper Ornithological Society, Allen Press, Lawrence, Kans.

Nakazawa, R., 1979, Incubation behaviour of the Kentish Plover, *Charadrius alexandrinus* with special references to the share of the sexes and of effect of ground surface temperature. Misc. Reports, Yamashina Inst., *Ornithology* **11:**54–63.

Nechaev, V. A., 1980, The Spotted Greenshank (*Okhotskii ulit*), *Priroda* (*Sofia*) **4:**102–106 (cited in Anderson, E. C., 1981, *J. Field Ornithol.* **52:**78).

Nettleship, D. N., 1973, Breeding ecology of turnstones *Arenaria interpres* at Hazen Camp, Ellesmere Island, N.W.T., *Ibis* **115:**202–217.

Nielsen, B. P., 1971, Migration and relationships of four Asiatic plovers *Charadriinae, Ornis Scand.* **2:**137–142.

Nisbet, I. C. T., 1961, Studies of less familiar birds, 113. Broad-billed Sandpiper, *Br. Birds* **54:**320–323.

North, M. E. W., 1937, Breeding habits of the Crested Wattled Plover (*Sarciophorus tectus latifrons*), *J. East Afr. Nat. Hist. Soc. Natl. Mus.* **13:**132–145.

Oring, L. W., 1973, Solitary Sandpiper early reproductive behavior, *Auk* **90:**652–63.

Page, G. W., and Whitacre, D. F., 1975, Raptor predation on wintering shorebirds, *Condor* **77:**73–83.

Page, G. W., Stenzel, L. E., and Wolfe, C. M., 1979, Aspects of the occurrence of shorebirds on a central California estuary, in: *Studies in Avian Biology No. 2* (F. A. Pitelka, ed.), pp. 15–32, Cooper Ornithological Society, Allen Press, Lawrence, Kans.

Parmelee, D. F., Greiner, D. W., and Graul, W. D., 1968, Summer schedule and breeding biology of the White-rumped Sandpiper in the central Canadian Arctic, *Wilson Bull.* **80:**5–29.

Phelps, W. H., Jr., 1975, Willet breeding in Los Roques Archipelago, Venezuela, *Auk* **92:**164–165.

Phelps, W. H., and Phelps, W. H., Jr., 1958, Lista de Las Aves de Venezuela con su distribucion. no Passeriformes, *Bol. Soc. Venez. Cien. Nat.* **19:**1–317.

Phillips, R. E., 1977, Notes on the behaviour of the New Zealand Shore Plover, *Emu* **77:**23–27.

Pienkowski, M. W., 1980, Aspects of the ecology and behaviour of Ringed and Grey Plovers *Charadrius hiaticula* and *Pluvialis squatarola*, Ph.D. thesis, University of Durham, U.K.

Pienkowski, M. W., 1981, How foraging plovers cope with environmental effects on invertebrate behaviour and availability, in: *Feeding and Survival Strategies of Estuarine Organisms* (N. V. Jones and W. J. Wolff, eds.), pp. 179–192, Plenum Press, New York.

Pitelka, F. A., 1950, Geographic variation and the species problem in the shore-bird genus Limnodromus, Univ. Calif. Berkeley Publ. Zool. **50:**1–108.

Pitelka, F. A. (ed.), 1979, *Studies in Avian Biology No. 2*, Cooper Ornithological Society, Allen Press, Lawrence, Kans.

Pitelka, F. A., Holmes, R. T., and MacLean, S. F., Jr., 1974, Ecology and evolution of social organization in Arctic sandpipers, *Am. Zool.* **14:**185–204.

Pitman, C. R. S., 1965, The eggs and nesting habits of the St. Helena Sand-Plover or Wire bird, *Charadrius pecuarius sanctae-helenae* (Harting), *Bull. Br. Ornithol. Club* **85**:121–129.

Portenko, L. A., 1972, *Ptitsy Chakotskogo polvostrova i ostrova Vrangelya,* Nauka, Moscow.

Powers, K., 1982, A comparison of two methods of counting birds at sea. *J. Field Ornithol.* **53**:209–222.

Prater, A. J., 1979, Shorebird census studies in Britain, in: *Studies in Avian Biology No. 2* (F. A. Pitelka, ed.), pp. 157–166, Cooper Ornithological Society, Allen Press, Lawrence, Kans.

Prater, A. J., 1981, *Estuary Birds of Britain and Ireland,* Poyser, Calton, England.

Pringle, J. S., and Cooper, J., 1977, Wader population (*Charadrii*) of the marine littoral of the Cape Peninsula, South Africa, *Ostrich* **48**:98–105.

Puttick, G. M., 1979, Foraging behaviour and activity budgets of Curlew Sandpipers, *Ardea* **67**:111–122.

Rand, A. L., 1936, The distribution and habits of Madagascar birds, *Bull. Am. Mus. Nat. Hist.* **72**:143–499.

Recher, H. F., 1966, Some aspects of the ecology of migrant shorebirds, *Ecology* **47**:393–407.

Recher, H. F., and Recher, J. A., 1969, Some aspects of the ecology of migrant shorebirds. II. Aggression, *Wilson Bull.* **81**:140–151.

Ridley, M. W., 1980, The breeding behaviour and feeding ecology of Grey Phalaropes *Phalaropus fulicarius* in Svalbard, *Ibis* **122**:210–225.

Ripley, S. D., 1944, The bird fauna of the West Sumatra Islands, *Bull. Mus. Comp. Zool. Harv. Univ.* **94**:307–430.

Ripley, S. D., II, 1961, *A Synopsis of the Birds of India and Pakistan,* Bombay Natural History Society, Bombay.

Robertson, H. A., and Dennison, M. D., 1979, Feeding and roosting behaviour of some waders at Farewell Spit, *Notornis* **26**:73–88.

Sauer, E. G. F., 1962, Ethology and ecology of Golden Plovers on St. Lawrence Island, Bering Sea, *Psychol. Forsch.* **26**:399–470.

Schneider, D., 1978, Equalisation of prey members by migratory shorebirds, *Nature* (*London*) **271**:353–354.

Schneider, D. C., and Harrington, B. A., 1981, Timing of shorebird migration in relation to prey depletion, *Auk* **98**:801–811.

Senner, S. E., 1979, An evaluation of the Copper River Delta as a critical habitat for migrating shorebirds, in: *Studies in Avian Biology No. 2* (F. A. Pitelka, ed.), pp. 131–146, Cooper Ornithological Society, Allen Press, Lawrence, Kans.

Serle, W., Morel, G. J., and Hartwig, W., 1977, *Field Guide to Birds of West Africa*, Collins, Glasgow.

Shephard, J. M., 1976, Factors influencing female choice in the lek mating system of the Ruff, *Living Bird* **14**:87–111.

Sibson, R. B., and McKenzie, H. R., 1943, Some observations on stilts in the Firth of Thames, *N.Z. Bird Notes* **1**:51–57.

Sinclair, J. C., 1978, The seabirds of a trawling voyage, *Bokmakierie* **30**:12–16.

Slater, P., 1971, *A Field Guide to Australian Birds,* Oliver & Boyd, Edinburgh.

Slud, P., 1964, The birds of Costa Rica, *Bull Am. Mus. Nat. Hist.* **128**:1–430.

Smith, S. M., and Stiles, F. G., 1979, Banding studies of migrant shorebirds in northwestern Costa Rica, in: *Studies in Avian Biology No. 2* (F. A. Pitelka, ed.), pp. 41–47, Cooper Ornithological Society, Allen Press, Lawrence, Kans.

Smythies, B. E., 1953, *The Birds of Burma*, Oliver & Boyd, Edinburgh.
Spaans, A. L., 1978, Status and numerical fluctuations of some North American waders along the Surinam coast, *Wilson Bull.* **90**:60–83.
Stenzel, L. E., Huber, H. R., and Page, G. W., 1976, Feeding behavior and diet of the Long-billed Curlew and Willet, *Wilson Bull.* **88**:314–332.
Steyn, P., 1973, African Jacana at last, *Bokmakierie* **25**:34–37.
Stidolph, R. H. D., 1973, Plumages of Variable Oystercatchers, *Notornis* **20**:311–313.
Strauch, J. G., and Abele, L. G., 1979, Feeding ecology of three species of plovers wintering on the Bay of Panama, Central America, in: *Studies in Avian Biology No. 2*. (F. A. Pitelka, ed.), pp. 217–230, Cooper Ornithological Society, Allen Press, Lawrence, Kans.
Summers, R. W., and Waltner, M., 1979, Seasonal variations in the mass of waders in southern Africa, with special reference to migration, *Ostrich* **50**:21–37.
Summers, R. W., Cooper, J., and Pringle, J. S., 1977, Distribution and numbers of coastal waders (Charadrii) in the South-western Cape, South Africa, summer 1975–1976, *Ostrich* **48**:85–97.
Sutton, G. M., and Parmelee, D. F., 1955, The Purple Sandpiper in southern Baffin Island, *Condor* **57**:216–220.
Thomas, D. C., 1969, Breeding biology of the Australian Spur-winged Plover, *Emu* **69**:81–102.
Thomas, D. G., and Dartnall, A. J., 1971, Ecological aspects of the feeding behaviour of two *Calidridine* sandpipers wintering in south-eastern Tasmania, *Emu* **71**:20–26.
Tree, A. J., 1979, Biology of the Greenshank in southern Africa, *Ostrich* **50**:240–247.
Tyler, S., 1978, Observations on the nesting of the Three-banded Plover *Charadrius tricollaris*, *Scopus* **2**:39–41.
Urban, E. K., 1978, *Ethiopia's Endemic Birds*, Artistic Printers, Addis Ababa.
Urner, C. A., and Storer, R. W., 1949, The distribution and abundance of shorebirds on the north and central New Jersey coast, 1928–1938, *Auk* **66**:177–194.
Vogt, W., 1938, Preliminary notes on the behavior and ecology of the Eastern Willet, *Proc. Linn. Soc. N.Y.* **49**:8–42.
Walters, J., 1979, Interspecific aggressive behaviour by Long-toed Lapwings (*Vanellus crassirostris*), *Anim. Behav.* **27**:969–981.
Weir, R. D., and Cooke, F., 1976, Autumn migration of shorebirds in the Kingston area of Ontario, 1964–1974, *Can. Field Nat.* **90**:103–113.
Wilcox, L., 1959, A twenty year banding study of the Piping Plover, *Auk* **76**:129–152.
Williams, J. G., 1964, *A Field Guide to the Birds of East and Central Africa*, Houghton Mifflin, Boston.
Williams, J. G., and Arlott, N., 1980, *Guide to the Birds of East Africa*, Collins, Glasgow.
Winterbottom, J. B., 1963, Comments on the ecology and breeding of sandplovers *Charadrius* in southern Africa, *Rev. Zool. Bot. Afr.* **67**:1–2.
Wishart, R. A., and Sealy, S. G., 1980, Late summertime budget and feeding behavior of Marbled Godwits (*Limosa fedoa*) in southern Manitoba, *Can. J. Zool.* **58**:1277–1282.
Wolfe, L. R., 1954, The Australian Snipe in Japan, *Emu* **54**:198–203.
Wolff, W. J., 1969, Distribution of non-breeding waders in an estuarine area in relation to the distribution of their food organisms, *Ardea* **57**:1–25.
Woodcock, M., 1980, *Collins Handguide to the Indian Subcontinent*, Collins, Glasgow.
Yamashina, Y., 1961, *Birds in Japan*, Tokyo News Service Limited, Tokyo.
Zwarts, L., and Drent, R. H., 1981, Prey depletion and the regulation of predator density: Oystercatchers (*Haematopus ostralegus*) feeding on mussels (*Mytilus edulis*), in: *Feeding and Survival Strategies of Estuarine Organisms* (N. V. Jones and W. J. Wolff, eds.), pp. 193–216, Plenum Press, New York.

Chapter 3

POPULATION DYNAMICS OF SHOREBIRDS

P. R. Evans and M. W. Pienkowski

Department of Zoology
University of Durham
Durham DH1 3LE, England

I. INTRODUCTION

In some parts of the world, shorebird populations have been monitored for more than 10 years by counts in their nonbreeding range (Prater, 1982). In particular, midwinter counts over the whole of western Europe have given good indications of the year-to-year changes in the populations of several sandpiper (Scolopacidae) species breeding in discrete Arctic areas such as western Siberia or Greenland. These counts indicate fluctuations within ± 50% of the mean, as in Red Knot *Calidris canutus* from Greenland/Ellesmere Island, apparently associated with annual variations in weather on the breeding grounds. Much less is known of the total populations, or fluctuations in these, of most plover (Charadrii) species, which are found in drier tropical regions. Those waders that are restricted to intertidal habitats in the nonbreeding season provide some of the few examples of smaller birds whose regional populations can be assessed relatively easily and accurately even though they contain many hundreds of thousands of individuals.

Densities of shorebirds in the nonbreeding season are usually much higher than on the nesting areas, especially for scolopacids. Unlike the winter surveys that have been made in some parts of the world, most long-term censuses of breeding areas have covered only a small proportion of possible habitats, and these chiefly in temperate latitudes, rather than in the Arctic where breeding numbers are highest. Studies of breeding performance over several years have been made chiefly by professional biologists, though some valuable contributions, mainly in single years, have resulted from expeditions including both amateur and professional ornithologists. Because studies in the breeding areas have been intensive rather than extensive, very little information is available on any changes in geographical range associated with changes in total population. Indeed, most information is concerned only with changes in population density in a defined locality. This must be accepted as a major limitation in the discussion of wader population processes given below, as it is known that reduction in the wintering population of Red Knot in western Europe has led to contraction in use of parts of the nonbreeding range, with markedly reduced numbers using French estuaries (R. Mahéo, unpublished); hence, it is possible that equally extensive year-to-year changes take place in the range occupied in the breeding season. Longer-term changes in range have been summarized for several species by Hale (1980). Meltofte (1976) reviewed previous semiquantitative assessments of breeding shorebird numbers in southern Peary Land, northern Greenland, and concluded that marked fluctuations in abundance had occurred in most species, but that different species reached peak numbers in different years.

Changes in the size of a geographically distinct population arise from changes in the balance between annual reproductive output and annual mortality. Even if the boundaries of the breeding area do not change from year to year, shifts may occur in the sites where breeding densities are highest. These are particularly likely if the breeding habitats are unstable, e.g., on coastal shingle that might be covered by blown sand (see Pienkowski, 1980) or on riverine sandbanks that might be moved by flood waters. In Arctic regions, the depth of winter snows lying over potential breeding sites will vary from year to year with the extent of precipitation and the wind direction. Snow depth will affect whether the sites become suitable for occupation in summer early enough for birds to establish there, and possibly whether sites become too dry for use by shorebird families in late summer (see Green *et al.*, 1977; Pienkowski, 1984b).

Adults of shorebird species nesting in temporarily suitable habitats inevitably show less site-fidelity from year to year than other species. How far this is genetically programmed and how far it occurs in direct response to changed conditions is not known. Pitelka *et al.* (1974) suggested the former for Curlew Sandpiper *C. ferruginea* and Pectoral Sandpiper *C. melanotus* in Alaska, and Hildén (1978b) for Little Stint *C. minuta* in northern Norway. Other species, whose nesting habitat changes less rapidly with time, show greater site-faithfulness, e.g., Dunlin *C. alpina* (Soikkeli, 1970) and Temminck's Stint *C. temminckii* (Hildén, 1979), but even the quality of the study areas for these two species deteriorated eventually, through changes in agricultural practice or eutrophication affecting the coastal vegetation (Soikkeli, personal communication; Hildén,

1978a). This appeared to depress recruitment of young birds rather than leading to change of site by older established breeders. It is in these site-faithful species that intensive studies of the effects of various factors on reproductive success are most relevant to an understanding of the reasons for changes in total population size. A chosen study area for such species may provide a representative subpopulation in which the effects of population density and habitat type on settlement of potential first-breeders, on proportion of breeders, on clutch size, and on hatching and fledging success can be measured and extrapolated to the whole population.

In opportunistic species, population density in an ephemeral habitat might determine not only whether potential first-breeders settle there but also whether older birds settle that have bred elsewhere in previous years. The density of birds finally settling in the site is likely to be affected not only by the number reaching the area but also by the quality of the habitat, a feature rarely measured in studies of breeding waders. Changes in reproductive output from year to year in a small intensively studied area are thus unlikely to be representative of those occurring over the whole of the range of an opportunistic species.

In this chapter, and with the above points in mind, we first examine published and unpublished information on reproductive output, from studies of breeding biology of a variety of shorebird species. [Work published before 1977 was summarized and reviewed by Goss-Custard (1981).] We then discuss mortality of adults on the breeding areas, on migration, and in the nonbreeding sites. We also summarize data on survival of immature birds until they reach breeding age. We close the chapter with a discussion of the evidence for regulation of shorebird numbers.

II. RECRUITMENT AND MORTALITY ON THE BREEDING GROUNDS

The aim of the first part of this section is to account for year-to-year variation in the total number of chicks that reach fledging from a certain specified area of breeding habitat. This number (N) is the product of four parameters, i.e., $N = c \times e \times h \times f$, where

c = No. of clutches laid
e = No. of eggs per clutch
h = Proportion of eggs hatching
f = Proportion of chicks fledging

The number of clutches laid in a given area is affected by the population density and the number of breeding attempts by each female present.

A. Population Density

Few waders breeding in temperate and Arctic latitudes spend the nonbreeding season close to their nesting areas. In many species, the males usually arrive first in spring, but in others, including Sanderling *Calidris alba* and phalaropes *Phalaropus* spp., the sexes arrive at the same time (see Myers, 1981). The population density of males settling on an area may be restricted by territorial behavior. Removal experiments have shown the presence of a "floating population" of Dunlin in Alaska (Holmes, 1966) and of Eurasian Oystercatchers *Haematopus ostralegus* on the small British island of Skokholm (Harris, 1970). Some of these birds took over vacant territories if their owners were removed at the appropriate season. In these studies, all males that defended areas managed to secure a mate. However, in Temminck's Stint, in several years, between 2 and 6 males of a total population of 20–30 birds defending territories failed to secure a mate (Hildén, 1978a). These males were chiefly 1-year-old birds that occupied larger than average sites, usually on the lakeside margins of the peninsula on which the study was made. Unmated territorial males were also recorded in the Western Sandpiper *Calidris mauri* by Holmes (1971). Little has been published on the determinants of territory size apart from the review of literature on *Calidris* species by Hildén (1979). Within a species, geographical variation in territory size is common, and Holmes (1970) argued that territory size of Dunlin in two parts of Alaska differed sixfold in response to differences in food density, as they were adjusted to allow enough food for adults in spring. Ratcliffe (1976) showed that densities of nesting Eurasian Golden Plovers *Pluvialis apricaria* within upland Britain were lower where food was scarcer. Pienkowski (1984c) suggested that territory size in Ringed Plovers *Charadrius hiaticula* might be related to potential ease of nest-finding by predators. Unfortunately, factors leading to greater nest cover tend to correlate with those leading to greater prey abundance. Not all waders defend breeding territories; e.g., the Redshank *Tringa totanus* breeds "semicolonially" (Hale, 1956, 1980), as does the Dunlin in some areas (Fuller, 1978; Symonds, 1981; Etheridge, 1982), so that other factors determine the nesting density.

In several species, whatever the spacing behavior of the males, it is the number of females settling in the area that determines the potential number of nests. In both Temminck's Stint (Hildén, 1978a) and Western

Sandpiper (Holmes, 1971), no conclusive evidence was found of the presence of nonbreeding females to parallel the presence of unmated males. This could be explained if females bred for the first time at a later age than males and spend their nonbreeding summers elsewhere than on the nesting areas. But the reverse seems to be true: males attempt to breed for the first time later than females (see below). Hence, the surplus of males on the breeding grounds defies easy explanation, unless there is differential mortality between the sexes in the first years of life, which seems unlikely.

B. Nesting Attempts

The most obvious factor influencing the number of breeding attempts by a female is the duration of the period in which breeding and rearing of young to fledging is possible. This period is influenced by climate, and tends to be correlated with latitude, such that the breeding season is longer in temperate areas than in the Arctic, e.g., in the Ringed Plover (Pienkowski, 1980, 1984c). Monogamous species, such as the Ringed Plover and Dunlin, may occasionally lay second clutches in temperate latitudes, after completing successful rearing of the young of the first clutch (Laven, 1940; Soikkeli, 1967), but they never do in Arctic areas. [Among the marshland and woodland waders such as snipe and woodcock, second clutches are regular in temperate areas (Hale, 1980).] Indeed, in years of late snow melt in Arctic areas, some wader species do not attempt to lay even a first clutch (Green *et al.*, 1977), for there would be insufficient time for the young to reach independence and prepare for migration before the weather deteriorates and food becomes scarce.

Females of some species, particularly those that have complex mating systems, may lay more than one clutch in quick succession at the beginning of the breeding season. Some of the additional clutches are reared by one or more males (see Pienkowski and Greenwood, 1979). Whether this occurs may depend on the regularity (predictability) of a high level of food supply at the correct time (see Emlen and Oring, 1977; Graul *et al.*, 1977). In the subarctic, females of species feeding on mass emergences of insects, such as Temminck's Stint (Hildén, 1975) and Northern Phalaropes *Phalaropus lobatus* (Hildén and Vuolanto, 1972; Raner, 1972), usually lay two clutches, and sometimes three. In Northern Phalaropes, extra clutches are laid in years when a surplus of males is present. The same is true of Dotterels *Eudromia morinellus* (Pulliainen, 1970; Nethersole-Thompson, 1973) and, at more southerly latitudes, of Spotted Sandpipers *Actitis macularia* (Hays, 1972; Oring and Knudsen, 1972). If the food supply varies from year to year, multiple clutches may be laid only in years of food abundance. This has been suggested for Mountain Plover *Charadrius montanus* in the central states of the U.S.A. (Graul, 1976), Sanderling in the Canadian Arctic (Parmelee and Payne, 1973), and Dotterel in northern Europe (Hildén, 1975).

Differences in number of clutches laid by a female may occur within a species in different geographical regions at the same latitude, e.g., female Sanderling in Greenland lay only a single clutch (Pienkowski and Green, 1976) whereas in the Canadian Arctic they lay two (Parmelee and Payne, 1973).

The reactions of females to losses of clutches also vary with climatic zone and food supply. The time of loss during the breeding cycle is also important—the later during incubation or guarding of the young, the less likely is renesting. In the Arctic, as exemplified by Alaska, the Dunlin replaces clutches lost to predators early in the breeding season (Holmes, 1966). Temminck's Stint does not normally re-lay after producing its multiple clutches, even in the most southern part of its range, i.e., subarctic Finland (Hildén, 1978a). Many species renest in temperate latitudes, but the food supply at the time the first clutch is lost may be all-important (Soikkeli, 1970). Individually marked female Ringed Plover in Britain have been shown to produce up to five clutches in one season in response to natural predation by foxes, crows, mustelids, etc. (Pienkowski, 1980, 1984c).

Meltofte *et al.* (1981) reported renesting by several species in northeastern Greenland in 1976, a year of severe nest predation by foxes; but renesting in the Arctic is probably restricted to replacement of nests lost before, or early in, incubation.

C. Clutch Size

The first clutches produced by most scolopacids usually contain four eggs, as do those of most plovers in Arctic and north temperate areas (see Maclean, 1972). However, in tropical and hot arid areas, and generally in the Southern Hemisphere, clutch sizes of wader species from all families are often only two or three. Less than four eggs may also be produced by those families or species of waders that feed their young, e.g., oystercatchers (Haematopodidae).

Even at latitudes where food abundance (for egg production) does not decrease markedly during the summer, replacement clutches in both species that do and those that do not produce genuine second layings may be reduced in size. In Ringed Plover, which usually lays a first clutch of four eggs in northeastern England, the clutch size of 35% of pairs decreased during re-layings, and none increased (Pienkowski, unpublished).

If replacement clutches are excluded from consideration, seasonal variation in clutch size has been shown only for Eurasian Oystercatcher (Harris, 1969) and does not occur during the short breeding season of Temminck's Stint (Hildén, 1978a) or even during the 5-month breeding season of the White-fronted Sandplover *Charadrius marginatus* in South Africa (Summers and Hockey, 1980). Ferns *et al.* (MS) found that average clutch size was less in areas of northeastern Greenland with a later start of breeding [following late snow clearance—see above and Green *et al.* (1977)] than in areas with an earlier start. Furthermore, decreased clutch size was associated with increased egg-size.

Suggestions that the multiple-clutch-laying species, e.g., Temminck's Stint and Sanderling, lay small eggs (e.g., Oring and Knudsen, 1972; Graul, 1973; Parmelee and Payne, 1973; Jenni, 1974) are not supported by the data (see Hildén, 1975; Pienkowski and Green, 1976).

D. Hatching Success

Published estimates of hatching success vary very widely from species to species, year to year, and even area to area. Among the lowest hatching success is that recorded for Ringed Plovers on beaches in eastern Britain, where at one site only 13% of eggs laid produced chicks in one year and 1% in another year; figures for two years in another site were 2 and 29%, respectively (Pienkowski, 1984c) Most of the losses arose from predation. On the salt marshes of the Ribble Estuary, England, in a series of four years, between 18 and 48% of eggs laid by Redshank produced chicks (Hale, 1980). Here, losses resulted chiefly from flooding by exceptionally high tides and trampling by cattle grazing on the marshes—hazards common to many waders breeding on salt marshes, a habitat found chiefly in temperate latitudes. Other coastal breeding waders may also suffer from the effects of unpredicted high tides, e.g., species nesting close to the strand line on shingle and sandbanks, such as the Ringed Plover. Their nests are also particularly susceptible to predation (as documented earlier). Many scavengers routinely search along the high-tide line and may thereby encounter nests or adults. Pienkowski (1984c) showed that clutches were significantly more likely to be lost on a day when other clutches were taken than would be expected by chance. Human disturbance on coastal beaches may also be serious, through direct crushing of nests and eggs, by causing desertion of nests by incubating birds, and aiding detection of nests by other predators. Dogs brought by human visitors may also cause destruction and desertion (Pienkowski, 1980; Senner and Howe, this volume). Pienkowski (1984c) found that distance from public access was the main correlate of area differences in hatching success of Ringed Plovers on the Lindisfarne National Nature Reserve in northeastern England.

Nests of waders breeding further from the water's edge may also be at risk from predators. Hatching success in Hildén's study population of Temminck's Stints varied among years from 58% down to 33%. Of the losses, three-quarters could be attributed to predators and less than one-sixth to individual eggs failing to hatch in otherwise successful clutches. Even the latter is a higher percentage than in most other species studied, e.g., 6% of eggs completing incubation failed to hatch for Ringed Plovers in Greenland and 8% in Britain (Meltofte, 1979; Pienkowski, 1980, 1984c), 5% for Dotterel (Pulliainen, 1970), 4% for Eurasian Golden Plovers (Ratcliffe, 1976) and 6% for Upland Sandpiper *Bartramia longicauda* (Higgins and Kirsch, 1975). Possibly the higher value (about 9%) for Temminck's Stint is a consequence of the mating system of this species, in which clutches may be left for some time before the start of incubation (Hildén, 1975; Pienkowski and Greenwood, 1979).

In Arctic areas, hatching success may be much higher. Holmes (1966, 1972) quotes figures of 70–90% of Western Sandpiper eggs and 75–80% of Dunlin eggs hatching in his Alaskan study areas. But Byrkjedal (1980) has shown that predation may be heavy in the early part of the breeding season in the Norwegian mountains when only restricted areas are free from snow and suitable for nesting Eurasian Golden Plovers. Meltofte *et al.* (1981) reported that, in an area of northeastern Greenland where the snow cover was up to 90% during early incubation, foxes simply walked from one small snow-free patch to another in search of nests of shorebirds. Pienkowski (1984c) found also that predation of nests decreased later in the season. As this was in a temperate breeding area (northeastern England), increasing availability of alternative prey for predators may be implicated, as well as increasing difficulty in nest finding.

No evidence of higher percentage losses of clutches to predators in areas of higher nest density has been obtained so far, although the tendency of a predator to visit an area once every few days to take eggs could provide a mechanism for this.

E. Fledging Success

After hatching, the chicks of most wader species feed themselves and one or both parents provide merely a warning or defense against predators, brood the young, and probably lead them to suitable feeding areas. In consequence, the chicks tend to be cryptically colored so that they can escape potential visual predators, including research biologists. Fledging success has thus proved one of the most difficult parameters to measure in studies of reproductive output of most waders. Those estimates that have been made range very widely within some species, e.g., in Eurasian Oystercatchers from 4.6% in northern Germany (Dircksen, 1932) to 86% on the island of Skokholm, Wales (Keighley, 1949), but not in others, e.g., Ringed Plover: 60% in northern Germany (Laven, 1940), 50% in Finland (Nordberg, 1950), 50% in Greenland, and 45% at Lindisfarne, northeastern England (Pienkowski, 1984a). However, at Lindisfarne, fledging success varied with hatching date in one year, being lower later in the summer (Pienkowski, 1984a).

Many possible causes of death of chicks are similar to those leading to losses of eggs, notably predation (see review by Goss-Custard, 1981), but starvation is an additional risk. This might arise not only through food shortage but also through meteorological factors restricting feeding success, e.g., periods of heavy rain or strong winds coinciding with the first few days of life. Heavy rain, strong winds, and, in Arctic areas, low temperatures decreased the food intake rate of Ringed Plover chicks in Britain and Greenland, and the birds stopped feeding in particularly severe conditions (Pienkowski, 1984a,b). Meltofte (1976) reported deaths of young shorebird chicks in northeastern Greenland during an unusually early snowstorm in early August. However, Pienkowski (1984b) argued that, despite the effects of weather on feeding rate in Ringed Plovers, losses of chicks of that species through starvation are rare after the first few days of life because potential feeding time in both temperate and Arctic areas appeared to be in excess. In Northern Lapwings *Vanellus vanellus* in southern England, however, Jackson and Jackson (1975) reported that prefledging mortality was directly related to the amount of rainfall in June.

In the Ringed Plover population at Lindisfarne, Pienkowski (1984a) found that most chick losses occurred by night, possibly to mammalian predators. Darkness does not, of course, occur during summer over most of the range of Arctic breeding shorebirds. Indeed, the longer summer nights at lower latitudes, and thus increasing possibilities of nocturnal predation, could be a factor contributing to the southern limit of breeding ranges. Half of the losses of Ringed Plovers at Lindisfarne occurred before the chicks were a week old and 95% before they were 2 weeks old (they fledge when 3 weeks old). Because predation is important in determining nesting success, it is tempting to try to relate the extent of chick losses to nesting density. However, in many species the chicks are led away from the immediate area of the nest by their parents soon after hatching, so that the density of family parties on the chick feeding grounds need not be related to the density of nests.

F. Adult Mortality during the Breeding Season

The most spectacular documented losses of waders on their breeding grounds have occurred in species nesting at high latitudes, especially in the high Arctic, where late snowstorms have occasionally led to deaths of large numbers of Ruddy Turnstones *Arenaria interpres* and Red Knot (Morrison, 1975) that had recently arrived on the breeding grounds (see also Meltofte, in press). Similar disasters may occur at lower latitudes but higher altitudes, e.g., in central Scandinavia where severe weather early one summer killed Northern Lapwings and Eurasian Oystercatchers on breeding sites (Marcström and Mascher, 1979). In the uplands of central England, a doubling of the normal annual mortality rate of Common Sandpipers *Actitis hypoleucos* coincided with exceptional and severe snowfall in late April and early May 1981, which also caused deaths of numerous other birds and mammals (Holland *et al.*, 1982, personal communication).

Losses of adults during incubation are often small. In Temminck's Stint, only 4 of 102 unsuccessful nests failed because of death of an adult (Hildén, 1978a). At Lindisfarne, although predation was the chief cause of loss of eggs and chicks, only one adult was known to have been taken by a predator during incubation. In contrast, about a quarter of the total annual mortality of adult Dunlin in southern Finland could be attributed to losses during incubation (Soikkeli, 1967). These differences could be due in part to differences in antipredator behavior. This varies greatly within and between species; indeed, the unpredictability may in itself be adaptive (see Meltofte, 1976, 1977, 1979; Green, 1978; Pienkowski, 1984c). There is some tendency, however, for birds of open habitats to leave their nests early if a potential predator approaches and to rely on egg-camouflage and luring the predator away, while those of more vegetated areas allow predators a close approach, relying for protection on the camouflage of the adult's plumage and, in heavily vegetated areas, nest concealment (see also Gochfeld, this volume).

G. Year-to-Year Variation in Breeding Output

Within a defined geographical area, the breeding output is the product of clutches laid × eggs per clutch × percentage of eggs hatching ×

percentage of chicks fledging. As territory size is reasonably consistent from year to year in a particular place, increases in the numbers of territories per hectare can be achieved only if there are poor habitats within the study area that may be settled in years when large numbers of males arrive, but not in other years. In years of high totals of territories, some males may fail to obtain a female [e.g., Western Sandpiper (Holmes, 1971), Temminck's Stint (Hildén, 1978a)]. Thus, the variation in number of breeding females on a site from year to year is often rather small. The greatest annual variation in potential reproductive output results from variations in the number of clutches laid. In temperate areas, this is usually in response to annual variations in losses of clutches, rather than to annual changes in the timing of breeding and length of breeding season, as rather few waders attempt to produce true second broods in a year. In high Arctic areas, however, total failure to breed in some years is of major importance in determining the magnitude of year-to-year variations in breeding output. Elsewhere, predation is the most obvious agent leading to such variations. At the egg stage, losses are chiefly of whole clutches; infertility and other causes of failure to hatch of single eggs are not important. At the chick stage, it is rare for whole broods to disappear, presumably through predation, simultaneously. Because predators may turn their attention to alternative prey as the birds' breeding seasons progress, it cannot be assumed that in years of heavy predation on eggs, predation on chicks will also be heavy. Furthermore, because in most situations shorebird eggs and young probably provide only a small component of a predator's diet, the behavior of the predator will depend on many other factors. Thus large local, seasonal, and annual fluctuations in intensity of predation on nests and young may occur (Ratcliffe, 1976; Pienkowski, 1980, 1984c).

III. ADULT SURVIVAL DURING THE NONBREEDING SEASON

For many waders, time spent on the breeding grounds is less than a quarter of their lives, particularly if they nest in the Arctic. As discussed above, mortality on the breeding areas is usually very low except in unseasonally severe weather at high latitudes or altitudes. Annual survival rates of adults from one breeding season to the next have been estimated for several species (Table I) on the assumption that all those marked in one summer will return to the same breeding area in the following summer if they are alive to do so. This assumption is probably valid for most

Table I. Average Minimum Annual Survival of Shorebirds Estimated by Return of Individually Marked Birds to Breeding Areas

Species	Survival	Place	Source
Eurasian Oystercatcher	94%	Friesian Islands	Schnakenwinkel (1970)
	89%	Skokholm, Britain	Harris (1967)
White-fronted Sandplover	91%	Cape Province, S. Africa	Summers and Hockey (1980)
Ringed Plover	59%	Friesian Islands	Laven (1940)
	70–75%	Friesian Islands	Bub (1962)
	80–90%[a]	Lindisfarne, Britain	Pienkowski (1980, 1984c)
Snowy Plover	75%	Friesian Islands	Rittinghaus (1956)
Ruddy Turnstone	78%	Aland Islands, Finland	Bergman (1946)
Common Redshank	75%[a]	Friesian Islands	Grosskopf (1959, 1964)
	78%	Ribble, England	Yates (1982)
Dunlin	73%	Finland	Soikkeli (1967, 1970)
	70–72%	Point Barrow, Alaska	Holmes (1966)
Temminck's Stint	81%[b]	Finland	Hildén (1978a)
Common Sandpiper	76%	Peak District, England	Holland *et al.* (1982)

[a] Increased with age.
[b] Decreased with age of bird; other estimates age-independent.

species that utilize habitats that shift little from year to year (see Section I). These estimates of survival are much higher than those calculated by Boyd (1962) from recoveries by the public of birds carrying numbered rings (bands). It is probable that rapid wear of the metal used for manufacture of such rings before the 1970s led to spurious apparent mortality. In addition, "general-purpose" small-scale ringing may have involved capture of those parts of a population more prone to predation, whereas studies based on color-ringing have often attempted to mark a large, and hopefully representative, sample of birds present in a particular area [see Swennen (1984) for evidence of segregation between birds in poor and good body condition on roost sites]. As indicated in Table I, survival rates of even the smallest waders are higher than would be expected for terrestrial songbirds of similar size.

A. Mortality during Migration

Mortality away from the breeding areas could occur on the long-distance autumn and spring migrations undertaken by many species, or while they are temporarily resident on molting or "wintering" sites. Very little is known of the extent of mortality on migration. Shorebirds fly fast

(50–80 km/hr), as revealed by radar observations (Lack, 1963; Evans, 1968; Richardson, 1979), and are thus less likely to be deflected from their chosen migration routes by adverse winds than are songbirds. Also, they store large loads of fat as fuel before departure for long oversea flights (Pienkowski and Evans, Vol. 6 of this series). Conversely, if they do utilize all their fat on such flights, they cannot necessarily land to refuel, e.g., between Greenland and Europe or northwestern Africa. The presence of newly arrived, exhausted, and emaciated Red Knot has been recorded from Mauritania by Dick and Pienkowski (1979) but it is not possible to estimate whether many birds perished before making a landfall. Some of those that did reach land, however, failed to recover, particularly juveniles. Emaciated Dunlin have been found in northeastern England after migration across the North Sea in one autumn. The possibility that juveniles on their first autumn migration may be more likely to die than adults that have flown the route before has not been tested. Although most of the emaciated birds found in Mauritania were juveniles, this could be a result of the dates of observations, for adults and juveniles of most wader species migrate at different periods (Pienkowski and Evans, Vol. 6 of this series). It is not even known whether the fat reserves carried by the two age groups, before departure for the same destination, are similar.

B. Mortality during Molt

The complete molt, of both flight and body feathers, of shorebirds normally takes place immediately after the breeding season but usually after the birds have left the immediate nesting area. Adults of many Western Palearctic breeding species travel considerable distances to a few major intertidal regions in western Europe to molt. The reasons for choice of these sites are not yet established, but may include safety from mammalian and other predators at a time when many of the molting birds are growing up to five or six primary feathers simultaneously in each wing, and therefore are presumably less able to take off and maneuver quickly. No estimates are available of mortality during this period of the annual cycle, nor during the molt into breeding plumage in the spring. However, the latter involves only replacement of body feathers, so that flight ability should not be impaired. In this respect the chances of survival should be no different from those at other times during the nonbreeding season. The provision of extra food for growth of new feathers or to compensate reduced insulation might require shorebirds to feed for longer each day during molt and lead to reduced alertness to predators. However, feeding time is known to be longer only during premigratory fattening and during winter for shorebirds staying in north temperate areas.

C. Mortality in the Nonbreeding Areas

Although shorebirds migrating long distances from the Arctic, or shorter distances from temperate latitudes, reach coasts and estuaries on the western seaboard of Europe that are normally free from ice and snow, in certain winters freezing conditions may persist for several weeks and large numbers of certain species may then die. For example, during the hard winter of 1962–1963, many Redshank, Dunlin, and Red Knot were found dead on the Wash in eastern England (Pilcher, 1964). Elsewhere in Britain, similar findings were made (e.g., Ash, 1964). However, it was not possible to determine what percentage of the birds of each species present on each estuary died, partly because counts before the hard weather began had rarely been made and partly because not all birds that died were available to be found by ornithologists. Many corpses probably were eaten by scavengers, such as foxes, soon after death.

To obtain more precise information on mortality and its variation from year to year, we began in 1975 to mark individuals of five species of shorebirds, at Teesmouth in northeastern England, with unique combinations of color rings. In subsequent winters we made regular, at least weekly, searches for the color-marked birds, in the expectation that most, if not all, adults that spent part of one winter on the estuary in one year would return to do so in later years. More recently, we have also organized searches for the color-marked birds on other estuaries in Britain and on the western European seaboard. Although a few birds marked at Teesmouth have indeed been seen elsewhere, these have been in addition to, rather than instead of, sightings of these individuals at Teesmouth in that winter. Hence, although the percentages of birds that return to Teesmouth in successive winters provide only minimal survival rates, we believe these to be very close to the true survival rates in four out of five species marked (Eurasian Curlew *Numenius arquata*, Sanderling, Ruddy Turnstone and Bar-tailed Godwit *Limosa lapponica*). In the fifth species (Black-bellied Plover *Pluvialis squatarola*) we now have evidence that a few individuals move as far west as Teesmouth only in years in which very cold weather reaches continental Europe as early as December. In other years, those individuals are believed to remain in the coastal Wadden areas of The Netherlands or northern Germany (Townshend, 1982b, in preparation). Thus, for this species, annual survival rates can be estimated accurately only for the large proportion of the population that visits Teesmouth each winter, and somewhat less accurate estimates are available by including all marked birds, as presented below.

The most detailed information has been obtained for three shorebird species that differ markedly in their winter ecology. Sanderling are restricted to open sea beaches at Teesmouth; there they feed chiefly upon small polychaetes (*Nerine cirratulus*) and crustaceans (*Bathyporeia pelagica* and *Eurydice pulchra*) (Evans *et al.*, 1980; Brearey, 1982). In most winters, about 600–800 birds, between 10–20% of the entire British wintering population, are found around the Tees Estuary, which lies close to the northern limit of their wintering range in Europe. One might expect, therefore, that winter survival would be lower here than in other major European wintering sites, if climate directly affects survival. In contrast, Ruddy Turnstone winter in large numbers on rocky shores much further north in Britain than at Teesmouth, and even in Iceland (Wilson, 1982) and Norway. Although, like Sanderling, they are entirely coastal in their winter distribution, they take a much wider range of invertebrate prey, chiefly from rocky shores. But they are also opportunistic feeders, often responding quickly to new food sources, such as spat of the mussel *Mytilus edulis*, washed off outlying rocks at Teesmouth by wave action and deposited on the shore; and adults and larvae of wrack-flies *Coelopa* spp., colonizing detached macroalgae thrown up on the strandline. About 300 Turnstone are found in midwinter at Teesmouth, only a few percent of the British wintering population. On the mudflats of the Tees Estuary are found Eurasian Curlew, feeding upon the large polychaete *Nereis diversicolor* and in autumn also on small shore crabs *Carcinus maenas*. In winter, some curlew, chiefly the shorter-billed males, move off the mudflats to feed on earthworms in coastal pastures; most of the longer-billed females stay on the estuary to feed (Townshend, 1981). Perhaps 1% of the Eurasian Curlew wintering in Britain are found at Teesmouth—some 500–600 birds. The estuary lies close to the northern limit of its wintering range. Until 1982, it was a lawful quarry species in Britain, but is now protected by the 1981 Wildlife and Countryside Act.

Annual minimum survival rates, measured as return rates to the wintering grounds at Teesmouth, of these three species are given in Table II. This shows that (1) survival varied considerably from year to year within each species, but not necessarily in parallel between species, (2) average annual survival was consistently high in Ruddy Turnstone, for which Teesmouth is well south of the northern limit of its wintering range [an exception occurred in one year (1980) in which summer mortality was heavy; see below], (3) average survival in curlew, the species formerly hunted, was lower than in the other two species, even though it is much larger (700–1000 g as opposed to 50–100 g), *contra* Boyd's (1962) general

Table II. Annual Variations in Survival between Successive Autumns of Adult Sanderling, Ruddy Turnstone, and Eurasian Curlew at Teesmouth, Northeastern England

Survival between autumns	Sanderling	Turnstone	Curlew
1975–1976	(100% (5))	97% (33)[a]	—
1976–1977	92 (49)[a]	90 (68)	82% (28)[a]
1977–1978	87 (45)	88 (77)	79 (61)
1978–1979	74 (53)	90 (69)	82 (51)
1979–1980	92 (49)	62 (69)	74 (42)
1980–1981	69 (45)	87 (45)	70 (33)
Average of annual rates	83%	85%	77%
No. of years	5	6	5
Adult survival (Boyd, 1962)[b]	56%	66%	75%

[a] Figures in parentheses are numbers of individually marked birds from which return rates were obtained.
[b] Estimated from recoveries of ringed (banded) birds by the general public.

finding that larger shorebirds survive better, and (4) survival rates of Sanderling and Ruddy Turnstone, species not hunted, are higher than those estimated by Boyd from recoveries of numbered metal-banded birds.

By regular observations of the color-banded birds, it has been established that the dates of departure of individuals from the estuary in successive years normally occur within the same 2-week period, irrespective of the winter conditions they have endured beforehand. Departure dates in spring 1981 and 1982 of those color-banded Sanderling that returned, after migration to their breeding grounds and back, are shown in Fig. 1. The winter of 1980–1981 was mild, but in December 1981 and January 1982 a 4-week period of exceptionally cold weather occurred. As may be seen, this did not affect the timing of spring departure. From these data, and similar observations for other species, e.g., Eurasian Curlew (Fig. 2), it has been possible to identify a date for each species representing the earliest normal date of departure of individuals that survived until the following autumn. For Sanderling and Ruddy Turnstone, this date has been taken as 16 March, for curlew as 1 March. This has allowed us to partition the annual mortality for each species into components representing mortality on the wintering grounds (birds seen in autumn, last seen *before* the normal date of earliest departure, and failing to return in the following autumn); and mortality on migration or on the breeding grounds (birds seen in autumn, last seen *after* the normal date of earliest

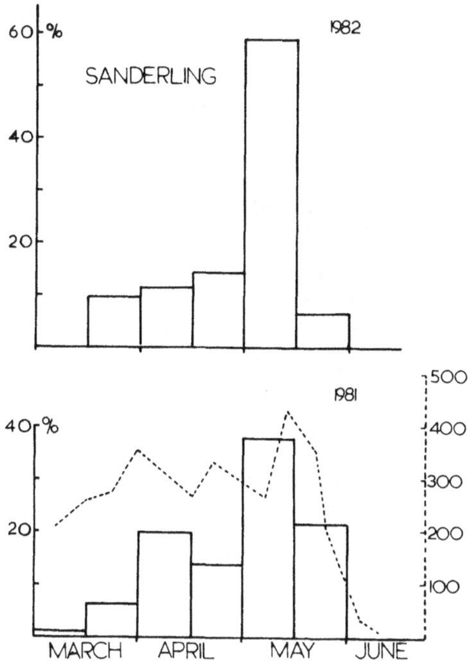

Fig. 1. Percentage of individually color-banded adult Sanderling *Calidris alba* departing in half-monthly periods from Teesmouth, northeastern England, in springs 1981 and 1982. All these individuals (1981, $n = 79$; 1982, $n = 97$) returned to the estuary in the following autumn. Counts of the total numbers present from March to June 1981 are also shown, emphasizing the use of this estuary also by passage migrants.

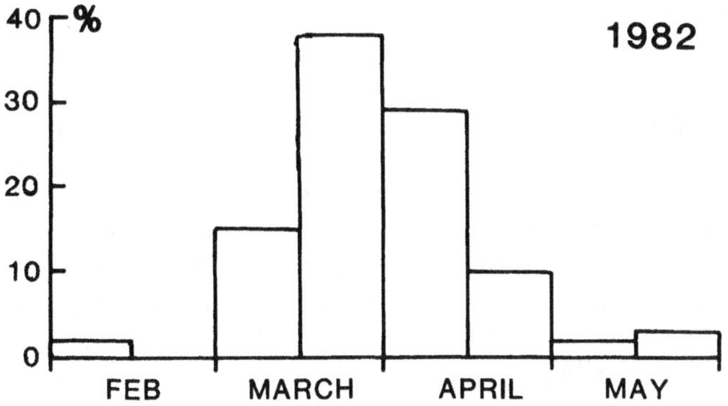

Fig. 2. Percentage of individually color-banded adult curlew *Numenius arquata* departing in half-monthly periods from Teesmouth, northeastern England, in spring 1982. All these individuals returned to the estuary in the autumn ($n = 58$).

Table III. Seasons of Disappearance of Adult Shorebirds, Color-Marked on Their Wintering Grounds at Teesmouth, Northeast England

No. of marked birds disappearing in:	Sanderling	Turnstone	Curlew
Winter 1975–1976	0	1	—
Summer 1976	0	0	—
Total for year	0	1	—
Winter 1976–1977	2	4	—[a]
Summer 1977	2	3	5
Total for year	4	7	5
Winter 1977–1978	4	6	5
Summer 1978	2	3	0
Total for year	6	9	5
Winter 1978–1979	10	4	8
Summer 1979	4	3	1
Total for year	14	7	9
Winter 1979–1980	2	12	7
Summer 1980	2	14	3
Total for year	4	26	10
Winter 1980–1981	7	4	8
Summer 1981	7	2	1
Total for year	14	6	9

[a] No birds marked until March 1977.

departure, and failing to return in the following autumn). Such a procedure may incorrectly assign a few birds that died in late spring on the wintering grounds to the "summer mortality" category, but not vice versa. Any bias present therefore tends to underestimate the contribution of winter mortality to the total annual mortality. Table III and Fig. 3 show that in spite of this possible bias, winter mortality (between arrival at Teesmouth in autumn and early or mid-March) is at least as important as mortality in all other periods of the year combined, and in some years much more so. These data also emphasize that although winter mortality varies from year to year in all three species, it does not do so in parallel. The effects of the severe winter weather in 1978–1979 led to an increase in mortality only of Sanderling [this was correlated with somewhat poorer food resources for Sanderling in that winter than in the previous one (Brearey, 1982)].

The data also indicate heavier than normal mortality in the summer (or on migration) in certain years. Again, these do not coincide between

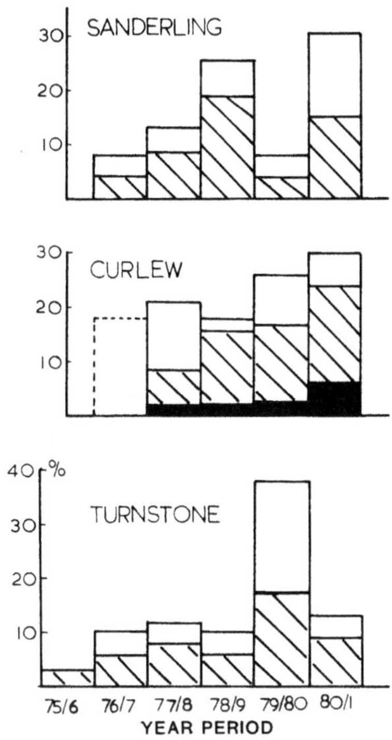

Fig. 3. Percentage mortality at Teesmouth, northeastern England, of Sanderling (top), Eurasian Curlew (center) and Ruddy Turnstone (bottom), partitioned into seasonal components. Mortality was estimated by the percentage of individually color-banded birds present in one winter that failed to return to the estuary in the following winter. For each year-period, e.g., 1975–1976, mortality is partitioned into a winter component (hatched), and a summer component (open rectangle) covering deaths on migration or on the breeding grounds. That part of mortality known to be attributable to shooting is shown as a black rectangle. For further details, see text.

the species. As the Ruddy Turnstone are part of the Greenland and Canadian Arctic breeding populations (Branson *et al.*, 1978), whereas the Sanderling are believed to include, and may consist chiefly of, the Siberian breeding population (Prater and Davies, 1978), and Eurasian Curlew may include both British and Scandinavian breeding birds, these differences in years of peak breeding season mortality are unremarkable. What is more significant is that the approach, if extended to other species, might allow monitoring of conditions on Arctic breeding grounds without the need for long-term studies there. For example, if Red Knot and Ruddy Turnstone wintering populations in Britain, both known to derive from breeding populations in Greenland and Canadian Arctic (Dick *et al.*, 1976; Branson *et al.*, 1978), show summer mortality that varies in parallel, even though the two species migrate at slightly different times and by somewhat different routes (see Wilson, 1981), then the summer mortality, measurable from observations on the wintering area, can be taken as indicating the suitability of conditions on the breeding grounds. Such information might be coupled with the information from satellite photographs of snow

Table IV. Return (Minimum Survival) Rates of Adult Bar-tailed Godwits Marked on the Tees Estuary

Between winters of	Survival
1977/78–1978/79	100% (5)
1978/79–1979/80	100 (5)
1979/80–1980/81	83 (6)
1980/81–1981/82	70 (20)
Average annual survival (over 4 years)	88%

[a] Unlike other species, most of the godwits caught at Teesmouth have been juveniles. All the adults included in Table IV were caught from August to October, during their molt.

conditions on the breeding areas to make predictions about success of breeding in a particular summer.

Less information is available about annual variation in survival of Bar-tailed Godwits, for they have proved more difficult to capture than other species at Teesmouth. Nevertheless, what information there is (Table IV) shows again the high annual survival rates typical of shorebirds, and the inadequacy of mortality rate estimation from recoveries, as opposed to resightings, of banded birds. Boyd (1962) obtained a survival rate of only 60% from birds ringed on passage in Norway.

Similar studies of survival have been made for Black-bellied Plovers at Teesmouth. The departure dates for this species (Fig. 4) occur much earlier in the year than those for the other two Arctic breeding waders, Ruddy Turnstone and Sanderling. Even in a spring of severe weather (1979), departures were not delayed by comparison with those in a mild spring (1980). The similarity in departure pattern arises because most of the individual birds that provided data for the 1979 spring also provided data in 1980, and the departure dates of individuals are remarkably consistent from year to year. In the analysis of the seasonal components of total annual mortality (Table V), deaths are attributed to the winter period if birds were last seen before the end of January and to the "summer" season if they were last seen after mid-February. Last sightings in early February are separated, as they are difficult to assign with confidence to either category. Table V shows that, as with the other species studied at Teesmouth, minimum survival rates varied markedly from year to year. However, the contribution of winter mortality to the total mortality was generally lower than in the other species, except in the severe winter of 1978–1979 and to a lesser degree in winter 1980–1981. The year-to-year

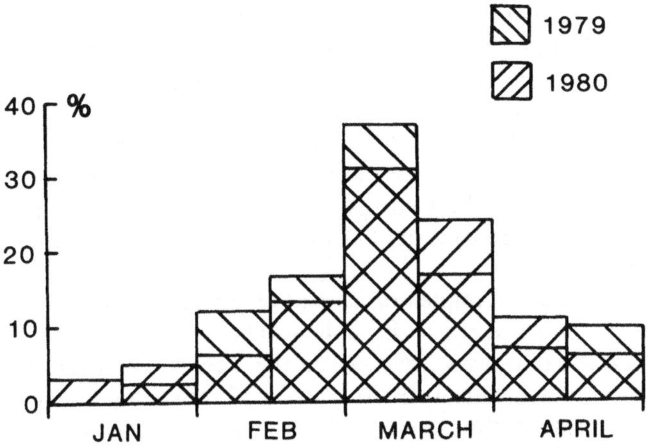

Fig. 4. Percentage of individually color-banded adult Black-bellied Plovers *Pluvialis squatarola* departing in half-month periods from Teesmouth, northeastern England, in springs 1979 and 1980. All these individuals returned to the estuary in subsequent autumns (1979, $n = 60$; 1980, $n = 62$).

pattern of winter mortality was similar to that found in the Sanderling, as was the pattern of summer mortality, except that many more Black-bellied Plovers disappeared in "summer" 1980. Both species are believed to migrate to breeding grounds in Siberia, but the plovers must spend some time in late March, April, and early May at a staging post en route

Table V. Return Rates of Adult Black-bellied Plovers to the Tees Estuary

Survival between winters of			No. of birds last seen in Winter	Early Feb.	Summer	% of total losses during winter
1975/76–1976/77	100%	(5)[a]	0	0	0	0
1976/77–1977/78	68	(19)	0	2	3[b]	0
1977/78–1978/79	100	(19)	0	0	0	0
1978/79–1979/80	78	(63)	9	0	4[b]	64
1979/80–1980/81	81	(107)	1	3	11	5
1980/81–1981/82	88	(83)	4	1	5	40
Average annual survival (over 6 years)	86%					

[a] Numbers in parentheses are sample sizes of color-marked birds.
[b] Remaining birds that failed to return departed in autumn, so season of death uncertain.

to the breeding grounds, probably in the Wadden areas of Holland, Germany, and Denmark (Pienkowski and Evans, Vol. 6 of this series), whereas many of the Sanderling stay at Teesmouth until they have molted into breeding plumage, and then migrate direct to the breeding grounds (as judged by the fat levels they carry before departure). Thus, the increase in summer mortality of plovers in 1980, above that expected if the summer mortality of the two species were to fluctuate in parallel, may have arisen because of mortality in late spring on the molting grounds in western Europe.

From the large samples of Black-bellied Plovers now color-marked, we hope in future years to be able to test for differences in mortality between adults that arrive to molt on the Tees in early autumn and then pass on to winter further south or west, and those that arrive on the Tees after molt, or stay after molt, to face whatever weather conditions the winters bring. Work is also in progress (Townshend, 1982a) to examine differences in survival between territorial and nonterritorial wintering birds. Finally, there is a portion of the plover population that does not reach Teesmouth until January, on the return movement toward the breeding grounds. Such birds spend 4–6 weeks on the estuary, normally at the most climatically unfavorable part of the winter. At the moment, this group is not separated from the earlier arrivals in calculations of annual mortality. However, as more data accumulate, it will be possible to examine their survival in comparison with that of birds that establish themselves on the Tees in late autumn. In this way, we intend to develop measures of the costs and benefits of different migration strategies within the species.

Unlike the other four species, no estimate of mortality of Black-bellied Plovers was available from Boyd's (1962) paper. The high survivals estimated for all five species in the present study update and correct figures given by Evans (1981) for the earlier years of the study.

A similar approach to measurement of mortality has been applied to Eurasian Oystercatchers in Britain by Goss-Custard *et al.* (1982). These authors were more interested in differences in survival with age, and so combined data from five winters to give average values of annual survival, as measured by rates of return of individually color-banded birds. Of the average annual disappearance rate of adults of 14%, only one-tenth occurred in "winter" (between August and the end of January), and the rest in spring and summer, away from the study site on the Exe Estuary, southwestern England. This picture differs from that for most of the species studied at Teesmouth for at least two possible reasons: (1) the Exe Estuary, on average, is one of the warmest in Britain, whereas Teesmouth is one of the coldest (see Pienkowski *et al.*, 1979) so that winter mortality would be expected to be lower at the Exe; (2) oystercatchers breed at temperate latitudes, those reaching the Exe in winter coming from nesting areas in Scotland, Faeroes, southwestern Norway, and The Netherlands (Goss-Custard *et al.*, 1982), whereas four of the five species studied at Teesmouth breed in the Arctic, where they are probably less vulnerable to predators and human disturbance (Pienkowski, 1984c).

It needs to be stressed that adults from a single breeding area may winter in widely different sites, with correspondingly differing chances of survival, and that birds at a single wintering site may originate from different breeding areas (as in the case of the oystercatcher).

IV. SURVIVAL OF IMMATURE BIRDS AND AGE OF FIRST BREEDING

A. Survival of Young Birds after Fledging

From his calculations of mortality, based on recoveries of individually numbered banded birds, Boyd (1962) concluded that survival was lower in the first year of a shorebird's life than later. Other methods of estimating survival have confirmed this conclusion, though absolute values for particular species have not necessarily coincided with those obtained by Boyd. In part, this results from biases inherent in use of information from metal-ringed birds, discussed earlier and leading to underestimation of survival; and in part from alternative sources of bias in other methods, but which also tend toward underestimation. For example, in the rather few studies of survival of individually color-banded chicks from fledging until the following breeding season, usually no more than a return rate of chicks to the natal areas has been measured. Such rates will underestimate true survival considerably if young birds disperse extensively from their birthplaces, as do Ringed Plovers in eastern Britain (Pienkowski, 1984a). Furthermore, as chicks are usually ringed well before they fledge, the survival rate estimated includes a component of prefledging mortality as well as postfledging mortality.

In a study of the estuarine Shelduck *Tadorna tadorna*, Pienkowski and Evans (1982) extrapolated from the rate of return to the natal breeding population (in this case at several years of age) of ducklings (29%), to estimate survival rate to recruitment age and settlement anywhere (54%). For this, they used records elsewhere on their molt migration route of birds older than 1 year but marked as ducklings. There should be no difference in the chance of sighting such birds whether they eventually settled at the natal area or elsewhere. Hence, to estimate the number surviving to recruitment age, the number of ducklings that eventually recruited to the natal site was increased in proportion to the total number of marked ducklings recorded elsewhere on the migration route, compared with the number of the latter settling later in the natal area. This method tends to overestimate survival, as birds seen in winter may not survive to the breeding season. However, together with the simple return rate, it gives maximum and minimum estimates for survival rates. Similar methods were used by Pienkowski and Evans (1982) to adjust estimates of survival rates for older birds (82% return giving an estimate of 93% survival). This approach is potentially useful for studies of other shorebirds, but may be more difficult as they tend to use a greater number of sites than do Shelducks, and their smaller marks are less visible.

An alternative approach to estimation of first-year survival has been to calculate what number of young per pair should have survived in order to replace adult deaths, if the population was to remain steady from year to year. This number can be compared with the number per pair known to have fledged, to give a percentage survival. This procedure depends on the accuracy of estimation of adult survival as well as on a correct estimation of number of young fledging.

Table VI summarizes some estimates of mortality during the first year of life. Not all refer to the period from fledging onwards; indeed, for some, survival is calculated from the first migration period.

At Teesmouth we have obtained a limited quantity of data on minimum survival rates of juveniles from their first to second autumns of life, and again from their second to third autumns. These data are summarized in Table VII, which includes a comparison with the appropriate survival rates for adults in the same year-periods. For three of the four species, apparent survival rates are lower for juveniles than for adults in the same year, but the differences are not all statistically significant. Even if they are, it does not prove that actual survival rates are lower between the first and second winters of life than in older birds, as there is insufficient information on the settlement patterns and movements of these younger birds. Quite possibly, those that settle on an estuary in their first autumn of life, but subsequently move on to another site in late autumn or early winter, may not return to the first estuary in later years. This seems likely if, as juveniles, they leave because they are forced out, e.g., by territorial behavior of adults arriving in late autumn after having molted elsewhere, as has been found for some Black-bellied Plovers at Teesmouth (Townshend, in preparation).

Table VII suggests that return rates of second-winter birds are not significantly lower (or higher) than those of adults in the same year-pe-

Table VI. Estimates of Shorebird Mortality from Fledging to 1 Year of Age

Species	% mortality	Method of estimation	Source
Eurasian Oystercatcher	36	Recoveries of banded chicks	Boyd (1962)
	40	Recoveries of banded chicks	Harris (1967, 1975)
Ruddy Turnstone	56	Recoveries of juveniles banded in autumn	Boyd (1962)
Eurasian Curlew	62	Recoveries of juveniles banded in autumn	Boyd (1962)
	50	Recoveries of banded chicks	Bainbridge and Minton (1978)
Dunlin	62	Recoveries of juveniles banded in autumn	Martin-Löf (1961)
	70	Proportion of 1-year-olds on breeding grounds	Holmes (1966)
Ringed Plover	61	Recoveries of banded chicks	Boyd (1962)
	41	Returns to natal area and sightings elsewhere of color-banded chicks known to have fledged	Pienkowski (1980, 1984a)

riods, even though the age of first breeding and/or acquisition of breeding plumage may be later in larger species (Eurasian Curlew and Black-bellied Plover as opposed to Ruddy Turnstone and Sanderling). In contrast, Pienkowski (1984a) found that the return rate of Ringed Plovers to a breeding area in northern Britain was similar between the first and second summers of life (57%) to that between fledging and the first summer (59%), but much lower than that of older birds (80+%). This could be due to continued dispersal, or to mortality associated with the first attempt at breeding (which occurs at 1 year of age in most individuals of this species). The latter explanation was suggested by Boyd (1962) to account for reduced survival of Eurasian Avocets in their fourth and fifth years of life, following first breeding.

The species for which survival rates of young and old birds most closely agree is Eurasian Curlew, for which adult mortality at Teesmouth was unexpectedly high, as discussed earlier. If the higher mortality of adults arises from hunting pressure, it is possible that juveniles, being smaller in size and therefore less desirable quarry, are less at risk. Thus, their survival from first to second autumns of life might indeed be lower than that of adults, if mortality due to hunting could be excluded. Nevertheless, in an analysis of recoveries of Eurasian Curlews ringed as chicks in Britain as a whole, Bainbridge and Minton (1978) found that mortality from the date of ringing to 1 year of age was 50%, double the average

Table VII. Survival of Young and Older Shorebirds between Successive Autumns at Teesmouth, Northeast England

Species/age	% survival	% survival of adults in same year	Year period
Ruddy Turnstone			
Juvenile	77 (13)	90 (68)	1976–1977
2nd year	100 (10)	88 (77)	1977–1978
3rd year	90 (9)	90 (69)	1978–1979
Sanderling			
Juvenile	82 (11)	92 (49)	1979–1980
2nd year	89 (9)	69 (45)	1980–1981
Eurasian Curlew			
Juvenile	87 (8)	82 (28)	1976–1977
	55 (11)	?	1981–1982
2nd year	100 (7)	79 (61)	1977–1978
Black-bellied Plover			
Juvenile	62 (13)	68 (19)	1976–1977
	61 (18)	78 (63)	1978–1979
	67 (9)	81 (107)	1979–1980
2nd year	100 (8)	100 (19)	1977–1978
	73 (11)	81 (107)	1979–1980
	100 (6)	88 (83)	1980–1981

annual mortality of adults, and they failed to find any difference between mortality rates calculated from returns of birds shot and those found dead.

Goss-Custard *et al.* (1982) showed that mortality of first-year and second-year Eurasian Oystercatchers during autumn and winter on the Exe Estuary was much higher than that of adults (12, 9, and 1.5%, respectively) during the same period. This might arise because young birds are less efficient at feeding by the specialized techniques required to open bivalves, or that they are prevented from feeding on occasions by their low competitive abilities on the feeding grounds.

Differences in feeding or competitive abilities are not the only factors leading to lower survival of first-year than adult shorebirds. Particularly in the Arctic breeding species, juveniles occur in a much wider range of habitats than adults when they first arrive on the nonbreeding areas in late summer or autumn. In particular, they may roost in more vulnerable places, and losses to mammalian predators may then be high. Foxes have killed several juvenile Black-bellied Plovers at Teesmouth in some autumns because the birds roosted at the edge of brackish pools (which were surrounded by rough grassland, through which foxes could approach unnoticed), rather than at the edge of open sand or mudflats where the adults congregated.

Table VIII. Average Age of First Recorded Nesting of Banded Shorebirds

Species/area	Average age (years)			Source
Ringed Plover	males	1.2		Pienkowski (1980;
(NE England)	females	1.2		1984a)
Dunlin				
Finland		2	(some 1)	Soikkeli (1970)
Alaska		1		Holmes (1966)
Temminck's Stint	males	1.8		Hildén (1978a)
(Finland)	females	1.0		
Ruddy Turnstone		2		Bergman (1946)
(Finland)				
Common Redshank		2	(some 1)	Grosskopf (1959)
(Friesian Islands)				
Eurasian Avocet		3		Boyd (1962)
Eurasian Oystercatcher				
Friesian Islands		3.3		Schnakenwinkel (1970)
Skokholm, Wales	males	5	(some 4)	Harris (1970)
	females	4	(some 3)	

B. Age of First Breeding

Smaller species of shorebirds breed for the first time when 1 or 2 years old, but larger species have a longer period of delayed maturity (Table VIII). In Temminck's Stint, age of first breeding in males was delayed in years when population density was high, as some 1-year-old males that defended territories failed to secure a mate.

Although no confirmation is available of age of first breeding of Black-bellied Plover, Bar-tailed Godwit, and Red Knot from studies of marked breeding populations, many individuals spend the summer on coasts and estuaries south of their breeding areas and within their "wintering" range. Many of these birds do not assume breeding plumage, though some do. It may be supposed that most do not breed until they are 2 or 3 years old. Some of the smaller shorebirds that migrate in their first autumn of life to Australia, South America, and South Africa also stay in these nonbreeding areas throughout the following calendar year and do not return toward their breeding areas until they are nearly 2 years old. These species include Curlew Sandpiper, and Rufous-necked Stint *Calidris ruficollis* (Thomas, 1970) in Australia, Semipalmated Sandpiper *Calidris pusilla* in Surinam, South America (Spaans, 1978), and Curlew Sandpiper in South Africa (Puttick, 1980).

It would be valuable to know if birds that do not return to the breeding range until they are 2 or 3 years old are predominantly of one sex.

V. ARE SHOREBIRD POPULATIONS REGULATED?

Regulation of an animal population implies that the numbers of a particular species, although fluctuating from year to year, tend to return to a steady average level. The most widely held view is that this can occur only by the action of density-dependent factors which affect either or both (1) reproduction, in such a way that this decreases as population density increases, and (2) mortality, so that this increases as density increases. Most population models assume that all individuals are similar. However, one of the hypotheses of behavioral regulation of populations recognizes that the genetic structure of the population changes as the population fluctuates and that the effects of density on dispersal, survival, and reproduction differ between genotypes (Chitty, 1967; Krebs and Myers, 1974). Most field studies of populations have focused not on the entire species or isolated breeding unit, but on changes in the numbers in a defined area, i.e., they have been studies of control of population density in part of an animal's range and not of population size in the whole distributional range.

Shorebird populations have been studied through extensive counts in some nonbreeding areas, but only by intensive surveys on most breeding grounds. There are, therefore, difficulties in integrating the information on winter-to-winter changes in total population size with information on changes in population density in very small parts of the breeding ranges. Nevertheless, it is possible to examine the evidence for the existence of density-dependent controls in both breeding and nonbreeding areas, to assess whether regulation of total population size in such highly mobile species as shorebirds is likely, or whether they should be considered as opportunistic species whose numbers fluctuate widely but usually well below the maximum sustainable capacity of their chosen habitats. The evidence will be considered below.

We recognize, following Enright (1976), that if the population of a species is indeed being regulated by density-dependent factors, acting with equal severity in all parts of its geographical range, the level about which regulation occurs may yet vary from area to area in response to variation in the severity of density-independent factors, such as climate, which cause mortality. This is illustrated in Fig. 5. Such variations may occur not only in space but also in time. If climate changes on the nonbreeding areas, mortality may also change and thereby alter the equilibrium level of the population; so also will a change in hunting pressure, unless hunting has merely removed birds that would have died anyway from other causes.

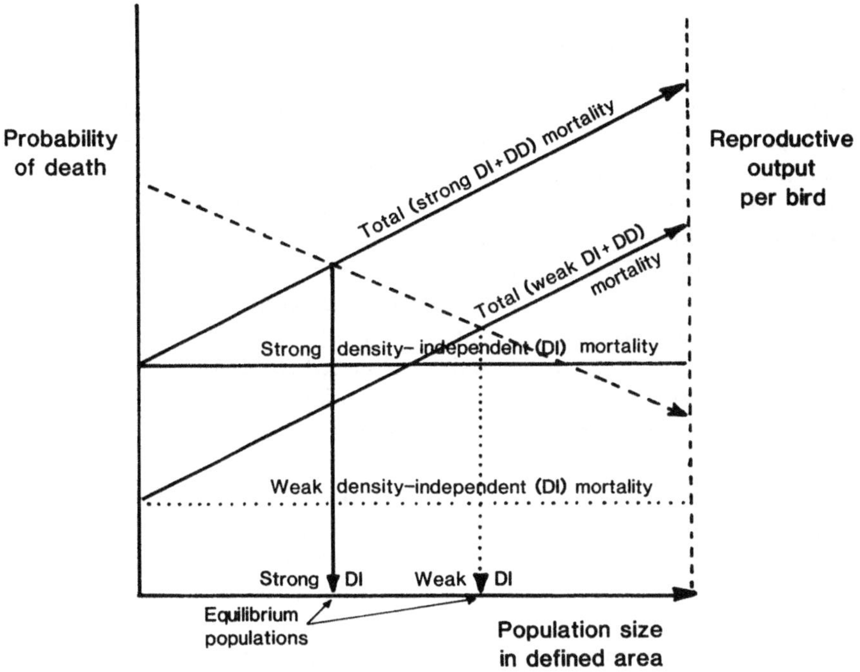

Fig. 5. Theoretical model of the effects on equilibrium population size of a change in the level of density-independent mortality (resulting from e.g., climatic change or change in hunting pressure). The model assumes that strengths of density-dependent effects on reproduction and mortality are not altered by the change in density-independent factors.

Because there may be considerable diversity in the breeding areas from which shorebirds of some species travel to a single wintering area, e.g., Eurasian Oystercatchers (Goss-Custard *et al.*, 1982), and because many individuals of some species shift their breeding sites from year to year (Hildén, 1979), it seems difficult to produce realistic models of shorebird population dynamics on anything less than a total population level.

A. Reproductive Output

In a stimulating review, prepared for a symposium of the International Waterfowl Research Bureau in 1977 but delayed in publication by several years, Goss-Custard (1981) considered the evidence, available from studies of breeding waders prior to 1977, that density-dependent factors might act in the breeding season to control reproductive output.

He could find no conclusive evidence of density-dependent effects on the numbers of chicks hatched or fledged per breeding pair in a range of species; but he argued that, as territoriality limited the density of breeding pairs of oystercatchers on the island of Skokholm (Harris, 1970) and Dunlin in Alaska (Holmes, 1970), and nonbreeding birds were present, production expressed per individual of the total population (breeders plus nonbreeders) probably varies in an inversely density-dependent way. This conclusion holds only if all available optimal habitat is occupied, so that changes in range of the population cannot occur, and if the nonbreeding birds present in the area are physiologically capable of breeding, as was certainly the case with at least some of the oystercatchers.

Although the Alaskan Dunlin that colonized Holmes' study area when he removed territorial birds were also physiologically mature (and bred), it is not known whether they would have established and bred elsewhere if the opportunity to establish in the study area had not been created. There are problems also in the interpretation of the Skokholm experiments, for the removal of 21 breeding birds and their replacement by others apparently had no effect on the numbers of nonbreeding pairs with territories (9) or in the nonbreeding flock (about 30). These numbers apparently did not fluctuate from year to year, and Harris could conclude only that "most of the individuals which replaced removed birds would probably otherwise have been non-breeders." But they might also have established elsewhere. Oystercatchers in Britain as a whole, and elsewhere in northwestern Europe, have spread inland as a breeding species during the present century (Buxton, 1962; Heppleston, 1972; Sharrock, 1976). Hence, it is difficult to argue that the total (or even the British) population is regulated, even though the number of pairs breeding on Skokholm appears to be limited by territoriality. As the numbers of breeding plus nonbreeding birds on Skokholm fluctuate very little from year to year, in the absence of the "floating" birds that take over territories experimentally made vacant, it is difficult to judge whether Goss-Custard's hypothesis—that production varies in an inversely density-dependent way when expressed per adult present in the study area—is tenable.

In contrast to the situation in Alaska, Soikkeli's (1967, 1970) studies of Dunlin in Finland established that breeding density was affected by the numbers of young produced 2 years previously, i.e., territory size was not immutable. Nevertheless, as Goss-Custard (1981) pointed out, as long as some birds are prevented from holding a territory and so lose the opportunity to breed, even a situation in which territory size is reduced when more birds attempt to settle could produce weak density-dependent restrictions on average production of young per adult present in the area.

Such weak density-dependent effects may be sufficient to regulate a population, but within fairly broad limits.

The argument so far has focused on the males; possession of a territory does not ensure a male's opportunity to breed, however. In Temminck's Stint (Hildén, 1979), Dunlin (Soikkeli, 1967), and Western Sandpiper (Holmes, 1971), some territorial males failed to secure mates in some years, usually when the density of males and therefore number of territories on the study area was highest. An "excess" of males in some years has also been recorded in Northern Phalaropes (Hildén and Vuolanto, 1972). This raises the question whether the average output of young per female is restricted by territoriality.

Adult females are less site-faithful than adult males, in Temminck's Stint (Hildén, 1979), Dunlin (Soikkeli, 1967, 1970), and indeed in most birds. Added to this, the age of first breeding in females of the smaller shorebirds is at 1 year of age, whereas that of males is often older (on average) in the same sites. Furthermore, nonbreeding physiologically mature females have not been recorded on the study sites of any species (other than oystercatchers) during years in which other females are breeding. All these factors point to a lack of restriction on numbers of females breeding in a population, attributable to territorial behavior, though it may restrict their density in any one area. It may be supposed that some adult females return first to former breeding areas, whereas others, and most 1-year-old females, disperse more widely until they find a suitable breeding site where they choose from males that go with it. Such dispersive behavior should be particularly pronounced and advantageous in species that utilize temporarily suitable habitats.

Such reasoning demands that, in most shorebird species, the value of a territory is only to the males: those that hold them securing the possibility of obtaining a mate; those that do not, forfeiting the possibility to breed. Territoriality is not found in some species, e.g., Redshank, or in all parts of a species' range, e.g., Dunlin. In these situations, it might be expected that the males would use other behaviors to secure a mate, e.g., more elaborate courtship displays. These presumably would not restrict the proportion of females that breed.

Territoriality in oystercatchers may have additional functions, at least in the island situation studied by Harris (1970). These shorebirds feed their young, and food is usually obtained from within the territory on Skokholm. Harris concluded that birds defended as large an area as they could, to ensure an adequate food supply for the young. In several other species that do not feed their young [e.g., Ringed Plovers (Edwards *et al.*, 1947; Pienkowski, 1980), Purple Sandpipers *Calidris maritima* (Bengtson, 1970)], females play at best some role in territorial establishment and defense from conspecifics. This implies some additional role of territories, but critical evidence as to its nature is lacking. Perhaps the most feasible possibilities are the spreading out of nests to reduce predation or the avoidance of being forced into suboptimal areas. Further work is needed to test these possibilities. For the second possibility, the existence of areas used only at high population size requires investigation. If such areas are identified, the breeding output and adult survival there relative to other sites need measuring: the latter will not be easy.

Our conclusion is that there is as yet no firm evidence for density-dependent restriction of reproductive output in those species breeding at higher latitudes: that all females that are capable of breeding, hormonally and nutritionally, are able to find a mate somewhere within the distributional range of the species, but not necessarily in the same site that was used in the previous breeding season. We further conclude that average reproductive output per female varies from year to year and from place to place, depending in particular on local variations in the intensity of predation. There is no evidence that such variations are determined by population density of the shorebirds, but many other factors have been implicated.

The major factor affecting reproductive output would appear to be weather at high latitudes. This determines whether or not breeding attempts are made and whether there is appreciable mortality of breeding adults. In neither case is the effect on reproductive output density-dependent, as was documented earlier, in Section II.B.

B. Mortality in the Nonbreeding Areas

Even if reproductive output and mortality on the breeding grounds are not affected by population size or population density, a shorebird population could still be regulated by density-dependent mortality on the nonbreeding area. This topic has been reviewed by Goss-Custard (1980), who concluded that the rates of food intake of at least some sections of shorebird populations are likely to be depressed by increasing bird density. In principle, this might lead to an increase in winter mortality, but feeding conditions might normally be so good that a large decrease in food intake rate might be permissible before mortality increased.

Not all possible feeding areas on an estuary are of equal quality. Some lie further from safe roost sites; some hold lower densities of food, or less preferred prey. Occupation of feeding areas in sequence has been shown for several shorebirds by Zwarts (1974), and for Eurasian Oystercatcher by Meire and Kuyken (1984), Red Knot (Goss-Custard, 1977), and Black-bellied Plover (Townshend, 1981), but only in the first-mentioned has it been shown that the most profitable feeding areas were filled first. Also in Eurasian Oystercatchers, Swennen (1984) has shown that mortality may be disproportionately higher in the poorer quality feeding areas. Thus, even though population density on a wintering ground need not increase as more birds arrive (if later arrivals occupy additional, but poorer quality, feeding sites), the percentage mortality of birds on that wintering ground may rise as the total numbers increase. This would produce an effect on the population similar to density-dependent mortality.

Direct tests of a possible relationship between winter mortality and density of birds on the feeding grounds are difficult to make, for the numbers of many species on a particular site fluctuate considerably during the winter period. While it is easy to follow these fluctuations, it is not usually easy to establish exactly when individuals die and therefore the appropriate bird density to which to relate mortality. Data presented earlier (Table V) indicate that the winter mortality of Black-bellied Plovers was much heavier in the severe winter of 1978–1979 than in the following mild winter, 1979–1980. The seasonal pattern of population change, however, was very similar in the two winters, apart from a large influx in February in the mild winter (Dugan, 1981). Loss of body condition of birds in the severe winter was proven only in January, when strong winds interfered with feeding behavior (Dugan *et al.*, 1981) and not when population density was highest, i.e., in December and February (count data from Dugan, 1981). Indeed, six of the nine birds that are presumed to have died in that winter were last seen in January. Taken together, these data indicate that mortality of Black-bellied Plovers in a wintering area near the northern limit of their nonbreeding range, if density-dependent at all, is not strongly so.

For many species, the effects of temperature, precipitation, and wind force on the availability of prey (Evans, 1976, 1979; Pienkowski, 1981, 1982, 1983a,b) lead to much wider fluctuations in the density of available prey from day to day during the winter than do the changes in density of the birds. This means that bird feeding rates fluctuate from day to day chiefly as a result of weather factors. Davidson (1981) summarized the weather conditions under which different species of shorebirds draw upon their fat reserves (and ultimately their muscles also) to avoid death from inadequate rates of food intake. He recognized four major groups of shorebirds: (1) those that normally feed inland but move to the shore when their usual feeding grounds become frozen; (2) the long-legged and/or visual-feeding estuarine species that have difficulty in foraging on open mudflats during gales; (3) the short-legged tactile-feeding estuarine species that can feed only at the tide edge or on exposed mud, and which are therefore affected chiefly by freezing conditions; and (4) open coastal species that are affected by wind strength through its power to create strong wave action that can interfere with feeding and remove sediments and prey (see also Evans, 1981). Davidson provided information on the levels of fat reserves carried by the different groups of birds and on the rate at which they are drawn upon under particular conditions. Visual feeders lay down most fat but tend to use it most rapidly when weather becomes severe.

Although most winter mortality seems to arise from prolonged periods of severe weather, weak density-dependent mortality of shorebirds might yet arise from differences in the rate at which food stocks are depleted, during the course of the winter, by different densities of birds (Goss-Custard, 1980). The effects of an increase in the density of birds on the reduction in absolute density of prey in an area would be greatest by the end of the winter, but the percentage availability of prey may be lowest, e.g., *Arenicola marina* (or even highest, e.g., *Mytilus edulis*) in midwinter, when temperatures are lowest. Hence, it is difficult to predict by how much the densities of available prey, and hence bird feeding rates, are affected by increases in bird density. Furthermore, it cannot be concluded that any such decreases in feeding rates necessarily lead to increases in mortality.

The possibility remains that in areas further south, where weather and feeding conditions are initially more favorable than those in the wintering areas at northern temperate latitudes, mortality might be affected by the density of birds. Indeed, it has been noted, in Mauritania, that densities of feeding waders are very high but food resources low compared to temperate areas, and the proportion of time spent feeding is high but intake low (Altenburg *et al.*, 1982). It has been suggested that some Bar-tailed Godwits and Black-bellied Plovers may leave the Banc d'Arguin, Mauritania (a major west African "wintering" ground for Palearctic waders), in the first 2 months of the year to avoid competition as food supplies decline there (Dugan, 1981), even though by doing so they have to return to the northern temperate wintering areas at a time of year when weather conditions are at their most severe and the percentage availability of estuarine foods at their lowest.

Predation on shorebirds on their nonbreeding grounds could also lead to density-dependent mortality, though there is no proof of this. The proportion of birds taken by predators on northern European estuaries is not known, but significant predation rates have been recorded in California by Page and Whitacre (1975). No estimates have been published of predation rates in tropical "wintering" areas.

C. Epilogue

If populations of shorebirds are regulated, or even limited, then either or both their breeding and nonbreeding areas should be fully utilized in some years. The breeding ranges of many species in the northern temperate zone have changed steadily this century: some expanding [e.g., Eurasian Oystercatcher, Little Ringed Plover *Charadrius dubius*, Eurasian Curlew, Redshank (Hale, 1980)], others contracting (e.g., Eurasian Golden Plover), and yet others shifting [e.g., Temminck's Stint decreasing in Finland but spreading southwards in Norway (Hildén, 1978a)]. In some cases, human activities have led to habitat modification (Hildén and Hyytiä, 1981), so that changes in range may not be equivalent to changes in total population size but be associated with changes in density. In others, e.g., Eurasian Oystercatcher, range expansion and increase in population have occurred in parallel. Areas that once were used by some species are now empty, even though the habitat appears not to have changed, e.g., Golden Plover nesting in Britain (Parslow, 1973). It is thus difficult to accept that all possible breeding areas are utilized at present.

Evidence from the wintering grounds is equally unhelpful. Numbers of birds reaching the furthest parts of migration systems fluctuate widely from year to year, e.g., in Tasmania (Thomas, 1970). Even within, rather than at the margins of, a nonbreeding range, birds may move further in some years than others, e.g., Black-bellied Plovers, so that the relative densities of a species in different parts of its range are not necessarily similar from year to year.

We wonder if the concept of population regulation in shorebirds is illusory; whether severe weather acting unpredictably to cause mortality on the wintering grounds and on the breeding grounds, where it may also prevent reproduction, serves to prevent shorebird populations from reaching the utopian levels of the carrying capacity of the habitats to which they are adapted.

ACKNOWLEDGMENTS

We thank our colleagues at Durham for discussions and information on many matters: Dr. D. M. Brearey, Miss R. H. W. Cooper, Dr. N. C. Davidson, Dr. P. J. Dugan, Mr. L. R. Goodyer, Mrs. A. Pienkowski, Dr. D. J. Townshend, and Mr. A. G. Wood. We are grateful to the following organizations for major financial support of our studies of shorebirds at Lindisfarne and Teesmouth: Nature Conservancy Council of Great Britain, Natural Environment Research Council, Science Research Council, Nuffield Foundation, British Ornithologists' Union, and E.E.C. Environment Programme.

REFERENCES

Altenburg, W., Engelmoer, M., Mes, R., and Piersma, T., 1982, Wintering waders on the Banc d'Arguin, Comm. No. 6 of the Wadden Sea Working Group, Leiden.

Ash, J. H., 1964, Observations in Hampshire and Dorset during the 1963 cold spell, *Br. Birds* **57**:221–241.

Bainbridge, I. P., and Minton, C. D. T., 1978, The migration and mortality of the Curlew in Britain and Ireland, *Bird Study* **25**:39–50.

Bengtson, S.-A., 1970, Breeding behaviour of the Purple Sandpiper *Calidris maritima* in west Spitsbergen, *Ornis Scand.* **1**:17–25.

Bergman, G., 1946, Der Steinwälzer, *Arenaria i. interpres* (L.) in seiner Beziehung zur Umwelt, *Acta Zool. Fenn.* **47**:1–144.

Boyd, H., 1962, Mortality and fertility of European Charadrii, *Ibis* **104**:368–387.

Branson, N. J. B. A., Ponting, E. D., and Minton, C. D. T., 1978, Turnstone migrations in Britain and Europe, *Bird Study* **25**:181–187.

Brearey, D. M., 1982, The feeding ecology and foraging behaviour of Sanderling and Turnstone at Teesmouth, N.E. England, Unpublished Ph.D. thesis, University of Durham, U.K.

Bub, H., 1962, Planberingungen am Sandregenpfeifer (*Charadrius hiaticula*), *J. Ornithol.* **103**:243–249.

Buxton, E. J. M., 1962, The inland breeding of the oystercatcher in Great Britain, 1958–59, *Bird Study* **8**:194–209.

Byrkjedal, I., 1980, Nest predation in relation to snow-cover—A possible factor influencing the start of breeding in shorebirds, *Ornis Scand.* **11**:249–252.

Chitty, D., 1967, The natural selection of self-regulatory behaviour in animal populations, *Proc. Ecol. Soc. Aust.* **2**:51–78.

Davidson, N. C., 1981, Survival of shorebirds (Charadrii) during severe weather: The role of nutritional reserves, in: *Feeding and Survival Strategies of Estuarine Organisms* (N. V. Jones and W. J. Wolff, eds.), pp. 231–249, Plenum Press, New York.

Dick, W. J. A., and Pienkowski, M. W., 1979, Autumn and early winter weights of waders in north-west Africa, *Ornis Scand.* **10**:117–123.

Dick, W. J. A., Pienkowski, M. W., Waltner, M. A., and Minton, C. D. T., 1976, Distribution and geographical origins of Knots *Calidris canutus* wintering in Europe and Africa, *Ardea* **64**:22–47.

Dircksen, R., 1932, Die Biologie des Austernfischers, der Brandseeschwalbe und der Küstenseeschwalbe nach Beobachtungen und Untersuchungen auf Norderoog, *J. Ornithol.* **80**:427–521.

Dugan, P. J., 1981, Seasonal movements of shorebirds in relation to spacing behaviour and prey availability, Unpublished Ph.D. thesis, University of Durham, U.K.

Dugan, P. J., Evans, P. R., Goodyer, L. R., and Davidson, N. C., 1981, Winter fat reserves in shorebirds: Disturbance of regulated levels by severe weather conditions, *Ibis* **123**:359–363.

Edwards, G., Hosking, E., and Smith, S., 1947, Aggressive display of the Ringed Plover, *Br. Birds* **40**:12–19.

Emlen, S. T., and Oring, L. W., 1977, Ecology, sexual selection and the evolution of mating systems, *Science* **197**:215–223.

Enright, J. T., 1976, Climate and population regulation: The biogeographer's dilemma. *Oecologia* (Berlin) **24**:295–310.

Etheridge, B., 1982, Distribution of Dunlin *Calidris alpina* nests on an area of South Uist machair, *Bird Study* **29**:239–243.

Evans, P. R., 1968, Autumn movements and orientation of waders in north-east England and southern Scotland, studied by radar, *Bird Study* **15**:53–64.

Evans, P. R., 1976, Energy balance and optimal foraging in shorebirds: Some implications for their distributions and movements in the non-breeding season, *Ardea* **64**:117–139.

Evans, P. R., 1979, Adaptations shown by foraging shorebirds to cyclical changes in the activity and availability of their intertidal invertebrate prey, in: *Cyclic Phenomena in Marine Plants and Animals* (E. Naylor and R. G. Hartnoll, eds.), pp. 357–366, Pergamon Press, Elmsford, N.Y.

Evans, P. R., 1981, Migration and dispersal of shorebirds as a survival strategy, in: *Feeding and Survival Strategies of Estuarine Organisms* (N. V. Jones and W. J. Wolff, eds.), pp. 275–289, Plenum Press, New York.

Evans, P. R., Brearey, D. M., and Goodyer, L. R., 1980, Studies on Sanderling at Teesmouth, N.E. England, *Wader Study Group Bull.* **30**:18–20.

Ferns, P. N., Green, G. H., and Pienkowski, M. W., MS, Clutch size and egg size in Arctic waders.

Fuller, R. J., 1978, Breeding populations of Ringed Plovers and Dunlins in the Uists and Benbecula, Outer Hebrides, *Bird Study* **25**:97–102.

Goss-Custard, J. D., 1977, The ecology of the Wash. III. Density-related behaviour and the possible effects of a loss of feeding grounds on wading birds (Charadrii), *J. Appl. Ecol.* **14**:721–739.

Goss-Custard, J. D., 1980, Competition for food and interference among waders, *Ardea* **68**:31–52.

Goss-Custard, J. D., 1981, Role of winter food supplies in the population ecology of common British wading birds, *Verh. Ornithol. Ges. Bayern* **23**:125–146.

Goss-Custard, J. D., Durell, S. E. A., Sitters, H. P., and Swinfen, R., 1982, Age-structure and survival of a wintering population of Oystercatchers, *Bird Study* **29**:83–98.

Graul, W. D., 1973, Adaptive aspects of the Mountain Plover social system, *Living Bird* **12**:69–94.

Graul, W. D., 1976, Food fluctuations and multiple clutches in the Mountain Plover, *Auk* **93**:166–167.

Graul, W. D., Derrickson, S. R., and Mock, D. W., 1977, The evolution of avian polyandry, *Am. Nat.* **111**:812–816.

Green, G. H., 1978, The census of waders, in: *Joint Biological Expedition to North East Greenland 1974: The Report of the Joint Expedition Comprising Wader Study Group N.E. Greenland Expedition 1974 and Dundee University N.E. Greenland Expedition 1974* (G. H. Green and J. J. D. Greenwood, eds.), pp. 84–109, Dundee University N.E. Greenland Expedition, Dundee.

Green, G. H., Greenwood, J. J. D., and Lloyd, C. S., 1977, The influence of snow conditions on the date of breeding of wading birds in north-east Greenland, *J. Zool.* **183**:311–328.

Grosskopf, G., 1959, Zur Biologie des Rotschenkels (*Tringa t. totanus*), II, *J. Ornithol.* **100**:210–236.

Grosskopf, G., 1964, Sterblichkeit und Durchschnittsalter einiger Küstenvögel, *J. Ornithol.* **105**:427–449.

Hale, W. G., 1956, The lack of territory in the Redshank *Tringa totanus, Ibis* **98**:398–400.

Hale, W. G., 1980, *Waders*, Collins, London.

Harris, M. P., 1967, The biology of Oystercatchers *Haematopus ostralegus*, on Skokholm Island, S. Wales, *Ibis* **109**:180–193.

Harris, M. P., 1969, Effect of laying date on chick production in Oystercatchers and Herring Gulls, *Br. Birds* **62**:70–75.

Harris, M. P., 1970, Territory limiting the size of the breeding population of the Oystercatcher (*Haematoqus ostralegus*)—A removal experiment, *J. Anim. Ecol.* **39**:707–713.

Harris, M. P., 1975, Skokholm oystercatchers and the Burry Inlet, *Rep. Skokholm Bird Observatory for 1974*, pp. 17–19.

Hays, H., 1972, Polyandry in the Spotted Sandpiper, *Living Bird* **11**:43–57.

Heppleston, P. B., 1972, The comparative breeding ecology of Oystercatchers (*Haematopus ostralegus* L.) in inland and coastal habitats, *J. Anim. Ecol.* **41**:23–51.

Higgins, K. F., and Kirsch, L. M., 1975, Some aspects of the breeding biology of the Upland Sandpiper in North Dakota, *Wilson Bull.* **87**:96–102.

Hildén, O., 1975, Breeding system of Temminck's Stint *Calidris temminckii, Ornis Fenn.* **52**:117–146.

Hildén, O., 1978a, Population dynamics in Temminck's Stint *Calidris temminckii, Oikos* **30**:17–28.

Hildén, O., 1978b, Occurrence and breeding biology of the Little Stint *Calidris minuta* in Norway, *Anser Suppl.* **3**:96:100.

Hildén, O., 1979, Territoriality and site tenacity of Temminck's Stint *Calidris temminckii, Ornis Fenn.* **56**:56–74.

Hildén, O., and Hyytiä, K., 1981, The population changes and present status of waders of Finland, *Proc. 2nd Nordic Congr. Ornithol.* **(1979)**:19–37.

Hildén, O., and Vuolanto, S., 1972, Breeding biology of the Red-necked Phalarope *Phalaropus lobatus* in Finland, *Ornis Fenn.* **49**:57–85.

Holland, P. K., Robson, J. E., and Yalden, D. W., 1982, The breeding biology of the Common Sandpiper *Actitis hypoleucos* in the Peak District, *Bird Study* **29**:99–110.

Holmes, R. T., 1966, Breeding ecology and annual cycle adaptations of the Red-backed Sandpiper (*Calidris alpina*) in northern Alaska, *Condor* **68**:3–46.

Holmes, R. T., 1970, Differences in population density, territoriality and food supply of Dunlin on Arctic and subarctic tundra, in: *Animal Populations in Relation to Their Food Resources* (A. Watson, ed.), pp. 303–319, Blackwell, Oxford.

Holmes, R. T., 1971, Density, habitat and the mating system of the Western Sandpiper (*Calidris mauri*), *Oecologia (Berlin)* **7**:191–208.

Holmes, R. T., 1972, Ecological factors influencing the breeding season schedule of Western Sandpipers (*Calidris mauri*) in subarctic Alaska, *Am. Midl. Nat.* **87**:472–491.

Jackson, R., and Jackson, J., 1975, A study of breeding Lapwings in the New Forest, Hampshire 1971–74, *Ringing and Migration* **1**:18–27.

Jenni, D. A., 1974, The evolution of polyandry in birds, *Am. Zool.* **14**:129–144.

Keighley, J., 1949, Oystercatchers, *Skokholm Bird Obs. Rep.* **1948**:6–9.

Krebs, C. J., and Myers, J. H., 1974, Population cycles in small mammals, *Adv. Ecol. Res.* **8**:268–369.

Lack, D., 1963, Migration across the southern North Sea studied by radar. Part 4. Autumn, *Ibis* **105**:1–54.

Laven, H., 1940, Beiträge zur Biologie des Sandregenpfeifers (*Charadrius hiaticula* L.), *J. Ornithol.* **88**:183–287.

Maclean, G. L., 1972, Clutch size and evolution in the Charadrii, *Auk* **89**:299–324.

Marcström, V., and Mascher, J. W., 1979, Weights and fat in Lapwings and Oystercatchers starved to death during a cold spell in spring, *Ornis Scand.* **10**:235–240.

Martin-Löf, P., 1961, Mortality rate calculations on ringed birds with special reference to the Dunlin *Calidris alpina*, *Ark. Zool.* **13**:483–491.

Meire, P., and Kuyken, E., (1984), Relations between the distribution of waders and the intertidal benthic fauna of the Oosterschelde, Netherlands, in: *Coastal Waders and Wildfowl in Winter* (P. R. Evans, J. D. Goss-Custard, and W. G. Hale, eds.), pp. 57–68, Cambridge University Press, Cambridge.

Meltofte, H., 1976, Ornithological observations in southern Peary Land, North Grønland, 1973, *Medd. Groenl.* **205**:1–57.

Meltofte, H., 1977, 'Fly-away trick' of some Arctic waders, *Br. Birds* **70**:333–336.

Meltofte, H., 1979, The population of waders Charadriidae at Danmarks Havn, north-east Greenland, 1975, *Dan. Ornithol. Foren. Tidsskr.* **73**:69–94.

Meltofte, H., 1984, Populations and breeding schedules of waders in high arctic Greenland, *Medd. Groenl. Bioscience* (in press).

Meltofte, H., Elander, M., and Hjort, C., 1981, Ornithological observations in northeast Greenland between 74°30′ and 76°00′ N lat., 1976, *Medd. Groenl. Biosci.* **3**:1–53.

Morrison, R. I. G., 1975, Migration and morphometrics of European Knot and Turnstone on Ellesmere Island, Canada, *Bird-Banding* **46**:290–301.

Myers, J. P., 1981, A test of three hypotheses for latitudinal segregation of the sexes in wintering birds, *Can. J. Zool.* **59**:1527–1534.

Nethersole-Thompson, D., 1973, *The Dotterel*, Collins, London.

Nordberg, S., 1950, Researches on the bird fauna of the marine zone in the Aland Archipelago, *Acta Zool. Fenn.* **63**:1–62.

Oring, L., and Knudsen, M. L., 1972, Monogamy and polyandry in the Spotted Sandpiper, *Living Bird* **11**:59–73.

Page, G., and Whitacre, D. F., 1975, Raptor predation on wintering shorebirds, *Condor* **77**:73–83.

Parmelee, D. F., and Payne, R. B., 1973, On multiple broods and the breeding strategy of Arctic Sanderlings, *Ibis* **115**:218–226.

Parslow, J., 1973, *Breeding Birds in Britain and Ireland: A Historical Review*, Poyser, Berkhamsted, England.

Pienkowski, M. W., 1980, Aspects of the ecology and behaviour of Ringed and Grey Plovers *Charadrius hiaticula* and *Pluvialis squatarola*, Unpublished Ph.D. thesis, University of Durham, U.K.

Pienkowski, M. W., 1981, How foraging plovers cope with environmental effects on invertebrate behaviour and availability, in: *Feeding and Survival Strategies of Estuarine Organisms* (N. V. Jones and W. J. Wolff, eds.), pp. 179–192, Plenum Press, New York.

Pienkowski, M. W., 1982, Diet and energy intake of Grey and Ringed Plovers *Pluvialis squatarola* and *Charadrius hiaticula* in the non-breeding season, *J. Zool.* **197**:511–549.

Pienkowski, M. W., 1983a, Surface activity of some intertidal invertebrates in relation to temperature and the foraging behaviour of their shorebird predators, *Mar. Ecol. Prog. Ser.* **11**:141–150.

Pienkowski, M. W., 1983b, The effects of environmental conditions on feeding rates and prey selection of shore plovers, *Ornis Scand.* **14**:227–238.

Pienkowski, M. W., 1984a, Behaviour of young Ringed Plovers *Charadrius hiaticula* and its relationship to growth and survival to reproductive age, *Ibis.* **126**:in press.

Pienkowski, M. W., 1984b, Development of feeding and foraging behaviour in young Ringed Plovers *Charadrius hiaticula* in Greenland and Britain, *Dan. Ornithol. Foren. Tidsskr.* **78**:in press.
Pienkowski, M. W., 1984c, Breeding biology and population dynamics of Ringed Plovers *Charadrius hiaticula* in Britain and Greenland: Nest predation as a possible factor limiting distribution and timing of breeding, *J. Zool.* **202**:83–114.
Pienkowski, M. W., and Evans, P. R., 1982, Breeding behaviour, productivity and survival of colonial and non-colonial Shelducks *Tadorna tadorna, Ornis Scand.* **13**:101–116.
Pienkowski, M. W., and Green, G. H., 1976, Breeding biology of Sanderlings in north-east Greenland, *Br. Birds* **69**:165–177.
Pienkowski, M. W., and Greenwood, J. J. D., 1979, Why change mates?, *Biol. J. Linn. Soc.* **12**:85–94.
Pienkowski, M. W., Lloyd, C. S., and Minton, C. D. T., 1979, Seasonal and migrational weight changes in Dunlin, *Bird Study* **26**:134–148.
Pilcher, R. E. M., 1964, Effects of the cold weather of 1962/3 on birds of the north coast of the Wash, *Wildfowl Trust Annu. Rep.* **15**:23–26.
Pitelka, F. A., Holmes, R. T., and MacLean, S. F., Jr., 1974, Ecology and evolution of social organization in Arctic sandpipers, *Am. Zool.* **14**:185–204.
Prater, A. J., 1982, Wader Research Group report, Debrecen, *Int. Waterfowl Res. Group Bull.* **47**:74–78.
Prater, A. J., and Davies, M., 1978, Wintering Sanderlings in Britain, *Bird Study* **25**:33–38.
Pulliainen, E., 1970, On the breeding biology of the dotterel *Charadrius morinellus, Ornis Fenn.* **47**:69–73.
Puttick, G. M., 1980, Energy budgets of Curlew Sandpipers at Langebaan Lagoon, South Africa, *Estuarine Coastal Mar. Sci.* **11**:207–215.
Raner, L., 1972, Förekommer polyandri hos smalnäbbad simsnäppa (*Phalaropus lobatus*) och svartsnäppa (*Tringa erythropus*)?, *Fauna Flora* **67**:135–138.
Ratcliffe, D., 1976, Observations on the breeding of the Golden Plover in Great Britain, *Bird Study* **23**:63–116.
Richardson, W. J., 1979, South-eastward shorebird migration over Nova Scotia and New Brunswick in autumn, *Can. J. Zool.* **57**:107–124.
Rittinghaus, H., 1956, Untersuchungen am Seeregenpfeifer (*Charadrius alexandrinus* L.) auf der Insel Oldeoog, *J. Ornithol.* **97**:117–155.
Schnakenwinkel, G., 1970, Studien an der Population des Austernfischers *Haematopus ostralegus* auf Mellum, *Vogelwarte* **25**:336–355.
Sharrock, J. T. R., 1976, *The Atlas of Breeding Birds in Britain and Ireland*, British Trust for Ornithology, Tring.
Soikkeli, M., 1967, Breeding cycle and population dynamics in the Dunlin (*Calidris alpina*), *Ann. Zool. Fenn.* **4**:158–198.
Soikkeli, M., 1970, Mortality and reproductive rates in a Finnish population of Dunlin *Calidris alpina, Ornis Fenn.* **47**:149–158.
Spaans, A. L., 1978, Status and numerical fluctuations of some North American waders along the Surinam coast, *Wilson Bull.* **90**:60–83.
Summers, R. W., and Hockey, P. A., 1980, Breeding biology of the White-fronted Plover *Charadrius marginatus* in the South-western Cape, South Africa, *J. Nat. Hist.* **14**:433–445.
Swennen, C., 1984, Differences in quality of roosting flocks of Oystercatchers, in: *Coastal Waders and Wildfowl in Winter* (P. R. Evans, J. D. Goss-Custard, and W. G. Hale, eds.), pp. 160–176. Cambridge University Press, Cambridge.

Symonds, F. L., 1981, A survey of breeding waders and wildfowl in Caithness, *Wader Study Group Bull.* **31**:9.

Thomas, D. G., 1970, Fluctuation of numbers of waders in south-eastern Tasmania, *Emu* **70**:79–85.

Townshend, D. J., 1981, The importance of field feeding to the survival of wintering male and female curlews *Numenius arquata* on the Tees Estuary, in: *Feeding and Survival Strategies of Estuarine Organisms* (N. V. Jones and W. J. Wolff, eds.), pp. 261–274, Plenum Press, New York.

Townshend, D. J., 1982a, The use of intertidal habitats by shorebird populations with special reference to Grey Plover (*Pluvialis squatarola*) and Curlew (*Numenius arquata*), Unpublished Ph.D. thesis, University of Durham, U.K.

Townshend, D. J., 1982b, The Lazarus syndrome in Grey Plovers, *Wader Study Group Bull.* **34**:11–12.

Wilson, J. R., 1981, The migration of high Arctic shorebirds through Iceland, *Bird Study* **28**:21–32.

Wilson, J. R., 1982, The wintering of shorebirds in Iceland, *Wader Study Group Bull.* **36**:16–19.

Yates, B. J., 1982, A population biology of breeding Redshanks *Tringa totanus*, Unpublished Ph.D. thesis, Liverpool Polytechnic, U.K.

Zwarts, L., 1974, *Vogels van het Brakke Getijgebied*, Amsterdam.

Chapter 4

BREEDING AREA FIDELITY, NATAL PHILOPATRY, AND THE SOCIAL SYSTEMS OF SANDPIPERS

Lewis W. Oring and David B. Lank

Department of Biology
University of North Dakota
Grand Forks, North Dakota 58202

I. INTRODUCTION

Sexual selection and kin selection theory have advanced with leaps and bounds in recent years, providing a framework for the study of mating system evolution and expression. Theoretical development has been enhanced by utilization of electrophoretic and various field techniques that aid in determination of parentage. At the same time, field studies of the long-term mating relationships of individually marked animals have shed new light on the genetic relationships within and between wild populations.

The genetic structure of free-living populations results from the interaction of dispersal (gene flow), size and spacing of groups (population subdivision), and mating systems (genetic combination) (Baker and Marler, 1980). These factors are not independent variables and must be studied in concert. In reviewing the predominant dispersal patterns of birds and mammals, Greenwood (1980) concluded that dispersal was, to a very large extent, coadapted with social structure. In mammals, where the predominant social system is based on female defense by males, dispersal typically is male biased, whereas in birds, where the dominant social system involves resource defense by males, dispersal typically is female biased.

Sandpipers (family Scolopacidae) include closely related species exemplifying most of the various types of polygynous, polyandrous, and monogamous mating systems (Emlen and Oring, 1977; Oring, 1982). Thus, examination of the relationships between dispersal patterns, mating and territorial systems, and other population parameters in this group may be particularly enlightening in evaluating the applicability of Greenwood's hypothesis.

Dispersal may be broken down into two components: (1) *natal dispersal*, dispersal of young prior to first breeding; and (2) *breeding dispersal*, dispersal of breeding adults between breeding efforts (Greenwood, 1980). In line with these definitions, *philopatry* refers to the breeding of young birds in the area of their birth, and *breeding area fidelity* refers to breeding of adults in the same location in successive years. While return rates provide only a partial look at dispersal, we rely upon them as a measure of dispersal because of the few data available on dispersal distances. All of the species discussed below are migratory, hence "dispersal" to wintering grounds occurs every year. However, as no breeding occurs in these areas, and as the birds are capable of precise homing from wintering to breeding areas, these movements are ignored.

This chapter reviews information on natal philopatry and breeding area fidelity in sandpipers from the standpoint of phylogenetic or species characteristics, sexual differences, population differences, and factors affecting individuals such as breeding experience. The more extensive data from a few intensively studied species are emphasized.

A. Why Return?

A bird that returns year after year to the vicinity of its natal or breeding site may reap advantages through: (1) familiarity with environmental factors such as food, cover, and predators; (2) familiarity with conspecifics, including mates, relatives, and others; and (3) enhanced competitive ability stemming directly from increased self-confidence on its home turf (Hinde, 1956). Environmental familiarity may improve the individual's ability to cope with stressful situations and/or facilitate the ease or speed of breeding. Familiarity with conspecifics may yield advantages from various forms of cooperative behavior with relatives and others. In addition, such tasks as mate acquisition and assessment of one's status relative to others in the population may be facilitated. Advantages reaped through environmental and conspecific familiarity may increase as a bird accrues experience with age. Where philopatry is common, individuals may benefit by breeding with others having similar "coadapted gene complexes" suited to local conditions (see Section III and Shields, 1982). Under most circumstances, these advantages would appear to be so great that there must be strong overriding reasons for birds not to return.

B. Why Disperse?

Animals may be expected to forego the advantages of returning to or remaining in a familiar area, only when these advantages are outweighed by disadvantages from (1) inbreeding (e.g., Greenwood *et al.*, 1978; see also Noordwijk and Scharloo, 1981) or (2) reduced chances for reproductive success due to a relative decrease in habitat quality. If adults are strongly site faithful, inbreeding depression should act as a selective force against philopatry, at least in one sex. The latter effect is most likely to occur due to: habitat alteration, causing a decrease in mate, food, or cover availability; extreme predation; or severe competition.

II. BREEDING AREA FIDELITY

Return rates for adult and young scolopacids are summarized in Table I. In studies lasting more than 2 years, the sample of returning adults counts some individuals more than once.

A. Species Differences

Species with the highest rates, i.e., males and females together averaging 70%+, are monogamous, have biparental care, and, in most, exhibit joint defense of breeding territories (Redshank, *Tringa totanus*; Willet, *Catoptrophorus semipalmatus*; Long-billed Curlew, *Numenius americanus*; Black-tailed Godwit, *Limosa limosa*; and Dunlin, *Calidris alpina schinzii*). The exception to this generalization is Temminck's Stint (*C. temminckii*). Although the Temminck's Stint mating system is polygamous (Hildén, 1975; Pienkowski and Greenwood, 1979), both males and females are territorial and provide parental care.

At the other extreme, Red Phalaropes (*Phalaropus fulicarius*), which do not defend territories and have uniparental care, have an adult return rate of only 8% for males and 0% for females (Schamel and Tracy, 1977). All three species of phalaropes exhibit lower site tenacity than most sandpipers, except where populations are isolated (Hildén and Vuolanto, 1972). The Buff-breasted Sandpiper (*Tryngites subruficollis*), another uniparental care species in which males defend only mating territories, and females are not territorial, also has very low breeding area fidelity. Bergman (1974) measured their population density in northern Alaska over 3 years and found it to vary from 1.0 to 10.0 birds/km^2.

Table I. Adult Breeding Site Fidelity and Natal Philopatry in Sandpipers[a]

Species	Percent of returning adults				Number of returning young				Mating system	Source
	n	M	F	?	*n*	M	F	?		
Tringa totanus	418	65	63		810			33	Monogamy	Grosskopf (1959, 1968)
	522	84	74		—	—	—		Monogamy	Hale and Ashcroft (1982)
Catoptrophorus semipalmatus[b]	107			74+	—	—	—		Monogamy	Howe (1982)
Actitis macularia	126		58		256	13	14		Resource-defense polyandry	Oring *et al.* (1983 and personal observation)
	178	54								
Numenius americanus	59		73		270	9	2		Monogamy	Redmond and Jenni (1982)
	30	67								
Limosa limosa	32		69		—	—	—		Monogamy	Jonas (1979)
	27	78								
Calidris pusilla	330	50	47		258	5	3		Monogamy	Gratto (personal communication)
C. mauri	104	58	49		187	3	4		Monogamy	Holmes (1971 and personal communication)

C. minutilla	50	65	38	—	—	—		Monogamy	Miller (1977)
C. temminckii	173	79	72	307	12	13		Rapid multiclutch polygamy	Hildén (1975, 1978, 1979)
C. alpina pacifica	—	—	—	34	2	0		Monogamy	Holmes (1966)
C. a. schinzii	—	—	—	448			16	Monogamy	Heldt (1966)
	323	77	71	511	30	27		Monogamy	Soikkeli (1970a,b)
Micropalama himantopus[b]	42	50	50	—	—	—		Monogamy	Jehl (1973 and personal communication)
Philomachus pugnax	30	40		47	0			Male dominance polygyny	Andersen (1948)
	11		27	43		2			
Phalaropus lobatus	—	—	—	108	8	2		Female access polyandry	Hildén and Vuolanto (1972)
	46	46	—	—	—	—		Female access polyandry	Rausch (1982)
P. fulicarius	30		0	93	1	0		Female access polyandry	Schamel and Tracy (1977)
	90	8							

[a] n = Number of "bird years" for adults, and number of individuals marked for young.
[b] These rates probably are lower than actually occurred; however, incomplete sampling should not have distorted the relative frequency of return of the two sexes.

Thus, species that are not strongly site faithful are characterized by uniparental care and little territorial behavior, whereas site-faithful species tend to have biparental care and be territorial. At present, however, we lack the ecological data to unravel the selective pressures responsible for observed interspecific variability in return rates. For example, almost nothing is known about overwinter mortality. We may speculate that the more highly philopatric species breed in more "stable" environments, but adequate tests of this hypothesis remain for the future.

B. Sex Bias

Data on return rates of males and females are available for 13 populations of 12 species (Table I). Only two of these show statistically significant sex differences: one of two studied populations of Redshank, and Least Sandpipers on Sable Island, Nova Scotia. In both, more males returned than females. Although not statistically different, eight of the remaining 11 populations had higher male return rates, while only two had higher return rates of females. Return rates were equal in one case. The two-tailed binomial probability of obtaining a 10:2 or greater ratio of differences in return is 0.04. It thus appears that males return to study sites at slightly higher rates than females in most species. The unweighted average from all 13 populations shows males returning at a 16% higher rate than females.

Greenwood (1980) concluded that acquisition of mates through control of territories should lead to greater philopatry of the resource-defending sex, and lesser natal philopatry and breeding area fidelity in the opposite sex. The scolopacid data suggest that Greenwood's generalization holds, but differences in male and female breeding area fideltiy are small. Detailed studies of breeding biology are needed to tease apart the proximal and ultimate factors responsible for sex biases in dispersal patterns.

Below we consider the behavior of each sex in monogamous, rapid multiclutch polygamous, polygynous, and polyandrous mating systems. The thesis is that differing intensities of sexual selection and different modes of mate acquisition (e.g., resource defense vs. intrasexual dominance) cause differences in the availability of breeding situations for the two sexes.

1. Monogamy

Monogamy and biparental care is the rule for most species of sandpipers. In this group, the intensity of sexual selection appears to be similar

(Emlen and Oring, 1977; O'Donald, 1980), as variance in the numbers of mates acquired by each sex is slight.

Monogamous species of sandpipers are usually territorial, but the nature of the territory, its size, and which sex defends it are highly variable. For example, in the Green Sandpiper (*Tringa ocrophus*) and Willet, both males and females defend large, multipurpose territories, and feeding areas and nest sites may be disjunct; whereas in Dunlin and Long-billed Curlew, males alone defend multipurpose territories. Despite this variation, it appears that in sandpipers with monogamous mating systems, a high degree of breeding area fidelity by both sexes is the rule (Table I).

The small male bias in return rates in these species may result from variation in the *territorial fidelity* of the two sexes. The best data are from Dunlin (Soikkeli, 1967, 1970a). Although there was only a 6% difference in the return rates of males and females (77% vs. 71%), females nested significantly further from former nest sites when switching mates. Greater dispersal by females than males might result in more females moving beyond the boundaries of a study site and hence not being recorded as returning. Female Redshank also undertake longer movements than males following repairing (Grosskopf, 1959). We hypothesize that the small sex biases in return rates seen in monogamous species are a result of the dispersal of a small percentage of females off the study site following failure to regain old territories and/or mates.

Why should females disperse further than males? Recall that males in these species usually return to the breeding grounds and reestablish themselves on territories before females. For both Dunlin and Western Sandpiper (*Calidris mauri*), female movements are sometimes caused by usurpation of territories and mates by earlier returning females (Soikkeli, 1970a; Holmes, 1971).

A late returning female may not only find her territory and mate taken, but she may also have to move some distance to find a new breeding situation. Her situation is more difficult than that of a male returning to find his territory occupied, for he is likely to face a single bird rather than a pair of territory defenders. Males may also be more willing to engage in agonistic encounters to regain old territories. Thus, the slight male bias in return rates, and the greater distances moved by females in monogamous species may be the proximal result of differences in arrival times of the two sexes to the breeding grounds (Oring and Lank, 1982).

2. Rapid Multiclutch Polygamy

In rapid multiclutch polygamy (RMP), a female may lay a clutch that is cared for only by a male, and then lay a second clutch, with the same or a different male, that she herself cares for. This system was discovered and well documented in Temminck's Stint (Hildén, 1975), and has been suggested for Little Stint (*Calidris minuta*) (Kistchinski and Flint, 1973) and for Sanderling (Parmelee and Payne, 1973). Emlen and Oring (1977) argued that sexual selection pressures in RMP systems should be little different from those in monogamous systems, as both sexes are polygamous [but see Pienkowski and Greenwood (1979) for a discussion of systems where birds change mates]. In Temminck's Stint, return rates and *within-study-site* dispersal distances are very similar to those recorded in monogamous species (Hildén, 1978, 1979), and these similarities may be due to basic similarities in the two types of mating systems. In both monogamous and RMP systems, males arrive on the breeding grounds before females, have stronger territorial fidelity than females, and females may fail to return to previous territories because of usurpers (Hildén, 1979). No data are available for other scolopacids with RMP systems.

3. Polygyny

In approximately six species of Calidrinae, 15 of Gallinagoninae, and at least four species of Scolopacinae, only females incubate, males defend territories, and males are potentially polygynous. Some of these species have resource-defense polygyny [e.g., White-rumped Sandpiper (*Calidris fuscicollis*) (Drury, 1961)], while others have male dominance polygyny or lek mating systems. In this latter group, males defend small mating territories, the most extreme forms being found in the Ruff (*Philomachus pugnax*) (e.g., van Rhijn, 1973) and Great Snipe (*Gallinago media*) (Lemnell, 1978). In all polygynous species, males are subjected to greater sexual selective pressures, and might be expected to show greater return rates and site-tenacity—especially high-status males. On the other hand, females in these systems can choose from among many males and nest sites. Thus, they may respond more quickly than males to changes in environmental quality and, therefore, be less site-faithful.

No data on return rates are available from any territorial polygynous species. Andersen (1948) recaptured 12 of 30 marked Ruffs and 3 of 11 marked Reeves (40% vs. 27%) in Denmark, but it is unclear whether both sexes were equally sampled, for he showed that a male's chances of being recaptured were related to his attendance rate at the lek the previous year (Andersen, 1951). Lemnell (1978) recaptured 5 of 21 male Great Snipes marked on a Swedish lek. These meager data provide little basis for discussion. The relationship between Ruff attendance at a lek from one year to the next and return is intriguing, however, for it suggests that the fidelity of males in polygynous systems may vary with their social status.

4. Polyandry

In Phalaropodinae (three species), females compete directly for access to males, and a few become polyandrous. There is no territorial defense and females compete as well on one pond as another, hence minimal site fidelity. Males, on the other hand, must weigh advantages of dispersing to a possibly superior habitat, with the disadvantage of losing familiarity with an area in which they will incubate and brood for nearly 40 days. Data available indicate low site tenacity on the part of males, and essentially none among females (Schamel and Tracy, 1977; Rausch, 1982; Table I). Whatever the ecological factors are that determine the site of phalarope breeding, they are of far greater importance than the advantages of site fidelity.

Spotted Sandpipers (*Actitis macularia*) have a resource-defense polyandrous mating system (Oring and Knudson, 1972; Hays, 1972), and females are subject to greater sexual selection pressure than males. Females attract males in large part on the basis of the quality of their multipurpose territory; and when dispersing to new, prime habitat, they face intense competition for territory acquisition. Males, on the other hand, can almost always breed, even after dispersing. All they have to do is wait for a female to become available after an initial nesting with a primary male, for eventually the primary male becomes involved with incubation and cannot defend his territory. Despite this, late arriving males may find their previous territory occupied by a new male, resulting in prolonged male–male chasing (Oring, Maxson, and Lank, unpublished).

In Spotted Sandpipers, then, the hypothesis that breeding area fidelity is related to the relative availability of breeding situations predicts greater female than male return. In addition to potential disadvantages associated with dispersal of females, the advantages of experience in a population appear to be greater for females. In Minnesota, returning females obtained significantly more mates, eggs, chicks, and fledglings than inexperienced females, whereas returning males had no more chicks hatch or fledge per year than inexperienced males (Oring *et al.*, 1983).

Overall return rates of Spotted Sandpipers did not show a significant sex bias (Table I), although females returned at slightly higher rates than did males. Sex differences in breeding site fidelity became evident, however, when the data were considered with regard to previous breeding experience of individuals, and their breeding success in a particular year. Figure 1 plots our data on return rates of Spotted Sandpipers to Little Pelican Island, Minnesota, as a function of "minimum age" (years since banding, or exact age, if a locally hatched chick), and whether or not they

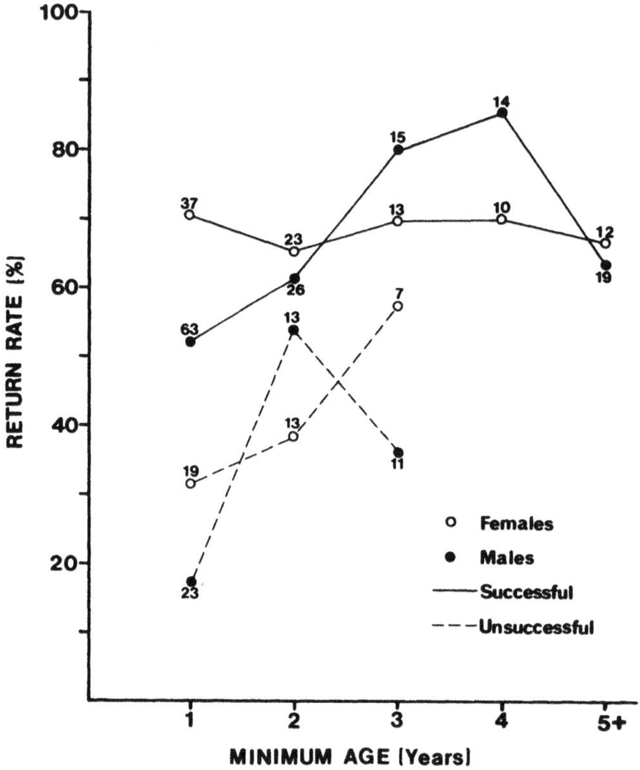

Fig. 1. Age-specific return rates of Spotted Sandpipers breeding on Little Pelican Island, Minnesota. The 5+ points combine information from 8 individual females and 13 individual males. Numbers indicate sample sizes (bird-years).

hatched chicks the previous year ("successful" or "unsuccessful" breeders).

Spotted Sandpiper males succeeding in their first breeding season, returned significantly less often than did successful males following subsequent breeding years (52% vs. 70%, $\chi^2 = 4.62$, 1 *df*, $p = 0.03$). Females, on the other hand, showed a consistent pattern after their first breeding season—returning if successful, at rates of 65–70%. In other words, females became locked in to the breeding site at the end of their first successful breeding season, while males were less site tenacious until after their third breeding season (Fig. 1). A direct comparison of return rates of successful first-year-breeding males to those of successful young females approached significance (52% vs. 70%, $\chi^2 = 3.08$, 1 *df*, $p = 0.08$). Unsuccessful birds of both sexes returned at similar rates (Fig. 1).

The timing of migration provides another way of looking at sex-related differences in return rates. The early migrating sex has more limited information on which to base a return–not return "decision." Information available is essentially limited to the quality of the site in past years plus physiographic assessment in the current year. But the later migrating sex, males in Spotted Sandpipers, can assess both territory quality and mate availability in the current year, for many females are already on territory when males arrive. In addition, the presence of these females is indicative of suitable reproductive conditions, for first-year-breeding females do not usually return following reproductive failure (Oring and Lank, 1982; see below). Thus, in Spotted Sandpipers, intense female sexual selection has led to early female arrival, and early female arrival provides males with a greater potential for assessing the quality of a breeding situation. Young males may be readily recruited into new populations by courting females; it seems less likely that females could be "recruited" as easily by suitable nesting habitat. This process may be partially responsible for the lowered site fidelity in young male Spotted Sandpipers, relative to females and older males.

The differential ease with which the sexes enter new populations is reflected in sex differences in philopatry following reproductive failure in birds (Greenwood and Harvey, 1976, 1982; Greenwood, 1980). In a number of cases, it has been shown that female birds change breeding sites more readily than males (Richdale, 1957; Catchpole, 1972). The pattern seen in Spotted Sandpipers is consistent with the hypothesis that the sex with fewer alternate reproductive opportunities shows increased site-fidelity. Thus, Spotted Sandpipers provide an important test of this hypothesis, as they run counter to the trend for greater male return rates found among birds in general (Greenwood, 1980), and substantiated here for scolopacids.

C. Effects of Age and Breeding Success

It has been a frequent finding that individual birds tend to move to new localities following breeding failure (Richdale, 1957; Catchpole, 1972; Harvey *et al.*, 1979; Thomas, 1980). The Spotted Sandpiper data presented in Fig. 1 showed that the probability of an individual returning varied with sex, age, and breeding success the previous year. Differences between the sexes were discussed above. Among failed breeders, experienced birds returned significantly more often than inexperienced ones (Fig. 1).

The comparatively high return rates of experienced birds after 1 year's failure reflect the fact that 81% of them had previously hatched chicks in the population. For this sandpiper species, we can formulate a set of rules that describe the observed breeding site fidelity.

1. If eggs hatch, return. This applies to all age and sex classes except for males breeding on the island for the first and possibly their second year. The behavior of these young males is not predictable. No other measure of reproductive performance (e.g., number of mates, number of young fledged) better predicts the probability of return.
2. If eggs do not hatch:
   a. and if breeding on the island for the first time, do not return, especially if a male.
   b. and if breeding in a second or later year, the decision is based on a composite of information including the recent failure, previous successes, and familiarity with alternate breeding opportunities. Hence, return rates are intermediate between those of successful breeding birds and birds failing in their first reproductive effort. Males may be less likely to return than females.

It is not surprising that these rules fail to account for all cases. Even events at other sites, including mortality and recruitment into other breeding groups, influence return as observed on the breeding ground. Nonetheless, these rules provide a mechanism that allows individuals to take into account changes in habitat suitability, both in terms of settling in potentially improved habitat, and dispersing from unsuitable areas.

The operation of these rules accounts for the large annual variation in adult breeding site fidelity seen on Little Pelican Island (Oring *et al.*, 1983). Nesting success varied greatly from year to year, and so did return rates, especially of inexperienced breeders. For example, following a "successful" year, 1977 (84% hatch), 11 of 14 inexperienced males and 5 of 6 inexperienced females returned. On the other hand, following the near-total breeding failure of 1975 (5% hatch), only 2 of 13 inexperienced males and 1 of 8 inexperienced females returned (Oring *et al.*, 1983).

Analysis of return rates of failed first-time breeders in years when most of the population was successful, versus years when most failed, did not show any "second-order" effects (Lank and Oring, unpublished). In other words, birds seem to be responding only to their individual success or failure and not to that of their neighbors.

From a population perspective, the degree of annual breeding site fidelity varies with the individual breeding success and the age structure of the breeders of each sex. Site fidelity of Spotted Sandpipers was examined, in less depth, in two other populations (Oring and Knudson, 1972; Hays, 1972), and overall return rates were similar to those on Little Pelican Island.

Do similar rules apply for individuals of other species? Long-term studies of other species show annual variation and/or the effects of previous breeding experience. Among male Northern Phalaropes (*Phalaropus lobatus*), Rausch (1982) found a significant relationship between breeding success and return, with 7 of 12 successful males returning to breed the following year, but 0 of 9 unsuccessful males doing so. Redmond and Jenni (1982) found that in Long-billed Curlews, hatching success in the year of capture was a significant predictor of return. For adults captured in 1977–1979, 80% (24/30) with successful nests were resighted the next year whereas only 42% (5/12) with unsuccessful nests were seen. There were indications that female curlews were more sensitive than males to both nest loss and disturbance due to capture. Despite the overall lack of sex difference shown in Table I, Redmond and Jenni (1982) concluded that under disturbed conditions (e.g., heavy predation), adult male curlews had greater site fidelity than females.

We saw earlier that a male Ruff's attendance at the lek, which may be a crude measure of reproductive success, was a predictor of its attendance in a subsequent year (Andersen, 1951). German Redshank returned at significantly lower rates (64.8%, $n = 418$) than British Redshank (79%, $n = 522$) (Grosskopf, 1959; Hale and Ashcroft, 1982), but there is no indication as to why this difference occurred. In Temminck's Stint, Hildén (1978) found that first-year females returned significantly less often than older birds (46% vs. 71%), but he did not state whether this reflected a differential response to nesting success by age class. Males showed no such first-year effect, their return rates declining steadily as they grew older (from 93% after the first year banded to 67% five years after banding), a fact Hildén attributed to mortality.

In Dunlin, breeding area fidelity was uniformly high and independent of reproductive success (Heldt, 1966; Holmes, 1966; Soikkeli, 1967, 1970a,b). Although Soikkeli did not present a breakdown of return data by years previously bred or by nesting success, there was no indication of a relationship between overall annual return rates and numbers of eggs hatched per adult (Soikkeli, 1967, 1970b).

In summary, while nesting success, age, and previous experience may be important predictors of breeding site fidelity in some species, they appear to have little effect in others. We argued elsewhere that we could understand interspecific variation in site fidelity for at least a few species of scolopacids in the context of other life history traits (Oring *et al.*, 1983).

Compared with Spotted Sandpipers, Dunlins, for example, lay fewer eggs per female per year, live longer, breed first at a later age, experience lower nest loss rates, and have more biparental care. This suite of life history characteristics is associated with higher, and possibly less flexible adult return rates. They are also the most philopatric scolopacids (see below), and probably the most inbred of the species discussed. We regard Dunlins as adapted for breeding in a more "stable" habitat, and Spotted Sandpipers as a more pioneering species that breeds in early successional habitat. Data on the responses of more species of scolopacids to changes in environmental quality are needed before the relationships between factors affecting breeding site fidelity can be more clearly understood.

III. NATAL PHILOPATRY

Baker and Marler (1980) state that "most young organisms return or stay close to their sites of origin." Nevertheless, natal dispersal is far more widespread than dispersal of breeding birds (Greenwood and Harvey, 1982). Why be philopatric? Shields (1982) reviews a range of explanations, both proximate and ultimate. Those most likely to apply to shorebirds are (1) the "somatic effects"—those that an individual gains through familiarity with a location, and (2) "ecogenetic" advantages of breeding with individuals having similar (but not *too* similar) genetic makeups. Ecogenetic advantages may play a role in maintaining local adaptation in, say, the timing of arrival at the breeding grounds, if such a trait were under partial genetic control.

A commonly stated explanation for natal philopatry is that the existence of a young individual is a good indication that the place that it came from is (or was) suitable for reproduction. However, as Shields' points out, the essential question is whether an individual does better by using this one piece of information than it would by incorporating additional information, especially information from conspecifics with breeding experience. As young commonly migrate later than adults, their best option may be to "follow the leaders." In species where successful breeders tend to return, and unsuccessful breeders do so to a significantly lesser degree, inexperienced breeders on average might benefit by attempting to breed where there are already large numbers of breeding birds. Among successful populations, this type of system may reach the point where density limits immigration (Oring *et al.*, 1983).

Accurate measurement of chick dispersal is very difficult, as one must mark large numbers of chicks, preferably determine which fledge, and then survey the breeding grounds and surrounding areas after chicks are sexually mature to see which breed. Even then, complications are considerable, as there may be marked between-year variations in overwinter survival, weather during migration, competitive exclusion and resource availability, and philopatry of chicks at ages older than minimum breeding age. Therefore, only studies continuing for a number of years beyond the time when chicks of the first few seasons are sexually mature have credibility in terms of determining philopatry rates. In species such as the Long-billed Curlew, where chicks first breed when 3+ years old (Redmond and Jenni, 1982), accurate determination of dispersal and philopatry is exceedingly difficult. The available data on scolopacid philopatry are summarized in Table I.

A. Species Differences

In most thriving populations of sandpipers studied to date, locally hatched chicks entering the breeding population do not account for most of the adults. For example, in a careful, long-term study of Dunlin along the German coast, Heldt (1966) recovered only 2.8% of the chicks ringed—just a fraction of the new breeders in the area. In years 8 to 10 of our study of Spotted Sandpipers, locally hatched chicks accounted for 31, 40, and 35% of the breeding females and just 21, 19, and 17% of the breeding males (Oring *et al.*, 1983, and unpublished). Here too, immigration of chicks hatched elsewhere accounted for a higher proportion of the new breeding birds than did locally hatched chicks. Among studies of eight populations of seven species, only in Finnish Dunlin did returning chicks provide most of the new breeding stock (Soikkeli, 1967, 1970a,b). Soikkeli's population was characterized by a very high chick return rate (approximately 15% seen; 11% found breeding) but also by low adult mortality and extreme population isolation. Furthermore, it is a real tribute to Soikkeli's thoroughness that most of the chicks were found at all, for 51% of the 57 chicks returning to breed were more than 2 km from their natal site, 77% over 0.5 km from their natal site, and 38 chicks shifted to the one other nearby population. It is also of interest that one of Soikkeli's chicks was seen as a migrant 280 km to the north, and that one of his breeding females was known to have hatched 280 km north of his study area.

In Temminck's Stint (Hildén, 1978), most philopatric chicks breed less than 300 m from their natal site, but two males were found breeding 2.5 km and 50 km from their hatch site, and a third chick of unknown sex was found breeding 170 km away. This study was also conducted on a geographically isolated population.

In the Spotted Sandpiper, the 27 returning chicks bred within 150 m of their natal site, that being the diameter of the island on which the study was conducted (Oring *et al.*, 1983). There was no significant relationship between hatch site and site of first breeding. Occasional sightings of 24 other locally hatched chicks at 2 to 6 years of age indicate that a significant number of chicks may have initially dispersed to breeding areas not far from the natal site. As with Dunlin and Temminck's Stint, some Spotted Sandpiper chicks may disperse long distances, but as yet this is undocumented.

Red Phalaropes, in contrast to the species discussed above, show practically no natal philopatry (Table I; Schamel and Tracy, 1977). In all three phalarope species, moderate to long-distance dispersal may be the rule rather than the exception.

B. Sex Bias

Greenwood (1980) predicted that most birds would have greater male than female philopatry because the dominant social system among birds is that of territorial defense by males. Through site familiarity obtained as juveniles, they gain more than females by being philopatric. Risk of inbreeding then becomes a selective factor for dispersal of the opposite sex.

In what is probably the most detailed study of natal dispersal of any bird species, it was found that in Great Tits (*Parus major*), sex bias in natal philopatry was manifested through differential response to population density. Males moved a greater number of territories at high density, whereas females moved a constant number of territories regardless of population density (Greenwood *et al.*, 1979). No comparable data are available for any scolopacid.

As with adult breeding site fidelity, the monogamous Dunlin shows no sex bias among returning young, at least in Finland (Table I). Not only are rates of philopatry similar for both sexes, but territorial fidelity (measured as distances from natal nest to first breeding nest) is similar in the two sexes (Soikkeli, 1970a).

Although sample sizes are small for most species, no sex bias in philopatry appears to be the rule for monogamous sandpipers (Table I). Temminck's Stints again show patterns similar to monogamous species, and Hildén (1979) shows that, as with Dunlin, there are no sex differences in within-study-site dispersal distances of returning young. The data from

Table II. Age Spotted Sandpiper Chicks Entered Their Natal Breeding Population[a]

	1	2	3	4	5	6
Female	10	3	1	0	0	0
Male	4	4	2	2	0	1

[a] Total number of local chicks seen as adults, both breeding and nonbreeding, was 51 (30 females, 21 males).

the polyandrous phalaropes hint that a degree of male philopatry may be present, and female philopatry is very rare.

Of 10 intensively studied species of scolopacids, only the Long-billed Curlew showed a significant male bias in return of chicks (Table I; Redmond and Jenni, 1982). Redmond and Jenni (1982) attributed this deviation from the sandpiper norm of no sexual bias in chick return to the long-lived nature of curlews, and advantages of kin cooperating in group mobbing behavior (see below).

If male bias in philopatry is expected in most sandpipers, one might predict a female bias in the Spotted Sandpiper. We have sighted 30 female and 21 male local chicks, in subsequent years, of which 14 females and 13 males bred (Table II). While there was no significant difference in the overall return rates of male and female chicks, females did return to breed significantly more often at 1 year of age, whereas males continued to enter the population for up to 6 years after hatching (Oring *et al.*, 1983; Table II). In interpreting these data, it should be recalled that the breeding sex ratio is not even, rather 75 females and 107 males have bred on Little Pelican Island over the years. Thus, fewer females than males were expected to establish themselves, the opposite of the observed pattern. Another way of looking at this is to look for a sex bias in the percentages of chicks in the breeding population given above; chicks represented a higher proportion of the breeding females than males during the last few years of the study. These results may be interpreted with reference to Greenwood's model. Females return because of the advantages received in intrasexual competition for territories. This may set up selective pressures against the return of males, which avoid possible sibling matings. But the patterns of arrival of sex and age classes on the breeding grounds (first experienced females, then experienced males and new females together, and finally new males; Oring and Lank, 1982) set up a more complex situation. Young females rarely breed with experienced males. Hence, at least in year one, a female is quite unlikely to breed with her site-faithful father. On the other hand, inexperienced males often are secondary mates of experienced females. Consequently, even in year one, there is a risk of mother × son mating. Thus, risk of inbreeding is greater for young males, and this may be an element behind the greater dispersal of young males as compared to young females.

While data from Long-billed Curlews and Spotted Sandpipers suggest that Greenwood's general hypothesis is applicable to these species, this does not appear to be the case for the other monogamous scolopacids, nor for Temminck's Stints. Does this group of shorebirds differ from birds in general in showing unbiased philopatry, and if so, why? Comparison of the data in Table I with Greenwood's summary for birds (Greenwood, 1980; Table II), and with his statement that "exceptions to the prevalent pattern of female biased dispersal in birds have been reported in only one family, the Anatidae" (p. 1144), gives the answer "yes"—these species are unusual in showing no sex bias in philopatry. Within the framework of Greenwood's model, this might occur because (1) reproductive opportunities are more evenly available to both sexes than is the case for other groups, (2) "somatic" (*sensu* Shields) effects are more important, and/or (3) "ecogenetic" effects are more important than in other groups. An indirect approach to this last point might be to compare the effective population sizes (N_e) of the species involved, as we would expect to find small deme sizes where such effects were important (Barrowclough, 1980; Shields, 1982). Unfortunately, no estimates of N_e are available for any scolopacid.

IV. LATE-SEASON MOVEMENTS OF SPOTTED SANDPIPERS

While many young Spotted Sandpipers return to breed in their natal population, the percentage of fledglings doing so is substantially lower than the number likely to be breeding at all sites combined. The greatest percentage of locally hatched breeding birds in any given year on Little Pelican Island was 40% for females and 30% for males, thus immigration from other areas was occurring. And yet Little Pelican Island is almost certainly exporting a considerable number of birds.

Over the years, many locally hatched chicks have visited Little Pelican Island, especially in early season, without making a concerted effort to establish a territory or form pair bonds. Others fought their way into the population, in some cases trying unsuccessfully to establish themselves in several different territories before succeeding. Still other chicks apparently bred elsewhere for years and then emigrated to Little Pelican

Island—one not until he was 6 years old, and two others not until their fourth year. Finally, some local chicks doubtless bred their entire lives elsewhere, without ever visiting Little Pelican Island after their natal year. Why this extreme variation in homing?

Almost as soon as chicks fledge, they depart Little Pelican Island. They are rarely seen again that summer, and yet we have observed fledged chicks, in late summer, resident on beaches 6 km distant. On the other hand, it is quite common for nonlocal fledglings to be resident on Little Pelican Island for days or weeks at a time in late summer. Similarly, it is common to have nonbreeding adults in residence on nearby spits, bars, and mini-islands, especially during late summer. These postbreeding juveniles and nonbreeding adults build up a set of experiences with potential breeding sites. From among banded nonbreeding adults, we have had 11 females and 6 males return in subsequent years to breed. There is no way that we can determine how many of the nonlocal fledglings that frequented the island in late summer returned to breed, as many of these birds were not individually marked.

The tendency of Spotted Sandpipers to home and to breed in the natal population lies intermediate between an extreme where young animals disperse to avoid crowding or inbreeding, or home to take advantage of a proven successful breeding site. The tendency to home is moderated by social experiences that inexperienced birds have, both in the time between fledging and migration in the natal year, and during northward migration preceding a first breeding year. As adults, Spotted Sandpipers may visit new areas after failed breeding attempts, or during the postbreeding period, and the experience they acquire in these new areas may influence their choice of a breeding site in subsequent years.

V. CONCLUSION

Site tenacity among breeding sandpipers is greatest in the species and sexes with the greatest competition for territories. It varies from situations where nearly all living adults return, regardless of previous breeding success, e.g., Dunlin and Willet (Soikkeli, 1970a,b; Howe, 1982), to situations with slight sex biases and strong effects of success and age, e.g., Spotted Sandpiper (Oring *et al.*, 1983), to situations where essentially no adults return, e.g., Red Phalaropes (Schamel and Tracy, 1977).

Most young sandpipers disperse to other populations, but some are philopatric. Chicks are less site tenacious than adults. The extent to which dispersal occurs depends in large part upon alternate reproductive opportunities. Dispersal is reduced in isolated populations; in species with no territories it may be greatest. Small monogamous species have no sex bias in natal philopatry—a situation unusual among birds. In territorial polygamous species, the sex facing the stiffest intrasexual competition may show greater natal philopatry, but data are available for only one species.

Redmond and Jenni (1982) speculate on the evolution and consequences of the male-biased philopatry seen in Long-billed Curlews. They feel that the presence of a sex bias suggests that curlews evolved in relatively small deme sizes, as these are the conditions under which dispersal of one sex or the other might be important to prevent inbreeding depression. They go on to suggest that the tendency of males only to engage in cooperative mobbing of predators in this species might have arisen through kin selection among philopatric males, and that other large shorebirds, which also mob (Sordahl, 1981), might also be expected to have sex-biased natal philopatry.

This scenario is plausible, and it will be interesting to see whether a pattern of sex-specific mobbing and biased natal philopatry occurs in other species of large shorebirds. This is another case where knowledge of the effective population sizes of different species would provide insight into the selective regimes involved. Unfortunately, we cannot determine N_e from the "return" data used in this chapter.

At present we see only part of the picture—the history of birds that remain in or enter populations, and the tabulation of those that do not return. But what induced immigrants to leave their home population and enter "our" study areas, and where did they come from? And are those that fail to return dead, or are they breeding elsewhere? As our potential for radio monitoring of birds over long distances and biochemical techniques improve, perhaps these questions will be answered. Even at current levels of technology, long-term studies of dispersal and mating relationships can go much farther than they have toward answering questions regarding the mechanisms by which population kin structures are established and maintained—a question at the very heart of modern evolutionary biology.

ACKNOWLEDGMENTS

The following researchers graciously provided us with unpublished data: C. L. Gratto, R. T. Holmes, J. R. Jehl, Jr., D. A. Jenni, E. H. Miller, R. L. Redmond, and M. J. Rausch. Preparation of this chapter was aided by NSF Grant DEB 79-11147 to L.W.O., and supported by the University of North Dakota Office of Research and Program Development.

REFERENCES

Andersen, F. S., 1948, Contributions to the biology of the Ruff (*Philomachus pugnax* L.), Part II, *Dan. Ornithol. Foren. Tidsskr.* **42:**125–148.

Andersen, F. S., 1951, Contributions to the biology of the Ruff (*Philomachus pugnax* L.), Part III, *Dan. Ornithol. Foren. Tidsskr.* **45:**145–173.

Baker, M. C., and Marler, P., 1980, Behavioral adaptations that constrain the gene pool in vertebrates, in: *Evolution of Social Behavior: Hypotheses and Empirical Tests* (H. Markl, ed.), pp. 59–80, Verlag Chemie, Weinheim.

Barrowclough, G. F., 1980, Gene flow, effective population sizes, and genetic variance components in birds, *Evolution* **34:**789–798.

Bergman, R. D., 1974, Wetlands and waterbirds at Point Storkersen, Alaska, Ph.D. dissertation, Iowa State University, Ames.

Catchpole, C. K., 1972, A comparative study of territory in the Reed Warbler (*Acrocephalus scirpaceus*) and Sedge Warbler (*A. schoenobaenus*), *J. Zool.* **166:**213–231.

Drury, W. H., Jr., 1961, The breeding biology of shorebirds on Bylot Island, Northwest Territories, Canada, *Auk* **78:**176–219.

Emlen, S. T., and Oring, L. W., 1977, Ecology, sexual selection, and the evolution of mating systems, *Science* **197:**215–223.

Greenwood, P. J., 1980, Mating systems, philopatry and dispersal in birds and mammals, *Anim. Behav.* **28:**1140–1162.

Greenwood, P. J., and Harvey, P. H., 1976, The adaptive significance of variation in breeding area fidelity of the Blackbird (*Turdus merula* L.), *J. Anim. Ecol.* **45:**887–898.

Greenwood, P. J., and Harvey, P. H., 1982, The natal and breeding dispersal of birds, *Annu. Rev. Ecol. Syst.* **13:**1–21.

Greenwood, P. J., Harvey, P. H., and Perrins, C. M., 1978, Inbreeding and dispersal in the Great Tit, *Nature (London)* **271:**52–54.

Greenwood, P. J., Harvey, P. H., and Perrins, C. M., 1979, The role of dispersal in the Great Tit (*Parus major*): The causes, consequences and heritability of natal dispersal, *J. Anim. Ecol.* **48:**123–142.

Grosskopf, G., 1959, Zur Biologie des Rotschenkels (*Tringa t. totanus*), II, *J. Ornithol.* **100:**210–236.

Grosskopf, G., 1968, *Die Vögel der Insel Wangerooge*, Institüt für Vogelforschung, No. 5, Mettcker & Söhne, Jever.

Hale, W. G., and Ashcroft, R. P., 1982, Pair formation and pair maintenance in the Redshank *Tringa totanus*, *Ibis* **124:**471–490.

Harvey, P. H., Greenwood, P. J., and Perrins, C. M., 1979, Breeding area fidelity of the Great Tit (*Parus major*), *J. Anim. Ecol.* **48:**305–313.

Hays, H., 1972, Polyandry in the Spotted Sandpiper, *Living Bird* **11:**43–57.

Heldt, R., 1966, Zur Brutbiologie des Alpenstrandläufers, *Calidris alpina schinzii*, *Corax* **1:**173–188.

Hildén, O., 1975, Breeding system of Temminck's Stint *Calidris temminckii*, *Ornis Fenn.* **52:**117–146.

Hildén, O., 1978, Population dynamics in Temminck's Stint *Calidris temminckii, Oikos* **30**:17–28.

Hildén, O., 1979, Territoriality and site tenacity of Temminck's Stint *Calidris temminckii, Ornis Fenn.* **56**:56–74.

Hildén, O., and Vuolanto, S., 1972, Breeding biology of the Red-necked Phalarope *Phalaropus lobatus* in Finland, *Ornis Fenn.* **49**:57–85.

Hinde, R. A., 1956, The biological significance of the territories of birds, *Ibis* **98**:340–369.

Holmes, R. T., 1966, Breeding ecology and annual cycle adaptations of the Red-backed Sandpiper (*Calidris alpina*) in northern Alaska, *Condor* **68**:3–46.

Holmes, R. T., 1971, Density, habitat, and the mating system of the Western Sandpiper (*Calidris mauri*), *Oecologia* (*Berlin*) **7**:191–208.

Howe, M. A., 1982, Social organization in a nesting population of eastern Willets, *Catoptrophorus s. semipalmatus, Auk* **99**:88–102.

Jehl, J. R., Jr., 1973, Breeding biology and systematic relationships of the Stilt Sandpiper, *Wilson Bull.* **85**:115–147.

Jonas, R., 1979, Brutbiologische Untersuchungen an einer Population der Uferschnepfe (*Limosa limosa*), *Vogelwelt* **100**:125–136.

Kistchinski, A. A., and Flint, V. E., 1973, A case of "double nesting" in the Little Stint (*Calidris minuta*), in: *Fauna and Ecology of Waders*, Vol. 1 (V. E. Flint, ed.), Moscow University Press, Moscow (in Russian).

Lemnell, A., 1978, Social behavior of the Great Snipe *Capella media* at the arena display, *Ornis Scand.* **9**:146–163.

Miller, E., 1977, Breeding biology of the Least Sandpiper, *Calidris minutilla* (Vieill.), on Sable Island, Nova Scotia, Ph.D. thesis, Dalhousie University, Halifax, Nova Scotia.

Noordwijk, A. J., and Scharloo, W., 1981, Inbreeding in an island population of the Great Tit, *Evolution* **35**:674–688.

O'Donald, P., 1980, Sexual selection by female choice in a monogamous bird: Darwin's theory corroborated, *Heredity* **45**:201–217.

Oring, L. W., 1982, Avian mating systems, in: *Avian Biology*, Vol. 6 (D. S. Farner, J. R. King, and K. C. Parkes, eds.), pp. 1–92, Academic Press, New York.

Oring, L. W., and Knudson, M. L., 1972, Monogamy and polyandry in the Spotted Sandpiper, *Living Bird* **11**:59–73.

Oring, L. W., and Lank, D. B., 1982, Sexual selection, arrival times, philopatry and site fidelity in the polyandrous Spotted Sandpiper, *Behav. Ecol. Sociobiol.* **10**:185–191.

Oring, L. W., Lank, D. B., and Maxson, S. J., 1983, Population studies of the polyandrous Spotted Sandpiper, *Auk.* **100**:272–285.

Parmelee, D. F., and Payne, R. B., 1973, On the multiple broods and the breeding strategy of Arctic Sanderlings, *Ibis* **115**:218–226.

Pienkowski, M. W., and Greenwood, J. J. D., 1979, Why change males?, *Biol. J. Linn. Soc.* **12**:85–94.

Rausch, M. J., 1982, Mating systems and the breeding biology of the Northern Phalarope (*Phalaropus lobatus*), B.S. honours thesis, Queen's University, Kingston.

Redmond, R. L., and Jenni, D. A., 1982, Natal philopatry and breeding area fidelity of Long-billed Curlews (*Numenius americanus*): Patterns and evolutionary consequences, *Behav. Ecol. Sociobiol.* **10**:277–279.

Richdale, L. E., 1957, *A Population Study of Penguins*, Oxford University Press (Clarendon), Oxford.

Schamel, D., and Tracy, D., 1977, Polyandry, replacement clutches, and site tenacity in the Red Phalarope (*Phalaropus fulicarius*) at Barrow, Alaska, *Bird-Banding* **48**:314–324.

Shields, W. M., 1982, *Philopatry, Inbreeding, and the Evolution of Sex*, State University of New York Press, Albany.

Soikkeli, M., 1967, Breeding cycle and population dynamics in the Dunlin (*Calidris alpina*), *Ann. Zool. Fenn.* **4**:158–198.

Soikkeli, M., 1970a, Dispersal of Dunlin *Calidris alpina* in relation to sites of birth and breeding, *Ornis Fenn.* **47**:1–9.

Soikkeli, M., 1970b, Mortality and reproductive rate in a Finnish population of Dunlin *Calidris alpina, Ornis Fenn.* **47**:149–158.

Sordahl, T. A., 1981, Predator mobbing behavior in the shorebirds of North America, *Wader Study Group Bull.* **31**:41–44.

Thomas, C. S., 1980, Certain aspects of the breeding biology of the Kittiwake (*Rissa tridactyla*), Ph.D. thesis, University of Durham, U.K.

van Rhijn, J. G., 1973, Behavioral dimorphism in male Ruffs, *Philomachus pugnax* (L.), *Behaviour* **47**:153–229.

Chapter 5

THE EVOLUTION OF POLYANDRY IN SHOREBIRDS

Sarah Lenington

Institute of Animal Behavior
Rutgers University
Newark, New Jersey 07102

I. INTRODUCTION

The existence of polyandry has posed a difficult problem for theories attempting to account for the evolution of mating systems. The theory of sexual selection predicts competition among males, greater female than male parental investment, and more sexual promiscuity on the part of males than females (Darwin, 1871; Fisher, 1930; Orians, 1969; Trivers, 1972; Emlen and Oring, 1977). Polyandrous mating systems, particularly in extreme cases such as the Northern Jacana (*Jacana spinosa*) (Jenni and Collier, 1972), seem to violate each of these predictions. Females are larger and more aggressive than males, parental care is performed only by males, and females copulate with several males whereas each male copulates with only one female.

Polyandry is extremely rare in higher vertebrates, being found only in birds (Jenni, 1974) and humans (Daly and Wilson, 1978). Many studies of avian polyandry have focused on shorebirds (Hays, 1972; Jenni and Collier, 1972; Oring and Knudson, 1972; Nethersole-Thompson, 1973; Kistchinski, 1975; Mayfield, 1978; Oring and Maxson, 1978; Maxson and Oring, 1980; Ridley, 1980), for most polyandrous species are found within the Charadriiformes (Jenni, 1974). As one might expect, it has been difficult to account for a behavioral system that deviates so radically from accepted theory. Indeed, it has even been suggested that chance may play a large role in the evolution of some behavior patterns leading to polyandry (Selander, 1972). However, as pointed out by Graul *et al.* (1977), it seems unlikely that behaviors having such drastic effects on fitness evolved by chance.

In this chapter I propose to examine several models for the evolution of polyandrous matings in shorebirds and I will attempt to evaluate the logic and evidence for each of these models. In addition, I will suggest some factors, not previously made explicit in models for polyandry, that may play a role in the evolution of this mating system in shorebirds.

II. FORMS OF POLYANDRY IN SHOREBIRDS

Classically, mating systems have been defined in terms of the number and duration of pair bonds formed by individuals of each sex. Recently, attempts have been made to define mating systems in terms of the variance in genetic contribution made by individuals of each sex to the next generation (Ralls, 1977; Gowaty, 1981). The limitations of these two definitions have been discussed by Faaborg and Patterson (1981) who propose defining mating systems solely in terms of the numbers of sexual partners with whom members of each sex attempt to raise a brood. Within this framework, polyandrous mating systems are those in which females attempt to raise broods with more than one male whereas males only attempt to raise broods with one female. This definition, therefore, excludes those species exhibiting "rapid multiple-clutch polygamy" (Emlen and Oring, 1977) and monogamous species in which individuals change mates between nesting attempts.

Polyandry has been reported or is suspected to occur in 13 shorebird species (Table I). Faaborg and Patterson (1981) divide polyandrous mating systems into three categories, two of which (sequential and simultaneous polyandry) are relevant to shorebirds. In sequential polyandry, a female may mate with two or more males in succession. Although she lays a clutch of eggs for each male, her association with each male is generally of short duration. In simultaneous polyandry, a female maintains a stable, long-term association with two or more males, each of whom incubates a separate clutch of eggs. The difference between sequential and simultaneous polyandry is probably one of degree rather than kind. In fact, both forms of polyandry may occur within a single species and even within a single population (Oring and Maxson, 1978; Faaborg and Patterson, 1981). The frequency of simultaneous polyandry tends to increase as resources become more abundant (Oring and Maxson, 1978).

Emlen and Oring (1977) coined the term "rapid multiple-clutch polygamy" for a mating system in which the female lays one clutch of eggs

Table I. Shorebird Species for Which Polyandry Has Been Reported or Is Suspected[a]

Family	Species
Charadriidae	*Charadrius morinellus*
Jacanidae	*Actophilornis africana*
	Hydrophasianus chirurgus
	Metopidius indicus
	Jacana spinosa
	J. jacana
Rostratulidae	*Rostratula benghalensis*
Scolopacidae	*Bartramia longicauda*
	Tringa erythropus
	Actitis macularia
	Phalaropus tricolor
	P. lobatus
	P. fulicarius

[a] Data from Jenni (1974), Rowan (1926), Higgens and Kirsch (1975), and Osborne and Bourne (1977).

for the male to incubate and then lays a second clutch of eggs for herself. However, the term "polygamy" is not appropriate in this case because individuals exhibiting this mating pattern may be either monogamous (both clutches fertilized by the same male) or polygamous [either male or female copulating with more than one individual (Graul, 1974)]. In this chapter I will refer to this pattern as a "rapid multiple-clutch system" and bypass the question of the number of sexual partners obtained by each individual.

Rapid multiple-clutch systems are often interpreted as a transitional stage between monogamy and polyandry. Species regularly exhibiting this mating pattern include the Mountain Plover (*Charadrius montanus*), Little Stint (*Calidris minuta*), and Temminck's Stint (*C. temminckii*) (Graul, 1973, 1974, 1976; Hildén, 1965, 1975). The Sanderling (*C. alba*) and Snowy Plover (*Charadrius alexandrinus*) may have rapid multiple-clutch systems in some locations (Parmelee, 1970; Parmelee and Payne, 1973; Warriner and Warriner, 1978), but engage in biparental care in others (Rittinghaus, 1956; Pienkowski and Green, 1976).

III. THEORIES FOR THE EVOLUTION OF POLYANDRY

Most theories for the evolution of polyandry agree that there probably has been an evolutionary sequence starting from monogamy with biparental care and proceeding either to monogamy with parental care performed by males alone or to a rapid multiple-clutch system. Once sex-role reversal has occurred to the extent that males perform all parental care, the next evolutionary step is for females to mate with and lay clutches for more than one male (Jenni, 1974; Emlen and Oring, 1977; Graul *et al.*, 1977; Oring and Maxson, 1978; Maxson and Oring, 1980; Faaborg and Patterson, 1981). Therefore, a critical evolutionary change in all models is the emancipation of females from parental care and the assumption of this behavior by males. However, the models differ in the reasons hypothesized for such a transition. Some theories give great emphasis to the role of predation; others emphasize the importance of food abundance.

Most of this chapter will be devoted to a discussion of the basis for male parental care, a crucial preadaptation for polyandrous mating systems. It seems to me that the key to understanding polyandry lies in explaining the transitional stages of female desertion of broods and the evolution of rapid multiple-clutch systems. Without first understanding the evolution of these early transitional stages, it may be impossible to understand the end point of fully developed polyandry.

A. Simple Predation Model

The first model relating polyandry to high predation rates, the simple predation model (SPM), assumes that if predation rates are high and food is abundant, it may be advantageous for females to lay clutches in two nests rather than one (Oring and Knudson, 1972; Emlen and Oring, 1977; Oring and Maxson, 1978; Faaborg and Patterson, 1981). This model is implicitly based on the assumption that in the face of environmental uncertainty it is better to "not put all your eggs in one nest" (Gillespie, 1974; Mock, 1983). In the SPM the evolutionary sequence is postulated as leading directly from monogamy with biparental care to a rapid multiple-clutch system, and from there to polyandry.

Although the SPM has a surface plausibility, it breaks down when examined more closely. The "choice" for a shorebird female is never to lay either a full clutch in one nest or half a clutch in each of two nests. Rather, typically in species with biparental care, a female lays a full clutch of 3–4 eggs in one nest. If the nest is destroyed, she lays a full replacement clutch. In species with rapid multiple-clutch systems, a full clutch is initially laid in each of two nests. If these are the choices and *if there is no additional cost associated with uniparental care*, it should always be advantageous for females initially to lay two clutches in two nests regardless of the predation rates (Gillespie, 1974). Even if predation rates are very low, the expected reproductive success will be higher for pairs laying two clutches in separate nests than for pairs that only lay a second clutch if the first is destroyed.

However, the assumption of no additional cost associated with uniparental care is unreasonable. Unfortunately, few data are available comparing fledging success of broods tended by only one parent with that of broods tended by two parents. Nevertheless, the widespread practice of biparental care, despite the fact that young are precocial, in itself suggests that uniparental care probably entails a substantial cost. Lenington (1980) found that in Killdeer (*Charadrius vociferus*), chicks were watched less when one parent tended the brood than when parental care was performed by two adults. It has been suggested that two parents may be more effective than one in protecting chicks from predators (Lack, 1968; Holmes, 1971; Lenington, 1980; Howe, 1982; Walters, 1982), preventing chicks from getting lost (Lenington, 1980; Walters, 1982), or in brooding chicks (Howe, 1982).

In the face of a cost associated with uniparental care, the effect of predation on the relative advantage of a rapid multiple-clutch system over biparental care is exactly the opposite of that hypothesized by the SPM. The process can be described by a very simple mathematical model.

Assume that (1) in a biparental system, a female always renests if her original nest is destroyed and she renests only once, and (2) in a rapid multiple-clutch system, both clutches are fertilized by the same male. Let

p = the probability a nest will be destroyed.
q = the probability a nest will not be destroyed.
X = the expected number of chicks raised to maturity if the nest is not destroyed and both parents participate in parental care.
aX = the expected number of chicks raised to maturity if the nest is not destroyed and only one parent tends the brood, $0 \leq a < 1$. Thus, a is the relative fitness of uniparental as compared with biparental pairs.
$1 - a$ = the cost of uniparental care.

In a biparental system, the expected reproductive success of the first nest $(R_1) = qX$. As a female always renests if her first nest is destroyed, the probability she will renest $= p$. Thus, the expected reproductive success of her second nest $(R_2) = pqX$. The expected reproductive success of a biparental pair for a single breeding season $(R_b) = R_1 + R_2 = qX + pqX$. In a rapid multiple-clutch system, the expected reproductive success of a nest incubated by a male $(R_♂) = qaX = R_♀$ (the expected reproductive success of a nest incubated by a female). Thus, the expected reproductive success of a uniparental pair for a single breeding season $R_u = 2qaX$. For a rapid multiple-clutch system to evolve, the following inequality must be satisfied: $R_u > R_b$, or $2qaX > qX(1 + p)$. This reduces to

$$a > \frac{1}{2}(1 + p)$$

From this it can be seen that even if predation rates are 0, a (the relative fitness of uniparental as compared with biparental pairs) can never be lower than $\frac{1}{2}$, if a rapid multiple-clutch system is to evolve. Furthermore, for a given value of a, the inequality is more likely to be satisfied as predation rates decrease rather than increase. This means that the cost of uniparental care $(1 - a)$ is only offset under conditions of low predation, when biparental parents produce chicks from one nest whereas uniparental pairs may produce chicks from two.

The preceding model is based on the assumption that a and p are independent. In reality, if two parents can protect a brood from predators better than one, a and p are likely to be inversely related. In such a situation, the inequality favoring a rapid multiple-clutch system is even less likely to be satisfied.

B. Replacement Clutch Model

A second theory for the evolution of polyandry combines the factors of high predation with those of energetic cost to the female of egg laying. In this replacement clutch model (RCM), it is hypothesized that when food availability is low, egg laying may constitute a severe energetic drain for females. If, in addition, the probability of predation is high, it is likely a female will also need to lay a replacement clutch. Yet incubating females can spend less time foraging than those that do not incubate. Under these conditions, it may be advantageous for a female to leave incubation and parental care to the male while she spends her time feeding. She may thereby increase her ability to lay replacement clutches should the necessity arise (Parmelee and Payne, 1973; Jenni, 1974; Emlen and Oring, 1977; Mayfield, 1978; Oring and Maxson, 1978; Ridley, 1980; Maxson and Oring, 1980). Once a system has evolved in which males are willing to perform all parental care, if food abundance subsequently improves, females could increase their reproductive success by laying a second clutch of eggs, fertilized by either the same or a different male. The female may then incubate the second clutch herself (rapid multiple-clutch system) or a second male may incubate the clutch (polyandry). Thus, in this model the first transition is from monogamy with biparental care to monogamy with uniparental care. The important difference between these two behavioral systems is that with biparental care, females, when faced with low food availability, may not be able to lay replacement clutches, whereas they may be able to do so if freed from parental care.

The foregoing model can be expressed mathematically as follows using the same terms as in the SPM and adding one further definition. Let

r = the probability a female will lay a replacement clutch, given that her first clutch has been destroyed.

Further, assume that in a uniparental system $r = 1$, whereas in a biparental system $0 \leq r < 1$. Then $R_b = qX + rpqX = qX(1 + pr)$, and $R_u = qaX + pqaX = qaX(1 + p)$. In this model, $R_u \neq R_u$ of the SPM, for the SPM assumed uniparental females always laid two clutches, whereas in the RCM uniparental females lay two clutches only if the first clutch is destroyed. Monogamy with uniparental care will evolve if

$$qX(1 + p) > qX(1 + pr)$$

or

$$a > \frac{1 + pr}{1 + p}$$

In this formulation, for a given value of a, the inequality is more likely to be satisfied as p increases and r decreases. Thus, theoretically, if the probability of predation is high and the probability an incubating female can produce a replacement clutch is low, female desertion could evolve. However, we are still left with the question of whether or not in reality this process is responsible for female desertion.

Table II presents a list of monogamous shorebirds in which females desert broods and the time at which desertion takes place. In most of these species, females desert after hatching and commonly they desert quite late in the breeding season. For example, Killdeer females desert only if a second brood is raised (Bunni, 1959; Lenington, 1980). They desert at or near the time the second broods hatch. In a study of the Thick-billed Plover (*Charadrius wilsonia*), Bergstrom (1981) found that during one breeding season, males incubated alone only at the last two nests of the season. Ashkenazie and Safriel (1979a) report that female Semipalmated Sandpipers (*Calidris pusilla*) desert broods sooner if they hatch later in the season, than if hatching is early.

Table II. Monogamous Species in Which Females Desert Broods and the Time at Which Desertion Takes Place

Species	Time of desertion	Reference
Vanellus vanellus	Information not available	Bannerman (1961)
Pluvialis squatarola	10 days after hatching	D. Hussell (personal communication)
Charadrius wilsonia	After egg laying for nests started late in the season	Bergstrom (1981)
C. vociferus	Near the time of hatching for second broods	Bunni (1959), Lenington (1980)
Numenius tenuirostris	Information not available	Dement'ev *et al.* (1969)
N. americanus	After hatching	Jenni (personal communication)
Tringa nebularia	10 days after hatching	Nethersole-Thompson (1951)
T. ocrophus	10 days after hatching	Bannerman (1961)
Catoptrophorus semipalmatus	2 weeks after hatching	Howe (1982)
Actitis hypoleucos	Information not available	Bannerman (1961)
Arenaria interpres	Near hatching	Nettleship (1973)
Calidris pusilla	2–8 days after hatching	Ashkenazie and Safriel (1979a)
C. mauri	After hatching	Holmes (1972)
C. alpina	1st week after hatching	Soikkeli (1967), Holmes (1966)
C. himantopus	1st week after hatching	Jehl (1973)

If female desertion were an adaptation to increase the probability of producing replacement clutches, desertion should occur early in the breeding season, rather than late when replacement clutches are not possible. Furthermore, despite the propensity of Killdeer females to desert second broods, replacement clutches are nearly always laid if the first clutch is destroyed (i.e., $r \approx 1$) (Lenington, unpublished data). Similarly, in other species with female desertion, replacement rates are high for clutches lost early in the season (Holmes, 1966; Soikkeli, 1970; Howe, 1982) and females lay replacement clutches *before* they desert the male. These data suggest that female desertion is not necessary to facilitate laying of replacement clutches. Therefore, although the need to obtain energy for replacement clutches could theoretically favor female desertion, it seems unlikely that this is the selective pressure actually producing such desertion.

Additional doubt is cast on both the SPM and the RCM by an examination of predation rates for several species of shorebirds. If high predation were a factor in producing polyandry, one would expect predation rates to be higher for polyandrous species or those with rapid multiple-clutch systems than for monogamous species (assuming no other correlated changes in antipredator strategies). Boyd (1962) published a list of hatching success rates (a rough indicator of predation) for several species of European Charadrii. Hatching success for monogamous species in his study was highly variable, ranging from 4.8% in the Northern Lapwing (*Vanellus vanellus*) to 96% in the Common Redshank (*Tringa totanus*). Table III presents data from additional studies not included in Boyd (1962), classified by mating system. Mean hatching success rate for monogamous species in Boyd (1962) and Table III is 65.57 ± 7.32% ($n = 41$). The mean rate for polyandrous species and those with a rapid multiple-clutch system is 62.66 ± 7.27% ($n = 10$). The difference is not significant. Typically, predation rates tend to be high for any ground-nesting bird (Lack, 1968). However, if predation rates do not differ between monogamous species and those with other mating systems, predation cannot be the selective factor producing the mating system.

C. Energetic Model

Both the energetic model (EM) and the RCM hypothesize that female desertion will be favored when food supply is low. The difference between the two models lies in the emphasis given to predation. The RCM suggests desertion will be favored when predation rates are high and energy for replacement clutches is limited. The EM assumes that the probability of female desertion is independent of the predation rate and related only to food abundance.

Several authors have suggested that polyandry has evolved in response to a highly variable food supply (Oring and Knudson, 1972; Nethersole-Thompson, 1973; Parmelee and Payne, 1973; Hildén, 1975; Graul *et al.*, 1977; Faaborg and Patterson, 1981). This model can best be explained by considering possible breeding strategies under two conditions: (1) when food supply is low, and (2) when food supply fluctuates markedly from year to year.

Female desertion of broods, although rare in other birds, is common in shorebirds. The list of species in Table II will probably grow as more is known about the breeding biology of other shorebird species. Shorebirds are also unusual in the amount of energy invested in a clutch of eggs. Clutch weight in monogamous shorebirds ranges from 40 to 104% of body weight ($\bar{x} = 69.5\%$, $n = 30$) (Lack, 1968; Graul, 1974), whereas in most avian species, clutch weight is a much smaller percentage of body

Table III. Hatching Success of Shorebird Nests[a]

Mating system	Species	No. of nests	% success	Reference
Monogamy	*Recurvirostra americana*	78	79.0	Gibson (1971)
	Vanellus spinosus	135	44.0–50.0	Thomas (1969)
	V. senegallus	6	33.3	Little (1967)
	Pluvialis apricaria	11	72.7	Parr (1979)
	Charadrius vociferus	20	55.5	Lenington (unpublished data)
	C. melodus	174	91.0*	Wilcox (1959)
		51	75.0*	Cairns (1977)
	C. alexandrinus	40	20.0	Hobbs (1972)
	C. bicinctus	40	20.0	Philips (1980)
	C. obscurus	30	40.0	Phillips (1980)
	Catoptrophorus semipalmatus	206	23.0	Howe (1982)
		81	31.5*	N. Sims (personal communication)
	Arenaria interpres	14	70.0	Nettleship (1973)
	Calidris alba	8	62.5	Pienkowski and Green (1976)
	C. mauri	215	83.7	Holmes (1972)
	C. alpina	215	59.0	Soikkeli (1970)
		42	81.0	Holmes (1966)
	C. himantopus	166	83.1	Jehl (1973)
Rapid multiple-clutch system	*Charadrius montanus*	101	62.1	Graul (1973)
	Calidris alba	9	55.5	Parmelee and Payne (1973)
	C. temminckii	193	55.0	Hildén (1975)
Polyandry	*Charadrius morinellus*	8	100.0	Pulliainen (1970)
	Actitis macularia	317	44.2	Oring *et al.* (in press)
		26	84.6	Hays (1972)
	Phalaropus fulicarius	55	61.8	Kistchinski (1975)
		41	43.9	Mayfield (1978)
		15	46.7	Ridley (1980)
	P. lobatus	66	72.8	Hildén and Vuolanto (1972)

[a] Values marked with an asterisk are based on the number of eggs that hatched chicks. Unlabeled percentages are based on the number of nests that hatch.

weight. Given the large amount of energy invested in a clutch, if food is scarce, a shorebird female may be severely stressed by egg production. Although this stress may or may not affect her ability to produce replacement clutches, it may well affect her lifetime fitness by decreasing her chance of survival to the next breeding season. Therefore, under conditions of food scarcity, it may pay a monogamous female to desert her brood, leaving all parental care to the male (Graul *et al.*, 1977; Ashkenazie and Safriel, 1979b; Lenington, 1980). Such desertion frees the female to spend all her time foraging.

If food supply is highly variable, females should desert in poor years after laying only one clutch. In good years, given that males are already preadapted to assume all parental care, a female's reproductive output could be increased by her laying and incubating a second clutch of eggs (rapid multiple-clutch system) or by laying a second clutch, incubated by another male (biandry) (Graul *et al.*, 1977). Thus, in this model, there are two crucial transitions: (1) from monogamy with biparental care to monogamy with male parental care, and (2) from monogamy with male parental care to polyandry or a rapid multiple-clutch system.

1. The Transition from Biparental to Male Parental Care

The advantage of monogamy with male parental care over monogamy with biparental care can be expressed as follows. Let

f = the probability a female lives to the next breeding season, given that she participates in parental care.

$f + h$ = the probability a female lives to the next breeding season, given that she does not participate in parental care. Thus, h = the amount a female's expectation of survival is improved if she deserts her brood. $h \leq 1 - f$.

E = expected reproductive success for each breeding season with biparental care.

aE = expected reproductive success for each breeding season with male parental care, $0 \leq a < 1$.

n = the maximum number of years a female reproduces.

If a female participates in parental care, her lifetime expected reproductive success from the time of her first breeding (L_b) is given by

$$E + fE + f^2E + \cdots + f^{n-1}E = E \sum_{i=0}^{n-1} f^i$$

If a female does not participate in parental care and only lays a clutch for one male each season, her lifetime expected reproductive success (L_u) is given by

$$aE + (f + h)aE + (f + h)^2 aE + \cdots + (f + h)^{n-1} aE = aE \sum_{i=0}^{n-1} (f + h)^i$$

For male parental care to evolve, L_u must be greater than L_b, or

$$a > \frac{\sum_{i=0}^{n-1} f^i}{\sum_{i=0}^{n-1} (f + h)^i}$$

From this it can be seen the extent to which male parental care is advantageous to females depends on h (the extent to which a female's probability of survival is improved by desertion) and n (the number of years during which a female can potentially reproduce). Note that in this model, unlike the preceding two models, a can be quite small [i.e., the cost of uniparental care $(1 - a)$ can be quite large] yet female desertion will still be favored.

However, a male's interests are not necessarily identical to a female's interests. Males may be willing to assume all parental care, rather than deserting themselves and leaving the female with the brood if:

1. Males have a high probability of remating with the same female in future years and their expected reproductive success is higher if they remate rather than finding a new mate. In this situation, males will be selected to behave in ways that favor a female's longevity (Emlen and Oring, 1977).
2. In the absence of mate fidelity, males should still be unwilling to desert if a female is less able than a male to successfully raise a brood to fledging and if males have no opportunities to find new mates during the rest of the breeding season. In this situation, male desertion would result in a lower expected reproductive success for males than would female desertion.

A rather large amount of indirect evidence favors the hypothesis that female desertion occurs under conditions of energetic stress. As mentioned earlier, shorebirds tend to invest an unusually large amount of energy in clutches as compared with other avian species. In addition,

Ashkenazie and Safriel (1979b) found that female Semipalmated Sandpipers (a species in which males perform most of the parental care) have a daily energy requirement 15% higher than that of males. By the time the young hatch, females suffer a marked energy deficit. The temporal pattern of female desertion, late in the season, is also consistent with the hypothesis that female desertion results from a declining food supply (Pitelka, 1959; Ashkenazie and Safriel, 1979a,b; Lenington, 1980; Bergstrom, 1981).

Only limited data are available on mate fidelity in shorebirds. However, these indicate that species vary considerably in the extent to which pairs remain mated for more than one season (Lenington and Mace, 1975; Oring and Lank, 1982). The pattern of variability is consistent with the EM. Six species in which females desert broods (*Numenius americanus, Calidris alpina, C. mauri, Micropalama himantopus, Catoptrophorus semipalmatus, Charadrius vociferus*) show a high degree of mate fidelity (Oring and Lank, 1982; Soikkeli, 1967, 1970; Holmes, 1971; Jehl, 1973; Howe, 1982; Lenington and Mace, 1975). In three of these species (*C. alpina, C. mauri, M. himantopus*), pairs that remate, after being paired in previous years, have a higher reproductive success than those that do not remate (Soikkeli, 1967; Holmes, 1971; Jehl, 1973). The extent of mate fidelity is not known for other species in which females desert broods. Mate fidelity, in contrast, is rare in two species (*Charadrius hiaticula, C. melodus*) in which females do not desert broods (Laven, 1940; Wilcox, 1959).

Evidence that the cost of uniparental care may be greater when females, rather than males, tend broods has been found in Baird's Sandpiper (*Calidris bairdii*). In this species, broods may be tended by either parent. Broods tended by males were heavier than were broods tended by females (Norton, 1973).

The major difficulty with the EM lies in the assumption that the stress of egg laying, in itself, is sufficient to account for female desertion. If the large amount of energy invested in a clutch were sufficient to account for female desertion, one would expect to see such behavior in other avian species that invest an equivalent amount of energy in egg production. Such a correlation, however, is not found. Ducks, like shorebirds, have precocial young. They also invest as much or more energy in each clutch of eggs (Lack, 1968; Ricklefs, 1974). For example, clutch weight in ducks of the genus *Anas* may be as much as 120% of body weight (Lack, 1968). Yet, unlike shorebirds, in many species of ducks all incubation and parental care are performed by females.

It is possible that ducks are less stressed by egg laying than shorebirds because of differences between the two groups in the amount of time over which eggs are laid. The energetic cost of egg laying is directly related to the rate at which the energy is expended (Ricklefs, 1974). Duck eggs are proportionately smaller than shorebird eggs and ducks lay larger clutches. Consequently, shorebird clutches are completed in 3–4 days whereas ducks may take a week or more to lay a clutch.

In addition, dietary differences between ducks and shorebirds may make it easier for ducks to obtain energy for egg production. The diet of ducks consists primarily of plant material whereas that of shorebirds consists of invertebrates. However, during the period of egg production, ducks, like shorebirds, primarily eat invertebrates (Krapu, 1974; Drobney, 1980; Kaminski and Prince, 1981). Thus, while eggs are being formed, there are no qualitative dietary differences between ducks and shorebirds, although there may well be quantitative differences in food abundance.

Probably the most important reason for the difference in energy balance between the two groups of birds is the amount of stored food reserves at the start of the breeding season. Energy for egg production comes from either dietary intake or stored reserves. Ducks arrive on the breeding grounds with substantial amounts of stored lipid. For example, as much as 10% of the body weight of a female mallard (*Anas platyrhynchos*) may be fat (Krapu, 1981). Although the dietary intake of invertebrates by ducks is important for producing egg proteins, much of the energy for egg production comes from these stored lipid reserves (Drobney, 1980; Krapu, 1981). In contrast, although shorebirds accumulate large amounts of body fat while on their wintering grounds (Holmes, 1966; Johnston and McFarlane, 1967; Summers and Waltner, 1979; Davidson, 1981), they probably arrive on the breeding ground with little or no stored food reserves (Holmes, 1966; Ashkenazie and Safriel, 1979b; Maxson and Oring, 1980). Egg production may therefore constitute a severe energetic stress for females because they have little excess body fat at the start of the breeding season. Consequently, female desertion and rapid multiple-clutch mating systems may not arise solely as a function of food supply on the breeding grounds, but rather from an interaction between feeding conditions on the breeding grounds and the physiological condition of the female at the start of the breeding season.

The physiological condition of the female is probably in part the result of migration distance and the time taken to complete migration. Several pieces of evidence are consistent with the hypothesis that long migratory distance may stress females and select for female desertion. Many species in Table II are Arctic breeders that often migrate long distances. Female desertion occurs in several sandpiper species breeding on the Alaskan tundra. Of these species, female desertion is more common in *Calidris pusilla* and *C. mauri*, both of which have long migrations, than *C. alpina*, which migrates a comparatively short distance (Ashkenazie and Safriel, 1979b). Willets (*Catoptrophorus semipalmatus*) breed in temperate areas, yet female desertion is a regular occurrence (Howe, 1982). However, this species overwinters in South America and has very long migrations. In other temperate breeding species whose migratory distances are shorter, female desertion is not a regular occurrence and tends to be seen only with those broods raised at the very end of the breeding season (Bunni, 1959; Lenington, 1980; Bergstrom, 1981). Female desertion has not been reported for any of the nonmigratory plovers or lapwings breeding in the Southern Hemisphere.

Clearly, many more data are needed on the energy budgets of shorebirds. The work of Ashkenazie and Safriel (1979b) and Maxson and Oring (1980) is a promising start in that direction. However, sorely needed is a more complete understanding of the energy requirements of a variety of species during the breeding season, knowledge of the physiological condition of individuals at the start of the season, and the energetic costs of migration at the end of the season. With these data, it may well be possible to develop a comprehensive explanation for differences in mating systems among species, as well as variability among populations or individuals in breeding strategy (Drent and Daan, 1980).

2. The Transition from Male Parental Care to Polyandry

If food supplies fluctuate, females may desert in years with poor food. In years when food is abundant, females may either remain with their original mate and brood, incubate a second clutch of eggs, or lay eggs for a second male. It will only pay a female to choose the second or third option if the cost of uniparental care is less than $\frac{1}{2}$. From the female's point of view, biandry is probably preferable to a rapid multiple-clutch system, because the energetic cost for her will be greater if she incubates. However, biandry may only be possible in populations with an excess of males (Hays, 1972; Nethersole-Thompson, 1973; Ridley, 1980). Although many polyandrous species or those with rapid multiple-clutch systems breed in habitats where food resources undergo considerable variation from year to year (Parmelee and Payne, 1973; Graul, 1974, 1976; Hildén, 1975; Ridley, 1980), at present there are no data indicating whether or not female incubation or biandry is more common in good food years than in years when food is scarce.

Once parental care is performed by males alone, it is possible for sexual selection to magnify differences between males and females. Extreme forms of polyandry, such as seen in the Northern Jacana, are probably the end result. At this point, however, the processes producing extreme polyandry are probably identical to those producing polygyny, except that male and female have completely changed roles (Graul *et al.*, 1977; Ridley, 1980).

IV. CONCLUSION

The evolution of polyandry, unlike the evolution of polygyny, is probably best explained as the outcome of a historical process. Typically, birds tend to be monogamous when food supply is low, and polygynous when food is abundant. Thus, polygynous matings tend to be explained as optimal adaptations to current, local breeding conditions (Orians, 1969; Emlen and Oring, 1977). Although polyandry is also hypothesized to occur under conditions of food abundance, it is probably the outcome of a process that first selected for male uniparental care. If a monogamous species were suddenly presented with unusually abundant resources, polygyny, rather than polyandry, would be the more probable result. However, once this selective constraint has been put on the behavior of males, a subsequent improvement in food availability may produce polyandry, in which sex roles are a mirror image of those found in the more commonly occurring polygynous systems.

ACKNOWLEDGMENTS

I wish to thank W. Graul for many hours of entertaining and profitable discussion about polyandry which provided the genesis for several ideas in this chapter. I am also grateful to M. Stephens and R. Phillips for their help with references and to D. Morrison and L. Abramovitch for their comments on ideas in this chapter.

REFERENCES

Ashkenazie, S., and Safriel, U. N., 1979a, Breeding cycle and behavior of the Semipalmated Sandpiper at Barrow, Alaska, *Auk* **96**:56–67.

Ashkenazie, S., and Safriel, U. N., 1979b, Time–energy budget of the Semipalmated Sandpiper, *Calidris pusilla*, at Barrow, Alaska, *Ecology* **60**:783–799.

Bannerman, D. A., 1961, *The Birds of the British Isles*, Oliver & Boyd, Edinburgh.

Bergstrom, P. W., 1981, Male incubation in Wilson's Plover (*Charadrius wilsonia*), *Auk* **98:**835–838.

Boyd, H., 1962, Mortality and fertility of European Charadrii, *Ibis* **104:**368–387.

Bunni, M. K., 1959, The Killdeer in the breeding season, Ph.D. thesis, University of Michigan.

Cairns, W. E., 1977, Breeding biology and behavior of Piping Plovers in southern Nova Scotia, Ph.D. thesis, Dalhousie University.

Daly, M., and Wilson, M., 1978, *Sex, Evolution, and Behavior*, Duxbury Press, Belmont, Calif.

Darwin, C., 1871, *The Descent of Man and Selection in Relation to Sex*, 1st ed., Murray, London.

Davidson, N. C., 1981, Survival of shorebirds (Charadrii) during severe weather: The role of nutritional reserves, in: *Feeding and Survival Strategies of Estuarine Organisms* (N. V. Jones and W. J. Wolff, eds.), pp. 251–260, Plenum Press, New York.

Dement'ev, G. P., Gladkov, N. A., and Spangenberg, E. P., 1969, *The Birds of the Soviet Union*, Vol. III, Israel Program for Scientific Translations, Jerusalem.

Drent, R. H., and Daan, S., 1980, The prudent parent: Energetic adjustments in avian breeding, *Ardea* **68:**225–252.

Drobney, R. D., 1980, Reproductive bioenergetics of Wood Ducks, *Auk* **97:**480–490.

Emlen, S. T., and Oring, L., 1977, Ecology, sexual selection, and the evolution of mating systems, *Science* **197:**215–223.

Faaborg, J., and Patterson, C. B., 1981, The characteristics and occurrence of cooperative polyandry, *Ibis* **123:**477–484.

Fisher, R. A., 1930, *The Genetical Theory of Natural Selection*, Oxford University Press (Clarendon), Oxford.

Gibson, F., 1971, The breeding biology of the American Avocet (*Recurvirostra americana*) in central Oregon, *Condor* **73:**444–454.

Gillespie, J. H., 1974, Natural selection for within-generation variance in offspring number, *Genetics* **76:**601–606.

Gowaty, P. A., 1981, An extension of the Orians–Verner Willson model to account for mating systems besides polygyny, *Am. Nat.* **118:**851–859.

Graul, W. D., 1973, Breeding adaptations of the Mountain Plover, *Charadrius montanus*, Ph.D. thesis, University of Minnesota.

Graul, W. D., 1974, Adaptive aspects of the Mountain Plover social system, *Living Bird* **12:**69–94.

Graul, W. D., 1976, Food fluctuations and multiple clutches in the Mountain Plover, *Auk* **93:**166–167.

Graul, W. D., Derrickson, S. R., and Mock, D. W., 1977, The evolution of avian polyandry, *Am. Nat.* **111:**812–816.

Hays, H., 1972, Polyandry in the Spotted Sandpiper, *Living Bird* **2:**43–55.

Higgins, K. F., and Kirsch, L. M., 1975, Some aspects of the breeding biology of the Upland Sandpiper in North Dakota, *Wilson Bull.* **87:**96–102.

Hildén, O., 1965, Zur Brutbiologie des Temminck-strandläufers, *Calidris temminckii* (Leisl.), *Ornis Fenn.* **42:**1–5.

Hildén, O., 1975, Breeding system of Temminck's Stint, *Calidris temminckii, Ornis Fenn.* **52:**117–146.

Hildén, O., and Vuolanto, S., 1972, Breeding biology of the Red-necked Phalarope, *Ornis Fenn.* **49:**57–85.

Hobbs, J. N., 1972, Breeding of Red-capped Dotterel at Fletcher's Lake, Dareton, NSW, *Emu* **72:**121–125.

Holmes, R. T., 1966, Breeding ecology and annual cycle adaptations of the Red-backed Sandpiper (*Calidris alpina*) in northern Alaska, *Condor* **68**:3–46.
Holmes, R. T., 1971, Density, habitat and the mating system of the Western Sandpiper (*Calidris mauri*), *Oecologia (Berlin)* **7**:191–208.
Holmes, R. T., 1972, Ecological factors influencing the breeding season schedule of Western Sandpipers (*Calidris mauri*) in sub-arctic Alaska, *Am. Midl. Nat.* **87**:472–491.
Howe, M., 1982, Social organization in a nesting population of eastern Willets (*Catoptrophorus semipalmatus*), *Auk* **99**:88–102.
Jehl, J. R., Jr., 1973, Breeding biology and systematic relationship of the Stilt Sandpiper, *Wilson Bull.* **85**:115–147.
Jenni, D. A., 1974, Evolution of polyandry in birds, *Am. Zool.* **14**:129–144.
Jenni, D. A., and Collier, G., 1972, Polyandry in the American Jacana (*Jacana spinosa*), *Auk* **89**:743–765.
Johnston, D. W., and McFarlane, R. W., 1967, Migration and bioenergetics of flight in the Golden Plover, *Condor* **69**:156–168.
Kaminski, R. M., and Prince, H. H., 1981, Dabbling duck activity and foraging responses to aquatic microinvertebrates, *Auk* **98**:115–126.
Kistchinski, A. A., 1975, Breeding biology and behavior of the Grey Phalarope, *Phalaropus fulicarius*, in east Siberia, *Ibis* **117**:285–301.
Krapu, G. L., 1974, Feeding ecology of Pintail Hens during reproduction, *Auk* **91**:278–290.
Krapu, G. L., 1981, The role of nutrient reserves in mallard reproduction, *Auk* **98**:29–38.
Lack, D., 1968, *Ecological Adaptations for Breeding in Birds*, Methuen, London.
Laven, H., 1940, Beitrage zur Biologie des Sandregenpfeifers (*Charadrius hiaticula* L.), *J. Ornithol.* **88**:183–287.
Lenington, S., 1980, Bi-parental care in Killdeer: An adaptive hypothesis, *Wilson Bull.* **29**:8–20.
Lenington, S., and Mace, T., 1975, Mate fidelity and nesting site tenacity in the Killdeer, *Auk* **92**:149–151.
Little, J. de V., 1967, Some aspects of the behaviour of the Wattled Plover, *Afribyx senegallus* (Linnaeus), *Ostrich* **38**:259–280.
Maxson, S. J., and Oring, L. W., 1980, Breeding season time and energy budgets of the polyandrous Spotted Sandpiper, *Behaviour* **74**:200–263.
Mayfield, H. F., 1978, Undependable breeding conditions in the Red Phalarope, *Auk* **95**:588–592.
Mock, D. W., 1983, On the study of avian mating systems, in: *Perspectives in Ornithology* (G. Clark, Jr., and A. Brush, eds.), Cambridge University Press, London.
Nethersole-Thompson, D., 1951, *The Greenshank*, Collins, Glasgow.
Nethersole-Thompson, D., 1973, *The Dotterel*, Collins, Glasgow.
Nettleship, D. N., 1973, Breeding ecology of turnstones, *Arenaria interpres*, at Hazen Camp, Ellesmere Island, NWT, *Ibis* **115**:202–217.
Norton, D. W., 1973, Ecological energetics of Calidridine Sandpipers breeding in northern Alaska, Ph.D. thesis, University of Alaska.
Orians, G. H., 1969, On the evolution of mating systems in birds and mammals, *Am. Nat.* **103**:589–604.
Oring, L. W., and Knudson, M. L., 1972, Monogamy and polyandry in the Spotted Sandpiper, *Living Bird* **11**:59–73.
Oring, L. W., and Lank, D. B., 1982, Sexual selection, arrival times, philopatry and site fidelity in the polyandrous Spotted Sandpiper, *Behav. Ecol. Sociobiol.* **10**:185–191.
Oring, L. W., and Maxson, S. S., 1978, Instances of simultaneous polyandry by Spotted Sandpipers, *Actitis macularia*, *Ibis* **120**:349–353.

Osborne, D. R., and Bourne, G. R., 1977, Breeding behavior and food habits of the Wattled Jacana, *Condor* **79:**98–105.

Parmelee, D. F., 1970, Breeding behavior of the Sanderling in the Canadian high Arctic, *Living Bird* **9:**97–146.

Parmelee, D. F., and Payne, R. B., 1973, On multiple broods and the breeding strategy of Arctic Sanderlings, *Ibis* **115:**218–225.

Parr, R., 1979, Sequential breeding by Golden Plovers, *Br. Birds* **72:**499–503.

Phillips, R. E., 1980, Behaviour and systematics of New Zealand plovers, *Emu* **80:**177–197.

Pienkowski, M. W., and Green, G. H., 1976, Breeding biology of Sanderlings in north-east Greenland, *Br. Birds* **69:**165–177.

Pitelka, F. A., 1959, Numbers, breeding schedule, and territoriality in Pectoral Sandpipers of northern Alaska, *Condor* **61:**233–264.

Pulliainen, E., 1970, On the breeding biology of the dotterel, *Charadrius morinellus, Ornis Fenn.* **47:**69–73.

Ralls, K., 1977, Sexual dimorphism in mammals: Avian models and unanswered questions, *Am. Nat.* **111:**917–938.

Ricklefs, R. F., 1974, Energetics of reproduction in birds, in: *Avian Energetics* (R. A. Paynter, ed.), pp. 152–192, Publ. Nuttall. Ornithol. Club, Cambridge, Mass.

Ridley, M. W., 1980, The breeding behavior and feeding ecology of Grey Pharlaropes, *Phalaropus fulicarius*, in Svalbard, *Ibis* **122:**210–226.

Rittinghaus, H., 1956, Untersuchungen am Seeregenpfeifer (*Charadrius alexandrinus* L.) auf der Insel Oldegoog, *J. Ornithol.* **97:**117–155.

Rowan, W., 1926, Notes on Alberta waders included in the British list, *Br. Birds* **20:**82–90.

Selander, R. K., 1972, Sexual selection and dimorphism in birds, in: *Sexual Selection and the Descent of Man* (B. Campbell, ed.), pp. 180–230, Aldine, Chicago.

Soikkeli, M., 1967, Breeding cycle and population dynamics in the Dunlin (*Calidris alpina*), *Ann. Zool. Fenn.* **4:**158–198.

Soikkeli, M., 1970, Mortality and reproductive rates in a Finnish population of Dunlin, *Calidris alpina*, Ornis Fenn. **47:**149–158.

Summers, R. W., and Waltner, M., 1979, Seasonal variation in the mass of waders in southern Africa with special reference to migration, *Ostrich* **50:**21–37.

Thomas, D. G., 1969, Breeding biology of the Australian Spur-winged Plover, *Emu* **69:**81–102.

Trivers, R. L., 1972, Parental investment and sexual selection, in: *Sexual Selection and the Descent of Man* (B. Campbell, ed.), pp. 136–179, Aldine, Chicago.

Walters, J. R., 1982, Parental behavior in lapwings (*Charadriidae*) and its relationship with clutch sizes and mating systems, *Evolution* **36:**1030–1040.

Warriner, J., and Warriner, R., 1978, Pajaro's plovers, *Point Reyes Bird Observatory* **45:**4–5.

Wilcox, L., 1959, A twenty year study of the Piping Plover, *Auk* **76:**129–152.

Chapter 6

COMMUNICATION IN BREEDING SHOREBIRDS

Edward H. Miller

Vertebrate Zoology Division
British Columbia Provincial Museum
and Biology Department
University of Victoria
Victoria, British Columbia V8V 1X4, Canada

I. INTRODUCTION

Shorebirds offer rich opportunities for comparative research because of their wide geographic distribution and their phylogenetic, ecological, and social diversity. There are more than 200 extant species, which occur in all zoogeographic regions of the world. Some species are resident and sedentary year-round, like the Shore Plover (*Charadrius novaeseelandiae*), which exists in small numbers only on the Chatham Islands east of New Zealand; others undertake phenomenal migrations each year, spending only 2 months or even less on the breeding grounds, with the remainder in migration or on wintering areas (e.g., certain Arctic species). Shorebirds inhabit the high mountains of Asia and of North and South America; they dwell in swamps and in hot, arid areas of Africa, Australia, and elsewhere; and they are found on muddy, rocky, or sandy seacoasts around the world (see Burger, this volume; Hale, 1980; Johnsgard, 1981). This wide geographic and ecological scope is paralleled by a great diversity of social systems, both on and off the breeding grounds. All these characteristics make shorebirds a likely group to enhance our understanding of the evolution and ecology of social behavior, a prediction supported by recent studies (e.g., Erckmann, 1981; Myers, 1981; Oring and Lank, 1982; Pitelka *et al.*, 1974; van Rhijn, 1983; Walters, 1980, 1982; and chapters in these volumes).

Studies on communicative behavior in shorebirds have lagged far behind those on social systems, and have contributed little to conceptual advances in the study of animal communication. A major reason for this is the difficulty in studying shorebirds. For example, in contrast to many species of oscines, on which most of our understanding of avian acoustic communication rests, shorebirds typically inhabit areas that are ecologically or geographically remote or difficult to work in, they often nest at low densities, are difficult to approach, offer no simple counterpart to oscine song, can be extremely mobile while in display, and are often hard to maintain or rear in captivity. Thus, though good descriptive studies on communication are possible for many species, few opportunities exist for rigorous field or laboratory experimental work on shorebirds like that routinely carried out on songbirds (e.g., see chapters in Kroodsma and Miller, 1982a).

Considering these facts, why should a chapter on shorebird communication be written, and what research potential do shorebirds offer? There are several answers to this question. First, it is important to begin to draw together our knowledge of shorebird communication, so that we can eventually understand its characteristics and its relationships to habitat, life history, and phylogeny. It is particularly urgent to expand and improve documentation of displays in certain populations and species, because they are uncommon, rare, or susceptible to extinction. The main offering of shorebirds to the study of animal communication lies in improved understanding of evolutionary pattern and process, through descriptive comparative studies, and in providing a data base against which to assess the generality of findings on more experimentally tractable taxa. For example, a major focus in communication research is ontogeny, which has implications for dialect formation, mimicry, duetting, and geographic variation. These areas are under intensive study in oscines, but are essentially unstudied and unknown in shorebirds, a group in which ontogeny of acoustic communication may be very different (see Section II.E.4).

This chapter begins with an overview of some external features of shorebirds that provide certain kinds of information about an individual's characteristics (including gender, age, etc.; see next paragraph). Following that, several studies on the organization and causation of displays are summarized. Acoustic components of displays are then reviewed, and their organization into repertoires, and functioning in several well-studied communication systems are discussed. Display characteristics and use are intimately related to social processes, so the relationship of social structure and ecology to communication is then considered. Finally, geographic variation and some evolutionary trends in displays, and the significance of displays to systematics, are treated.

Current concepts and theories of animal communication are diverse and divergent (Green and Marler, 1979; Hailman, 1982). As a consequence, there is no simple, single way to discuss communication even within a taxonomic group, without admitting strong conceptual biases. Two examples illustrate this point. Consider first the question of whether animals communicate honestly. A paradigm of honest communication would predict that breeding males of many species would accurately transmit information in their songs about their distance from other males. This could be effected through physical characteristics of song that degrade predictably with distance, thereby permitting listeners to accurately estimate distance from the singer (Richards and Wiley, 1980; Wiley and Richards, 1982). A contrasting paradigm might suggest that a singer could deceive listeners to his advantage by concealing his distance; this might lead to listeners expending time and energy in trying to locate a singer, or might keep them at great distances from him. Song features that undergo little degradation or attenuation over distance would thereby evolve. These simple paradigms provide different explanations for observed song characteristics, and lead to different predictions about evolutionary trends in song structure.

A second example deals with the kinds of information encoded in signals, assuming honest communication. Smith (1977) proposes that the messages encoded in signals are fixed in number and in what they refer to; a signal's meaning to a receiver results from his appraisal of many external circumstances in the immediate context of signaling, plus his prior experience, the signaler's identity and characteristics, etc. A converse view has been expressed by Beer (1975): why can't a particular type of display have a variety of messages, at different times and in different contexts?

The above contrasts serve to illustrate the current lack of wide acceptance of any particular paradigm for animal communication. In this chapter, I follow Smith (1977) in treating displays as though they make various kinds of information available about a signaler's behavioral and nonbehavioral states, even though there is often no evidence that such information is communicatively significant. This approach has the virtue of specifying the potential range of information available to receivers, and the important parts of that range can be narrowed through observation or experiment. Further, I consider that displays and display behavior cannot be viewed just from an adaptationist perspective, for their characteristics arise or are influenced also by other factors (notably phylogeny).

II. COMMUNICATION IN BREEDING SHOREBIRDS

A. External Characteristics and Appearance

In this section, I outline some important attributes of shorebirds that are apparent from external features: gender, population origin, individuality, age, and physiological state.

Many shorebird species are cryptically patterned and colored, and follow a widespread trend in being darker above than below (countershading), and in having the dorsal plumage approximate substrate color (chromatic matching) (see Graul, 1973a; Hailman, 1977). However, even the most unassuming species have distinctive features of potential communicative significance. For example, the Least Sandpiper (*Calidris minutilla*), the smallest of the 24 species in the Calidridini, is inconspicuously colored and patterned in the breeding season. Notable features are the yellowish legs and fairly dark appearance, especially on the breast (Fig. 1a; see Prater *et al.*, 1977). As in other monogamous species of calidridines, males are smaller than females (see Jehl, 1970). Male nuptial plumage is dark and gray, in contrast to the warmer, browner tones of females, and it is easy to distinguish the sexes by this plumage difference when paired birds are together, often even when in flight. This sexual difference is paralleled in the White-rumped Sandpiper (*Calidris fuscicollis*; Sutton, 1932) and the Upland Sandpiper (*Bartramia longicauda*; E. H. Miller and D. R. M. Hatch, unpublished observations). Jehl (1973) reports that ventral barring in male Stilt Sandpipers (*Calidris himantopus*) tends to be darker than in females. In brightly colored subspecies of the Dunlin (*Calidris alpina*), the hindneck of males "appears as a pale whitish collar," but in females is spotted, streaked, and only "slightly lighter in tone than . . . the crown and mantle" (Ferns, 1981, p. 94). Also, the upper breast of females has a more intense buffy wash (Ferns, 1981). These observations could be multiplied, but suffice to emphasize that a broad trend exists even in drab species of Scolopacidae for strong plumage differences that reliably signal gender. Sexual differences characterize many other taxa, of course, particularly those with extensive black in the plumage like *Vanellus* and *Pluvialis*. Similarly, males and females in many *Charadrius* species and in the Least Seedsnipe (*Thinocorus rumicivorus*) differ in the extent and intensity of black coloration in facial markings

Fig. 1. (a) Least Sandpiper (*Calidris minutilla*) in nuptial plumage [photograph by P. A. Johnsgard]. (b) African Black Oystercatcher (*Haematopus moquini*) showing the distended naked eye-ring [photograph by A. E. Burger]. (c) Northern Jacana (*Jacana spinosa*) in a display in which the striking primaries and carpal spurs are exposed [photograph by T. R. Mace].

and breast bands. Polyandrous species are typified by females being brighter, darker, or more intensively/contrastingly marked than males (e.g., *Phalaropus*, Rostratulidae). In general, it is likely that sexual dimorphism in nuptial plumage of most shorebird species is probably great enough to be a reliable and communicatively significant static-optical indicator of gender.

Sexual dimorphism in shorebirds is also pronounced in other morphological features: males of the Ruff (*Philomachus pugnax*) are adorned with ruffs and naked patches on the face, lacking in females (Fig. 2c; Glutz von Blotzheim *et al.*, 1975; Hogan-Warburg, 1966); male oystercatchers have bright, pink bills, in contrast to the dull, orange bills of females (Miller and Baker, 1980); the iris of female Eurasian Avocets (*Recurvirostra avosetta*) tends to be brown, that of the male red-brown (Witherby *et al.*, 1941); the bill of female American Avocets (*R. americana*) is much shorter and more strongly upcurved than that of males (Hamilton, 1975); and so on. Many sexual differences in morphology have resulted from selective pressures surrounding feeding ecology; energetics of egg-laying, incubation, and display; intersexual competition; etc. Such differences have not always arisen or been modified to signal a bird's gender; nevertheless, they often make that important information reliably available to receivers.

Other classes of nonbehavioral information made available through morphology include species and subspecies status. Species differences need not be discussed here; most such differences are obvious, but are not reinforced or accentuated through reproductive interactions between species (see Miller, 1982). An example for shorebirds concerns two sibling species of plovers (*Pluvialis dominica*, *P. fulva*) which breed sympatrically in northwestern Alaska; they exhibit no divergent shift in plumage characters there, despite their similar appearance (Connors, 1983). Differences between subspecies may facilitate segregation during late winter and early spring, and Ferns (1981) and Ferns and Green (1979) document many potentially important differences in plumage of Dunlin subspecies at that season. Subspecies identity is unambiguous in the Masked Plovers (*Vanellus miles miles* and *V. m. novaehollandiae*) of Australia, because of their strongly differing plumages and facial characteristics (van Tets *et al.*, 1967; see also Bock, 1958). The extent to which this information is used in social behavior between the subspecies is not known, though they interbreed in a recently formed overlap zone in Queensland (van Tets *et al.*, 1967).

Individual distinctiveness is not obvious in the plumage and soft parts of many species of shorebirds. Some species show great interindividual variation, however. In the Ruddy Turnstone (*Arenaria interpres*), the

Fig. 2. (a) Variation of facial markings in the nuptial plumage of four male Ruddy Turnstones (*Arenaria interpres*) [after Fig. 1 of Ferns (1978)]. (b) Threat display between breeding male Ruddy Turnstones [after Fig. 117 of Glutz von Blotzheim *et al.* (1977) and Fig. A of Cramp *et al.* (1983)]. (c) Variation of wattles and ruffs in the nuptial plumage of six male Ruffs (*Philomachus pugnax*) [after Fig. 94 of Glutz von Blotzheim *et al.* (1975)].

complex facial markings differ greatly among members of both sexes (Fig. 2a; Ferns, 1978), and these markings are prominently displayed in certain kinds of aggressive encounters (Fig. 2b). Similarly, wattle color in male Ruffs occurs in various shades of yellow and red, and in concert with color and markings of the ruff, allows individuals on lekking hills to be easily distinguished by human observers (Fig. 2c; Hogan-Warburg, 1966; Shepard, 1975). Extreme variation also occurs among Black-winged Stilts (*Himantopus himantopus*), in which the face and upper neck may vary from pure white to mainly black (Goriup, 1982). Some of the differences among male Ruffs are due to age, for the size of ruffs and ear tufts of captives has been noted to increase in successive years (Hogan-Warburg, 1966). Another feature that changes with age in this species is leg color, which is lead-gray in nestlings, and successively becomes greenish-gray, reddish-green, then reddish or orange, in the course of seasons (Andersen, 1951; Hogan-Warburg, 1966). Changes in leg color with age also occur in the American Woodcock (*Scolopax minor*; Shissler *et al.*, 1982). Birds in their first summer have darkly pigmented legs (and bills), while those at least 1 year old tend toward pinkish legs and bills. In the Red-breasted Dotterel (*Charadrius obscurus*), the breast band of males is deep reddish-brown, and that of the female is pinkish-red; Stead (1932) suggests that this coloring and that on the head increase in successive years (but see Reed, 1981). Sheathbills (Chionididae) owe their name to the horny culmen sheath at the base of the bill. This becomes separated from the culmen and increases in depth with age in the Lesser Sheathbill (*Chionis minor*), and is effectively displayed in certain kinds of short-range visual threat (Figs. 3, 4; Burger, 1979, 1980a,b,c).Other facial features also change with age: the head crest becomes more prominent, the black caruncles cover a larger part of the face anterior to the eyes, and the fleshy eye-rings thicken and change from very pale to bright pink (Fig. 3a; Burger, 1980a,c). The significance of morphological features that provide information about age may lie in the simultaneous transmission of information about two age-dependent qualities: experience and social status. There is no available evidence on this point, though plumage darkness, fighting success, and territory quality are positively correlated in Greater Golden Plovers (*Pluvialis apricaria*; Edwards, 1982), and birds appear to mate positive-assortatively according to extent of blackness in plumage (Parr, 1980). Plumage blackness does not appear to change with age in this species (Parr, 1980). However, age-related differences in breeding success arising from prior experience and foraging efficiency provide a possible mechanism for the evolution of enhanced discriminatory behavior, based on external features (see Cadman, 1980; Gratto *et al.*, 1981; Groves, 1978; Heppleston, 1971).

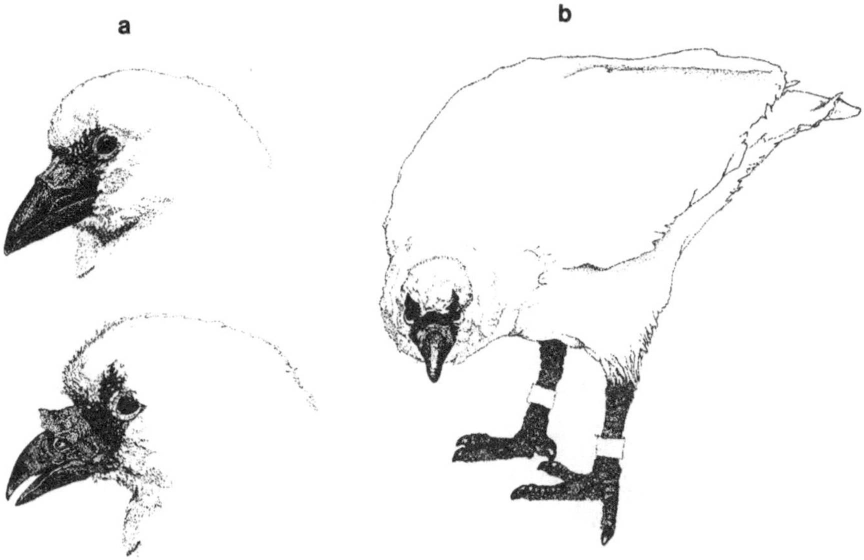

Fig. 3. (a) Changes with age in the facial features of the Lesser Sheathbill (*Chionis minor*). A 4-month-old juvenile is shown above, and a 4-year-old adult below [drawings by A. McClung, after photographs by A. E. Burger]. (b) Forward (threat) display by a Lesser Sheathbill [drawing by A. McClung, after photograph by A. E. Burger].

Short-term physiological changes can be reflected in the appearance of soft parts, thereby providing other kinds of information to receivers. Pupillary contraction and dilation are probably important sources of information in short-range interactions in species with brightly colored irides, including Burhinidae, Haematopodidae (Fig. 1b), Magellanic Plovers (*Pluvianellus socialis*), etc. In the latter species, "vivid iris coloration is featured prominently in courtship and territorial displays . . . or when a bird is excited . . . the eye is opened wide and the pupil contracts, giving added prominence and brightness to the expanded coral iris. At these times, the inner rim of the iris is yellow" (Jehl, 1975, p. 46). Alteration of blood flow can also alter the appearance of soft parts. In the Comb-crested Jacana (*Irediparra gallinacea*), the cephalic shield can change quickly between rich yellow and orange-red (Hindwood, 1934; Potter, 1934). Most soft parts apparently cannot undergo such short-term transformations, though seasonal changes in color, size, or distension to signal breeding condition may be widespread. Possible candidates include tarsi, wattles, and fleshy eye-rings (Fig. 1b). I know of no published observations on this matter.

Fig. 4. Male Lesser Sheathbills (*Chionis minor*) in "aggressive upright" postures during a pause in fighting. The arrows point to the black carpal spurs. [Drawing by A. McClung, after photograph by A. E. Burger.]

To summarize, there are many kinds of nonbehavioral and some behavioral (physiological) information made available, in a static-optical way, by the external appearance of shorebirds. It is likely that many external features that passively encode and emit such information owe their characteristics to selective pressures favoring (or disfavoring) communication. It is most tempting to invoke adaptive interpretations for features that are displayed, exaggerated, or modified through stereotyped motor patterns, of course. Thus, the carpal spurs of the Northern Jacana (*Jacana spinosa*) are rich yellow in color, with a naked area of the same color around their base. In several types of agonistic display with the wings extended, the spurs and the striking yellow primaries are clearly

Fig. 5. Male Upland Sandpiper (*Bartramia longicauda*) in various stages of display upon landing [drawing by A. McClung, after photographs by E. H. Miller and D. R. M. Hatch].

visible (Fig. 1c). Similarly, the black carpal spurs of the Lesser Sheathbill contrast with the surrounding white plumage, and are displayed when the wings are lifted out from the sides (Fig. 4; Burger, 1980b,c). Numerous examples could be added, of features that are hidden except when in display. Particularly notable are the striking patterns on individual feathers, and on surfaces of wings and tails in many species, including *Vanellus*, Burhinidae, Rostratulidae, Egyptian Plover (*Pluvianus aegyptius*), *Gallinago*, some Haematopodidae, etc. (Fig. 5).

Many optical characteristics can be altered through modification of the plumage, or by adoption of suitable postures or orientation to receivers. In many species of *Charadrius* plovers, an upright posture is adopted in which the bird stretches vertically and fluffs the plumage, thereby increasing the extent of the breast bands (Cairns, 1977, 1982; Cunningham, 1973; Glutz von Blotzheim *et al.*, 1975; Maclean, 1977; Mundahl, 1977; Simmons, 1953a,b; Smith and Hosking, 1955). In the Wrybill (*Anarhynchus frontalis*), "a resident bird drives out an intruder by running at it in a horizontal posture. The run [can end] in an upright display in which the breast is expanded. In the horizontal run the chest stripe is expanded . . ." (Hay, 1983, p. 11). Comparable upright postures in which apparent size is increased through vertical stretching and fluffing of the plumage occur in the Magellanic Oystercatcher (*Haematopus leucopodus*) (Miller and Baker, 1980) and the Magellanic Plover (Figs. 5, 6, and 8 of Jehl, 1975); the latter also shares with Haematopodidae, Jacanidae, and Burhinidae the trait of increasing the size of the nape during such displays (Fig. 6 of Jehl, 1975; Maclean, 1972). Fluffing of the plumage occurs in postlanding displays of the Upland Sandpiper (Figs. 5a,c,d), and in this species the throat is distended and strikingly white while calling (Figs. 5b,e). The same effect occurs in the Least Sandpiper, and may have provided the impetus for evolutionary elaboration of the throat for distension in display, as in the related Pectoral Sandpiper [*Calidris melanotos* (Holmes and Pitelka, 1961; Myers, 1982; Pitelka, 1959)].

B. Studies on Display Organization

Several researchers have analyzed shorebird displays based on simple models of causation. This approach has generally assumed that behavioral structure reflects a particular balance of a few underlying "tendencies" or neurophysiological states. Behavioral characteristics that have been viewed in this light include the component structure of displays, quantitative changes in display components, and the structure of sequences, as related to different social circumstances. The premises upon which such research is based are not widely accepted, but the analytic procedures and results provide useful information about the composition of shorebird display behavior. In this section, I outline a few examples of detailed research along these lines.

1. Ground Display in the Black-tailed Godwit

Black-tailed Godwits (*Limosa limosa*) show a number of distinctive visual displays during short-range agonistic interactions on the ground. Lind (1961) has described and analyzed them in detail, and this section discusses his findings. He recognized five major display types used in the context mentioned [see Figs. 3 and 9a of Lind (1961), Figs. A–E of Cramp (1983), and Fig. 40 of Glutz von Blotzheim *et al.* (1977)]:

Upright. Attitude upright; legs stretched; carpal joints slightly exposed; neck usually extended vertically; dorsal plumage sleeked to slightly ruffled; tail often lowered, never spread [note: in this and other displays, the tail is rotated "round the longitudinal axis and always in such a way that the upper part (= surface) faces the opponent" (Lind, 1961, p. 26); this degree of rotation varies with spreading]; bill 0–60° below horizontal.

Upright-II. Attitude upright; legs stretched; carpal joints not exposed, but wing tips slightly elevated; neck extended vertically; dorsal plumage ruffled; tail lowered and maximally spread; bill horizontal.

Forward. Attitude almost horizontal; legs not stretched; carpal joints not exposed; neck extended obliquely forward; dorsal plumage usually very ruffled; tail lowered and spread (seldom maximally); bill generally forms an angle with neck of 90°, but may be kept horizontal.

Crouch. Attitude horizontal; legs strongly bent; carpal joints slightly exposed only when dorsal plumage sleeked (it varies from ruffled to sleeked); neck withdrawn; tail may be lowered and spread (seldom maximally); bill position varies.

Tilt. Attitude tilted, with breast close to the ground; legs strongly bent; carpal joints not exposed (?); neck withdrawn; dorsal plumage sleeked; bill "most often points forward" (Lind, 1961, p. 26).

The relative frequencies of occurrence of the display types are summarized in Table I, across columns that are approximately ordered from high "attack drive" (left-most column) to high "escape drive" (right-most column). Overall, there is no striking correspondence of most display types with attack/retreat tendencies, though Forward occurred in 35% of Retreat contexts vs. only 16% of Attack contexts, and Upright-I occurred in 33% of Attack contexts vs. only 24% of Retreat contexts (based on data in Lind's Table 2). Yet Upright-II, which is similar in many components to Upright-I, is very dissimilar in distribution, occurring in 5% of Retreat contexts and 4% of Attack contexts. The heterogeneous distribution of Crouch across columns further emphasizes the complex relationship of display type to attack/retreat tendencies.

The distribution of display components shows comparably complex trends, for no component simply becomes more or less common as contexts move from Attack to Retreat (Table II). Ruffling of the dorsal plumage, for example, is about twice as common in Crouch postures during Retreat as during Attack (100 vs. 52%), but occurs in only 5% of Upright-I postures during Retreat—about a third the incidence of Upright-I during

Table I. Relative Frequency of Occurrence (in Percent) of Five Display Postures Used during Agonistic Ground Encounters by Black-tailed Godwits (*Limosa limosa*)[a,b]

	Context					
	I. Attack		II	III. Retreat		
Posture	During	Preceding	Intermediate	Preceding	Direct orientation	Orientation away
Upright-I	25	44	34	33	18	18
Upright-II[c]	5	1	3	2	0	18
Forward	21	9	11	32	45	26
Crouch	44	41	29	26	28	38
Tilt[d]	5	6	23	6	10	0

[a] Data from Table 2 of Lind (1961).
[b] Social contexts are arranged from attack (left-most column) to retreat with orientation away (right-most column).
[c] Termed "Hostile tail-display" by Lind (1961).
[d] Termed "Tail-up" by Lind (1961).

Table II. Relative Frequency of Occurrence (in Percent) of Certain Components in Agonistic Ground Encounters by Black-tailed Godwits (*Limosa limosa*)[a]

		Display component		
Posture	Context[b]	Dorsal plumage ruffled	Tail lowered and/or spread	Bill 0–30° below horizontal
Upright-I	Attack (I)	14	86	31
	Retreat (III)	5	72	76
Forward	Attack (I)	54	73	48
	Retreat (III)	86	95	0
Crouch	Attack (I)	52	67	63
	Retreat (III)	100	100	0
Tilt	Attack (I)	0	70	90
	Retreat (III)	0	38	46

[a] Data from Table 3 of Lind (1961).
[b] See column headings, Table I.

Attack. Tail lowering and/or spreading, and the position of the bill relative to the horizontal, likewise show no systematic relationship to context. The bill is held near the horizontal most often during Attack contexts in which the bird is in Forward, Crouch, or Tilt; but this strong trend is reversed for Upright-I.

Lind treats these and selected data in detail, in discussing the motivational basis for godwit displays. Here it is sufficient to summarize by emphasizing that the displays and their components do not map simply onto social contexts. Rather, each occurs in various contexts, and (for components) in various combinations. The rough correspondences that do exist suggest that the displays and their components encode information of a graded or probabilistic sort, and there must be many more classes of information than just tendencies to approach or retreat. These interpretations are supported by other of Lind's data: in Attack contexts when birds were separated by less than 2 m, Upright-I comprised 30%, Forward 20%, and Crouch 33% of the displays; at more than 4 m, they comprised 0, 6, and 65%, respectively.

2. Aerial Display of the Northern Lapwing

Many species of shorebirds have aerial displays (see Section II.E.2), and the first to be described in detail was the "ceremonial flight" of the Black-tailed Godwit (Lind, 1961). That study documented the remarkable stereotypy of the display sequence, as well as some of the external factors

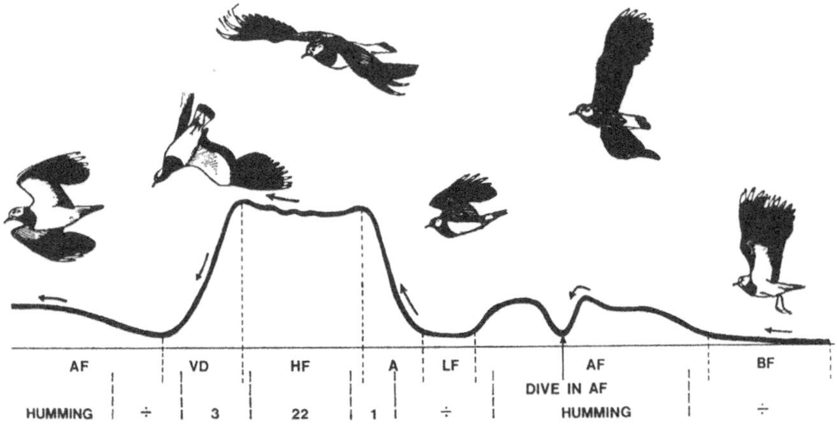
AF
VD
HF
A
LF
AF
BF
DIVE IN AF
HUMMING
÷
3
22
1
÷
HUMMING
÷

♂
♀

that can modify its structure. More detailed research was carried out on the Northern Lapwing (*Vanellus vanellus*) by Dabelsteen (1978), along the same lines.

Aerial display in the Northern Lapwing occurs in the following main contexts: (1) apparently spontaneous; (2) in response to predators; (3) on returning to the territory (e.g., early in the season, males are absent from the territory all night, and return in the morning); (4) in response to other males; and (5) in response to females.

The basic unit of aerial display in the Northern Lapwing has several distinct phases (Fig. 6, upper part). (1) The male takes flight with slow, deep wing-beats ("butterfly flight," BF). (2) During the last few beats of BF, flight speed and wing-beat frequency increase, and the bird revolves about its long axis every four wing-beats or so. "The revolving . . . is made by one wing-beat, and takes place in the upstroke, whereas the downstroke is made with the opposite wing tip pointing downwards" (Dabelsteen, 1978, pp. 139–140). When the bird revolves, it changes course to the same side, resulting in a zigzag flight course. This display component ("alternating flight," AF) is always accompanied by "humming" sounds, made by the modified outer three (and to some extent the fourth) primaries, which vibrate during the downstroke (Fig. 6, lower part). (3) The bird stops the AF and proceeds with slow wing-beats of very small amplitude just above the ground ("low flight," LF). (4) Ascent, during the end of which motif 1 of song is uttered. (5) Following ascent, the bird flies straight in "high flight" (HF) and emits the end of motif 1 plus motif 2, and the start of motif 3 of song. Toward the end of HF, the bird revolves on its long axis so that the inclination of the wing plane exceeds 90°, and slows down during a slight rise. (6) "Vertical dive" (VD) follows HF. During one or two wing-beats following HF, the bird turns on its back with its back down more or less vertically toward the ground. During descent, the bird often makes a few wing-beats while revolving back on its long axis. VD is accompanied by motif 3 of song. Other units can follow VD.

The preceding description conceals the numerous variations that can occur. Nevertheless, there is strong sequential dependency of the components: Takeoff is followed by BF nearly 70% of the time, BF by AF

←

Fig. 6. (Upper) Diagram of a typical aerial display in the Northern Lapwing (*Vanellus vanellus*). BF, butterfly flight; AF, alternating flight; LF, low flight; HF, high flight; VD, vertical dive. Song consists of motifs 1, 2, and 3; "÷" = silence. See text for further details. [From Dabelsteen (1978, Fig. 1).] (Lower) Typical wing shapes of male and female Northern Lapwings, showing the sexual difference in length of the primaries, the outermost three or four of which produce the "humming" sound in the downstroke of AF. [From Glutz von Blotzheim *et al.* (1977, Fig. 54).]

Table III. Transition Matrix for Components of Aerial Display in the Northern Lapwing (*Vanellus vanellus*)[a]

Preceding component	Following component							Row totals
	BF	AF	LF	A	HF	VD	Landing	
Takeoff	106	27	2	19		—	—	154
Butterfly flight (BF)	—	48	4	26			8	86
Alternating flight (AF)		—	111	157			120	388
Low flight (LF)		11	—	103			1	115
Ascent (A)				—	465		7	472
High flight (HF)					—	504		504
Vertical descent (VD)		299	14	38		—	82	433
Column totals (*N*)	106	385	131	343	465	504	218	2152

[a] Data from Fig. 6 of Dabelsteen (1978). Certain impossible transitions are indicated by dashes.

nearly 60%, HF is always followed by VD, and so on (Table III). Much of the variation results from differences in display form according to context. For example, the number of components per flight display varies significantly according to context. Fewest components occur in displays elicited by females, and upon returning to the territory, and most occur in displays elicited by other males. The larger number of components in the latter context is accompanied by a great variety of deviations from "normal" display. For example, song motifs may be repeated or omitted, revolutions may be increased or absent, etc. In all, nearly half of the initial part of aerial display involving other males shows deviations of some sort, as compared with only 15% in displays elicited by females, and 19% in spontaneous displays. Thus, when aerial display occurs in response to other males, it is apparently less stereotyped and more frequently modified than when emitted in response to other females, or spontaneously. This tendency is clearly shown by dives, which may be oriented toward males some distance away. Then "they are performed like attack dives . . . with the ventral side turned toward the adversary, and are often repeated" (Dabelsteen, 1978, p. 164). Dives may also be oriented toward females, but "they are less violent and are rarely repeated" (Dabelsteen, 1978, p. 164). In Ascent, males often rise in one another's direction, then fly in HF parallel to or away from one another, orientations that are rarely shown to females.

The flight display of the Northern Lapwing is a complex but remarkably stereotyped sequence of optical and acoustic components. It is given only by males, so is influenced by endocrinological factors, and various of its characteristics (including overall incidence) vary according to date, stage of the breeding cycle, weather, time of day, and other nonsocial factors (e.g., they are generally very brief in the morning). The display contains components of ritualized attack, approach, and other tendencies, and variation in underlying physiological states affects display form. Finally, variations in form result from proximate social factors. The large number of known or probable sources of variation make it likely that the communicative significance of the display and its components varies greatly for different classes of receivers and for different contexts of occurrence. In this way it resembles the agonistic ground displays of the Black-tailed Godwit.

3. Ground Display by Lekking Male Ruffs

The preceding accounts of godwits and lapwings focused on general aspects of display organization and variation, but did not consider contributions to variation by individual or class differences. One of the clearest cases where socially distinct classes of individuals exist is on leks of the Ruff. There, males fall into two classes based on their behavior and plumage: "independent" males include territorial and potentially territorial individuals, which tend to have dark or colored ruffs and ear tufts; "satellite" males are nonterritorial but often associate with residents, and tend to have white ruffs and tufts (see Fig. 2c; Hogan-Warburg, 1966; Shepard, 1975). The behavioral differences between members of these two social classes have been studied intensively by van Rhijn (1973), whose work is summarized here (see also van Rhijn, 1983).

Main behavioral units recognized by van Rhijn are shown in Fig. 7A. They are based mainly on posture, relative position of the head and bill to the body, and erection of the ruff and head tufts. These units occur with different frequencies for satellites and residents. Thus, Upright comprised about 9% of 395 postural displays by residents, but did not occur in 428 displays by satellites. Conversely, Tiptoe and Strut accounted for about 18% of satellite displays, but none of the residents' (see marginal totals, Table IV). The organization of behavioral sequences also differed strongly. Squat was followed by Forward far more often in residents than satellites, and was followed by Half-squat far less frequently (tabled figures of +75% and −71%, respectively). This difference is related to the higher frequency of occurrence of Forward in residents, particularly when satellites were present: residents spent 13% of their time in Forward when females were with them, and 34% when satellites were also present. On these sorts of evidence, van Rhijn proposed a simple model for causation of the displays (Fig. 7B). At least two variables (H, V) are needed to explain resident behavior, and at least one for satellites (and females).

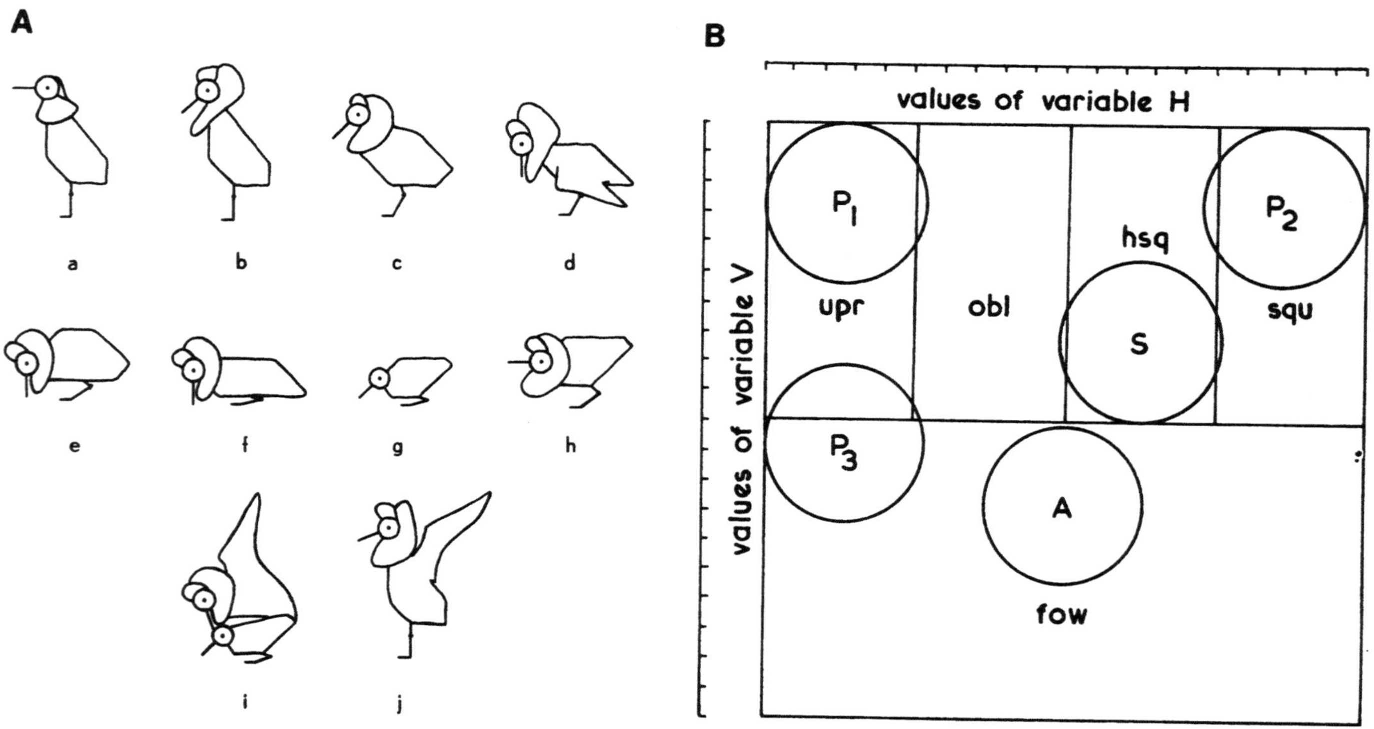

Fig. 7. (A) Major behavioral units recognized by van Rhijn in his study on Ruffs (*Philomachus pugnax*). a, Upright (upr); b, Tiptoe; c, Oblique (obl); d, Strutting; e, Half-squat (hsq); f, Squat (squ); g, Crouching; h, Forward (fow). [After van Rhijn (1973, Fig. 1).] (B) Causation and function related to major behavioral units recognized by van Rhijn. A, behavior in aggressive situations; S, behavior in sexual situations; P_1, protective behavior (flying away); P_2, protective behavior (lying flat); P_3, protective behavior (after fights). Variables H and V are postulated underlying variables. [From van Rhijn (1973, Fig. 8).]

Table IV. Transition Matrix for Postural Displays of Resident and Satellite Male Ruffs (*Philomachus pugnax*) on Leks[a]

Preceding component	Following component							
	Obl	Hsq	Squ	Fow	Upr[b]	Tip[c]	Str[c]	Totals[d]
Oblique (Obl)	—	−34	N	+56	(+24)	(−46)	(−1)	86, 140
Half-squat (Hsq)	−37	—	+9	+28	(+1)	(−1)	(−1)	67, 142
Squat (Squ)	−4	−71	—	+75	(N)	(N)	(N)	34, 52
Forward (Fow)	+6	−17	+5	—	(+5)	(N)	(N)	171, 4
Upright (Upr)[b]	(+81)	(N)	(N)	(+19)	—	—	—	37, 0
Tiptoe (Tip)[c]	(−94)	(−1)	(−1)	(N)	—	—	(−4)	0, 83
Strutting (Str)[c]	(N)	(−57)	(−29)	(N)	—	(−14)	—	0, 7
Totals[d]	88, 143	80, 129	94, 80	102, 3	31, 0	0, 67	0, 6	395, 428

[a] Data after Table 1 of van Rhijn (1973). Each cell entry shows the percent occurrence of a particular transition, for residents relative to satellites, based on row totals. For example, the Obl–Hsq transition occurred 34% less in residents than in satellites, and the reverse (Hsq–Obl) occurred 37% less (N = no observations).
[b] Upr was given only by resident males, in the sample.
[c] Tip and Str were given only by satellite males, in the sample.
[d] Totals are shown first for residents, then for satellites (total samples are thus 395 and 428, respectively).

Satellite behavior is intermediate between that of residents and females, and two unique components to it (Tiptoe, Strut) were judged to be similar to the more generalized displays, Upright and Oblique.

The general significance of van Rhijn's study lies in the documentation of substantial communicative differences among social classes, differences that can be explained by a model that assumes a common causal structure. In addition, it is unnecessary and misleading to simply label the differences as due to variation in "aggressiveness," "submissiveness," etc.

C. Acoustic Communication

1. Review

Vocal and nonvocal sounds of shorebirds range from soft, short-range calls like those between adults and young, to loud nuptial aerial song that may carry for miles. Such variation even occurs within species so it is perhaps not surprising that few comprehensive studies on acoustic communication in shorebirds have been made (Table V). In addition, most studies have focused on a particular aspect of communication. In the American Woodcock (*Scolopax minor*), for example, nearly all research

Table V. Summary of Available Information on Sound Analyses in Shorebirds[a]

Family and species[b]	Comments[c]	Analysis[d]	References[e]
Burhinidae (9)			
Burhinus oedicnemus	B, DD	s	10
B. senegalensis	B, D	s	10
Charadriidae (63)			
Anarhynchus frontalis	B, DD	s	54
Charadrius alexandrinus	B, DD	s	10, 16
Ch. bicinctus	B, DDD	s	54
Ch. dubius	A, B, DDD	s	10, 16, 64
Ch. hiaticula	A, B, DD	s	10, 16, 63, 66, 67
Ch. montanus	B, C, DDD	s	18, 44
Ch. morinellus	B, D	s	10
Ch. novaeseelandiae	B, C, DDD	s	53
Ch. obscurus	B, DD	s	54
Ch. semipalmatus	D	s	44
Ch. vociferus[f]	B, C, DDD	s x	10, 19, 21, 50, 52
Chettusia (= *Vanellus*) *leucura*	B, D	s	10
Ch. (= *Vanellus*) *gregaria*	D	s	10
Hoplopterus (= *Vanellus*) *spinosus*	B, D	s	10
Pluvialis apricaria	B, D	s	10
P. dominica	D	s o x	19, 44, 66
P. squatarola	A, B, D	s	10, 44, 64
Vanellus chilensis	D	s	44
V. vanellus	A, B, DD	s	10, 11, 72
Chionididae (2)	No published analyses	—	—
Dromadidae (1)	No published analyses	—	—
Glareolidae (16)			
Cursorius cursor	D	s	10

[a] Only species whose sounds have been analyzed graphically in some way are included. See the text for discussion.
[b] Number of species in each family is listed in parentheses.
[c] A = analytical; B = with behavioral observations; C = comprehensive survey; D = descriptive (D = few calls, . . . , DDD = many calls); E = experimental.
[d] s = sound spectrograms; o = oscillograms; p = power spectra; a = amplitude profiles; x = other.
[e] 1, Adret (1982); 2, Ailes (1976); 3, Baker (1974); 4, Baker (1982); 5, Beightol (1972); 6, Beightol and Samuel (1973); 7, Bourgeois (1977); 8, Bourgeois and Couture (1977); 9, Bursian (1971); 10, Cramp (1983); 11, Dabelsteen (1978); 12, Dowsett *et al.* (1977); 13, Ferdinand (1966); 14, Forsythe (1967); 15, Forsythe (1970); 16, Glutz von Blotzheim *et al.* (1975); 17, Glutz von Blotzheim *et al.* (1977); 18, Graul (1974); 19, Greenewalt (1968); 20, Grudzien (1976); 21, Heckenlively (1972); 22, Heidemann and Oring (1976); 23, Hirons (1977a); 24, Hirons (1977b); 25, Holmes (1973); 26, Howe (1972); 27, Howell (1979); 28, Jenni *et al.* (1974); 29, Knudson (1972); 30, Kroodsma and Miller (1983b); 31, Labutin *et al.* (1982); 32, Lemnell (1978); 33, Mace (1981); 34, Mal'chevskii (1981); 35, Marcström (1974); 36, Miller (1977); 37, Miller (1979a); 38, Miller (1979b); 39, Miller (1979c); 40, Miller (1982); 41, Miller (1983a); 42, Miller (1983b); 43, Miller (1983c); 44, Miller (this chapter); 45, Miller and Baker (1980); 46, Miller *et al.* (1983); 47, Myers (1982); 48, Nemetschek (1977); 49, Nethersole-Thompson and Nethersole-Thompson (1979); 50, Nol (1980); 51, Oring (1968); 52, Phillips (1972); 53, Phillips (1977); 54, Phillips (1980); 55, Reddig (1978); 56, Samuel and Beightol (1972); 57, Samuel and Beightol (1973); 58, Skeel (1976); 59, Skeel (1978); 60, Sordahl (1979); 61, Sordahl (1980); 62, Thönen (1968); 63, Tikhonov and Fokin (1979); 64, Tikhonov and Fokin (1980); 65, Tikhonov and Fokin (1981a); 66, Tikhonov and Fokin (1981b); 67, Tikhonov and Fokin (1982); 68, Veprintsev and Zablotskaya (1982); 69, Warham and Bell (1979); 70, Weir (1979); 71, Weir and Graves (1982); 72, Wollemann (1978); 73, Wollemann and Olaszy (1977); 74, Zvonov and Tikhonov (1981).
[f] Additional sound spectrograms are in Gaunt (1983b; *Ch. vociferus*) and Mal'chevskii (1982; *S. rusticola*).

Table V. (*Continued*)

Family and species[b]	Comments[c]	Analysis[d]	References[e]
Glareola pratincola	D	s	10
G. nordmanni	D	s	10
Pluvianus aegyptius	B, C, DD	s	10, 27
Rhinoptilus cinctus	D	s	12
Haematopodidae (9)			
Haematopus ater	A, B, D	s	37, 44, 45
H. chathamensis (= *H. unicolor*)	B, D	s	3
H. leucopodus	B, D	s	45
H. ostralegus	A, B, DD	s	3, 10, 63–65, 67
H. palliatus	B, D	s	44, 45
H. unicolor	B, D	s	3
Ibidorhynchidae (1)	No published analyses	—	—
Jacanidae (8)			
Jacana spinosa	A, B, C, DDD	s	28, 33
Pedionomidae (1)	No published analyses	—	—
Pluvianellidae (1)	No published analyses	—	—
Recurvirostridae (13)			
Himantopus himantopus	B, DD	s	10, 72
H. mexicanus	B, D	s	61
Recurvirostra americana	B, D	s	61
R. avosetta	A, B, DD	s	1, 10, 63, 64, 67, 73
Rostratulidae (2)	No published analyses	—	—
Scolopacidae (90)			
Actitis hypoleucos	A, B, DD	s	10, 17, 65, 66
A. macularia	B, D, E	s	10, 22, 29, 44
Arenaria interpres	A, B, D	s	10, 63–65, 67
Bartramia longicauda	B, D	s	2, 10, 17, 44
Calidris alba	B, DD	s	10, 16
C. alpina	A, B, DD	s o p x	4, 10, 16, 19, 42, 44
C. canutus	B, DD	s	10, 16
C. ferruginea	B, D	s	10, 16
C. fuscicollis	B, D	s	16, 44
C. himantopus	B, D	s o p	29, 42
C. maritima	B, DD	s	10, 16, 44
C. mauri	B, D	s	25
C. melanotos	A, B, D	s	16, 47, 64
C. minuta	B, DD	s	10, 16, 65, 66
C. minutilla	A, B, D	s o p	10, 29, 36–39, 41, 43, 44
C. pusilla	B, D	s o p	42, 44
C. temminckii	A, B, DD	s	10, 16, 64, 66
Catoptrophorus semipalmatus	B, C, DDD	s	60
Coenocorypha aucklandica	B, D	s	69
Gallinago gallinago	A, B, DD, E	s	10, 17, 20, 55, 62
G. media	A, B, D	s	10, 13, 17, 32
G. stenura	B, D	s	10, 31
Limicola falcinellus	B, D	s	10, 16
Limnodromus griseus	B, D	s o	19, 43, 46

(*Continued*)

Table V. (*Continued*)

Family and species[b]	Comments[c]	Analysis[d]	References[e]
Limosa lapponica	B, DD	s	10, 17
L. limosa	B, DD	s	10, 17, 73
Lymnocryptes minimus	B, D	s	10
Numenius americanus	B, C, DDD	s o	14, 15, 19, 60
N. arquata	B, DDD	s	10, 17
N. minutus	B, C, DD	s	30, 68
N. phaeopus	B, C, DDD, E	s	10, 17, 58, 59
Phalaropus fulicarius	B, DD	s	10, 44
Ph. lobatus	B, D	s	10
Ph. tricolor	B, C, DDD	s	26
Philomachus pugnax	B, D	s	10, 66
Scolopax minor	A, B, DDD	s o p x	5–8, 19, 40, 56, 57, 70, 71
S. rusticola[f]	A, B, D	s o a	10, 17, 23, 24, 34, 35, 48
Tringa erythropus	B, DD	s	10, 17
T. flavipes	B, D	s o	10, 19, 44
T. glareola	A, B, DD	s	10, 17, 64, 66
T. (= *Heteroscelus*) *incanus*	D	s	44
T. melanoleuca	D	s o x	19
T. nebularia	A, B, DDD	s	10, 17, 44, 49, 65, 66, 74
T. ochropus	B, C, DDD	s	10, 17, 51, 65, 66
T. solitaria	B, C, DDD	s	10, 51
T. stagnatilis	B, DD	s	10, 17
T. totanus	B, DDD	s	10, 17, 72, 73
Xenus cinereus	B, D	s	10
Thinocoridae (4)	No published analyses	—	—

has concerned individuality in the "peent" call of males as a tool facilitating estimation of population size (Table V references; see also Thomas and Dilworth, 1980). Other common, widespread, and easily studied species remain unstudied or understudied. There has been no general study of acoustic signaling in the Common Snipe (*Gallinago gallinago*), an abundant breeding species in Europe, Asia, and North and South America; none of the nine species of oystercatchers has been adequately studied, despite their wide geographic distribution, conspicuousness, and approachability; and no repertoires of calidridine sandpipers (Holarctic, 24 species) have been well documented. Sound analyses are completely unavailable for certain key taxa, as well: Chionididae, Dromadidae, Ibidorhynchidae, Pedionomidae, Pluvianellidae, Rostratulidae, and Thinocoridae. Finally, relatively few studies have been quantitative, only one has involved experimentation (playback experiments; Refs. 22, 29), and none has used captive birds.

In light of the major gaps in our knowledge, it would be premature to review many facets of acoustic communication by shorebirds, but certain widespread trends in structural characteristics are suggested by available studies (see references in Table V). Many sounds are long and of narrow bandwidth, and are spectrally simple with weak or no overtones and little frequency modulation (Figs. 8, 9, 12, 14, 17). Many kinds of sounds, including those just mentioned, exhibit sudden frequency shifts, with the call continuing unbroken at much lower or higher frequency (Figs. 8E, F, 12A, 14). Simple harmonic overtones occur commonly (Figs. 8C, E, I–L and 17E; see Gaunt, 1983a). Single brief calls are widespread; these often have simple patterns of frequency and amplitude modulation (e.g., simple increase then decrease), and are highly variable even within species. Such brief sounds, plus various sorts of trills, are abundant in shorebirds (Figs. 8, 12, 17). Structurally complex sounds with strong sequential organization occur in many species, and are equivalent to primary song of many oscines in their beauty and presumed functions. No song repertoires have been reported for any species of shorebird.

It is not possible to judge how closely the structure or sequential organization of shorebird sounds matches Morton's (1977, 1982) generalizations about links between behavioral tendencies and acoustic morphology ("motivation-structural rules"). However, the varied structure of calls uttered by shorebirds even in a single context, such as when a potential predator is near their brood, suggests that any correspondence is likely to be complex (Fig. 8; see also Sections II.C.2 and II.E.1). For example, brief calls occur in most species, "noisy" (unpatterned broadband) calls occur in others, and strong rhythmic frequency modulation characterizes some call types; but there are no obvious common characteristics. Similarly, the structure of soft calls by adult birds brooding young or tending them at close quarters varies greatly among species (Fig. 17; see Section II.E.1). The relationship of sound features to the physical and social environment is considered in Section II.D.

2. Repertoire Characteristics and Functional Classes

Characterization and classification of signals are essential preliminary stages in the study of communication, regardless of how much subsequent refinement may be necessary. Signals are usually classified within a structural–functional framework simultaneously. No particular conceptual or theoretical significance can be attached to such a classification, however, unless there is prior reason to expect structural and functional classes to correspond. Correspondence is probably closest between physical characteristics of certain stereotyped, long-distance sounds, for which adaptations to combat environmental degradation and frequency-dependent attenuation are likely. Such a generalization applies to many vocalizations used by males to attract females or to deter competitors of the same sex. Here, sound classes are often discrete and context-specific, but most sounds defy simple structural or functional classification, as can be illustrated by a few examples.

Oring (1968) studied two species of Tringini, the Green Sandpiper (*Tringa ocrophus*) of the Palearctic and the Solitary Sandpiper (*T. solitaria*) of the Nearctic (see also Oring, 1973). Based on call structure and use, Oring identified seven sound types common to adults of the two species, plus three more for juvenile *T. ocrophus* and six for juvenile *T. solitaria*. The juvenile sounds varied greatly, and their classification was more difficult than for adult sounds. The main functional classes for the latter were sexual-agonistic (two "song" types, plus "epigamic chatter" and "epigamic long whistle"), "alarm-attack," "alarm-flee," and "contact." The general contexts of occurrence for the latter three are suggested by Oring's names for them.

A significant point that emerges from Oring's study is that apparently simple sound types can serve more than one function, and several sound types can overlap in apparent functions. Thus, "alarm-flee" calls are given by either species when danger is imminent, but also "when a bird takes off spontaneously . . . as a bird flies by or from its nest prior to the start of incubation . . . and . . . when a strange pair attempts to establish itself in the territory of another pair" (Oring, 1968, p. 409). In the first context, there is overlap with the "alarm-attack" call, which, however, is used "when danger is apparent but not imminent" (Oring, 1968, p. 406) as well as in other, very different contexts.

Sordahl's (1979) fine study of Willet (*Catoptrophorus semipalmatus*) vocalizations revealed similar trends. "Thus *kyah-yah* serves as a greeting and contact call, an indication of flight intention, and a flight enticement call" (Sordahl, 1979, p. 563). At the other extreme, Sordahl heard two call types ("klik" and "dik" calls) only during copulation attempts. In the Whimbrel (*Numenius phaeopus*), "low trill calls" are used commonly by both sexes, as when gliding in to land, by incubating birds calling in response to their mates, and during changeovers at the nest (Skeel, 1976, 1978; see also Cramp, 1983). This species also has at least three calls used in response to potential predators: "short predator alarm trill," "long predator alarm trill," and "scolding trill call" (Skeel, 1976, 1978).

Examples of multiple functions of single sound classes and common functions by different sound classes could be multiplied (e.g., Adret, 1982; Graul, 1974). Some of the apparent problems can be resolved by studying the structure and function of calls very closely. Thus, the "alarm-flee" call of the Green Sandpiper had a median number of three repetitions when an adult or its brood was directly threatened, but only one in response to a strange intruding pair (Oring, 1968), suggesting that finer classification using structural features might yield closer correspondence with function. Similarly, Sordahl (1979) noted that "kleep" calls of Willets in response to predators tended to occur at a more rapid rate during aerial than terrestrial harassment. In the Spotted Sandpiper (*Actitis macularia*), "song" used in aggressive contexts is of higher amplitude, higher frequency, and faster tempo than in reproductive contexts (Heidemann and Oring, 1976). Conversely, Skeel (1976, 1978) points out that the Whimbrel's "short predator alarm call" is used only in chasing aerial predators, indicating that greater attention to contextual features can improve understanding of structural and functional correspondence.

Elaboration of classification schemes using finer structural and functional criteria will inevitably lead to improved congruence between them. However, the generality of such schemes decreases correspondingly, and the crucial questions about why general structural classes span diverse functions, and why general functions involve more than one sound type, become more remote. A solution to this problem can come from determining which factors are held in common by different contexts, along lines suggested by Smith (1977). Why should incubating Eurasian Avocets utter rhythmic call series when another bird, be it a conspecific or nonpredatory heterospecific, passes or flies over the nest (Adret, 1982)? The appropriate focus in assessing the message in the call is at a level that merges all contexts, so that common features can be extracted. Communication is egocentric, according to Smith's paradigm, so that a caller signals something about its behavioral state, or its readiness or likelihood to engage in various behaviors. Often this information indicates a bird's inclination to locomote, interact, etc.—information that can be appraised and placed in appropriate contexts by receivers in imbuing the messages of signals with more specific and specialized meanings. An example using calls of adult Calidridini attending their young can illustrate this approach.

Various species of calidridines use two distinct classes of sounds when they are attending their broods, and are approached by humans (and presumably by other kinds of predators). These are: (1) a relatively brief, frequency-modulated call; and (2) a trill. The two call types are emitted in similar circumstances in all the species shown in Fig. 8 (I–L): trills are most common on the ground, or at a great distance from the nest or brood; brief frequency-modulated calls are most commonly uttered as a parent flies toward or near an intruder, or when close to the nest or brood. The latter call type is also the one normally given when a parent jumps into flight. In general, then, the call types reflect a parent's "excitement."

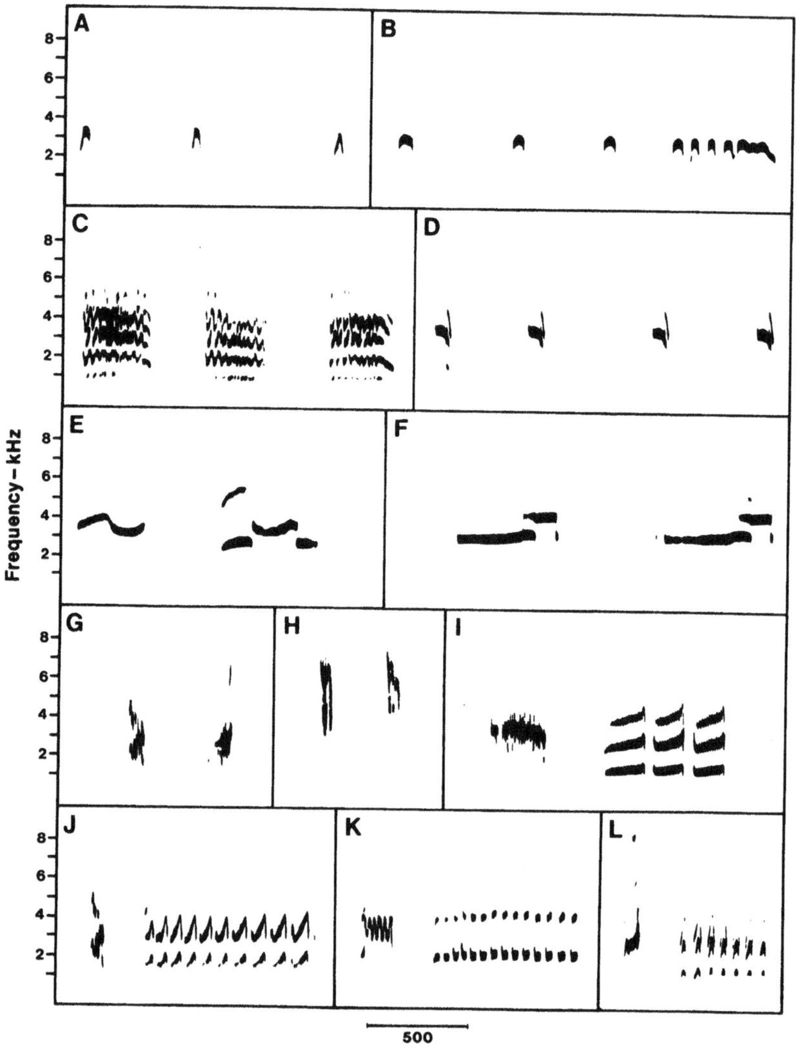

Fig. 8. Sound spectrograms of calls by adults near their nests or young, in response to human intrusion [to *Larus* gull in (E)]. (A) Sequence of ground calls by American Oystercatcher (*Haematopus palliatus*; Punta Tombo, Argentina). (B) Sequence of ground calls by Blackish Oystercatcher (*Haematopus ater*; Punta Tombo, Argentina). (C) Sequence of flight calls by Southern Lapwing (*Vanellus chilensis*; near Rio Gallegos, Argentina). (D) Sequence of ground calls by Semipalmated Plover (*Charadrius semipalmatus*; near Churchill, Manitoba). (E) Two flight calls by Black-bellied Plover (*Pluvialis squatarola*; Coats Island, Northwest Territories). (F) Two ground calls by American Golden Plover (*Pluvialis dominica*; near Churchill, Manitoba). (G) Two flight calls by male Red Phalarope (*Phalaropus fulicarius*; Coats Island, Northwest Territories). (H) Ground and flight calls by female White-rumped

The two call types are thus associated predictably with locomotory mode, and provide reliable information about a signaler's ongoing and (short-term) future behavior. This is a fairly simple outline of what actually goes on, which I have studied intensively only in the Least Sandpiper. In that species, the relative frequency of occurrence of trills and short calls changes even as a bird alters direction in flying toward or away from an intruder, and structurally intermediate sounds to trills and short calls occasionally occur. The calls are given in response to human intruders, to dogs, horses, and gulls—in brief, to almost any large animal whether or not it is a potential predator. It seems most straightforward to attempt to characterize these sound types functionally by the contexts in which they occur, in respect to the caller's behavior, and to determine the main classes of receivers (chicks?, mate?, predator?, etc.). A striking feature of this communication system is how individually distinctive different birds are, both in the physical characterisitics of their calls and in how often the two types are used. One male I studied rarely uttered trills, and another almost never gave short sounds; individuality would thus seem to be a crucial part of context when receivers evaluate meaning in such a system of sounds.

Smith's approach is useful in specifying a repeatable, objective method for determining many kinds of information that are available to receivers, and it avoids interpretive terms like "mobbing sounds," "alarm calls," "warning calls," etc. In our present state of knowledge, the approach does not fully explain the existence of two sound types in the contexts outlined, which also occur in the distantly related Red Phalarope (*Phalaropus fulicarius*; Fig. 8G), several oystercatcher species (Fig. 8B; Miller and Baker, 1980), Black-winged Stilt (Wollemann, 1978), and Eurasian Avocet (Adret, 1982), but apparently not in the closely related White-rumped Sandpiper (Fig. 8H), or in Charadriidae (Fig. 8C–F) (also see Oring, 1968; Skeel, 1976, 1978, Sordahl, 1979). *A priori*, it would seem as reasonable for "design" of such a communication system to encode information in graded features (as in Charadriidae?), rather than in different call types, especially as grading is apparently built into many of

---

Sandpiper (*Calidris fuscicollis*; Coats Island, Northwest Territories). (I) Two calls (not in sequence) illustrating two call types in Dunlin (*Calidris alpina*; near Churchill, Manitoba). (J) Two calls (not in sequence) illustrating two call types in Semipalmated Sandpiper (*Calidris pusilla*; Coats Island, Northwest Territories). (K) Two calls (not in sequence) illustrating two call types in Least Sandpiper (*Calidris minutilla*; Sable Island, Nova Scotia). (L) Two calls (not in sequence) illustrating two call types in Purple Sandpiper (*Calidris maritima*; Coats Island, Northwest Territories).

Analyzing filter bandwidth throughout, 300 Hz. Time marker is in msec.

these sound types anyway (see remarks above about the Willet's "kleep" call).

Because of the complexities surrounding congruence between structure and function, and because our current knowledge is so meager, it seems pointless to attempt any interpretive synthesis of repertoire size in shorebirds at present. In addition, studies to date suggest that complete repertoires have not been documented for many species, if any: why should juvenile Solitary Sandpipers have twice as many call types as juvenile Green Sandpipers (Oring, 1968), or young Long-billed Curlews (*Numenius americanus*) have twice as many as young Whimbrels (Forsythe, 1967, 1970; Skeel, 1976, 1978)? Why should adults of a highly visual, colonial species like the Eurasian Avocet have 10 distinct sound classes in the perinatal phase alone, while cryptic, dispersed breeders like Whimbrels have the same number for the entire breeding period (Adret, 1982; Skeel, 1976, 1978)? And why should Common Greenshank (*Tringa nebularia*) chicks have at least 10 structurally different sound types, when Willet chicks have only one (Nethersole-Thompson and Nethersole-Thompson, 1979; Sordahl, 1979)? These comments are not meant as criticism, but only to stress the inadequacy of current documentation; it is not surprising that the most diverse repertoire known (several times the size of any other) belongs to the most intensively studied species (Nethersole-Thompson and Nethersole-Thompson, 1979).

3. Graded Communication

An area crucial to further understanding of vocal communication in shorebirds concerns structural variation within sound classes (Klingholz and Meynhardt, 1979; Maier, 1982; Marler, 1976). Is such variation disorderly and meaningless, or does it encode slight shifts in probabilities of subsequent behavior, or reveal minor behavioral changes in signalers? An example of variation of this nature is in the long rhythmic series of simple calls that characterize aerial displays of unpaired Least Sandpipers (Miller, 1979a, 1983a). Major breaks in the rhythm of calling usually co-occur with obvious structural deviations of calls (Fig. 9). More subtle, systematic variations occur during unbroken rhythms, including a correlation between call duration and the presence of frequency modulation of the carrier frequency (Fig. 9A). The regularity of such trends may impart communicative significance to them, but whether they are significant or simply adventitious remains speculative. The bounds of such structural variation, their relationship to signaler behavior, and their meaning to receivers are badly in need of research (see Section II.B.2).

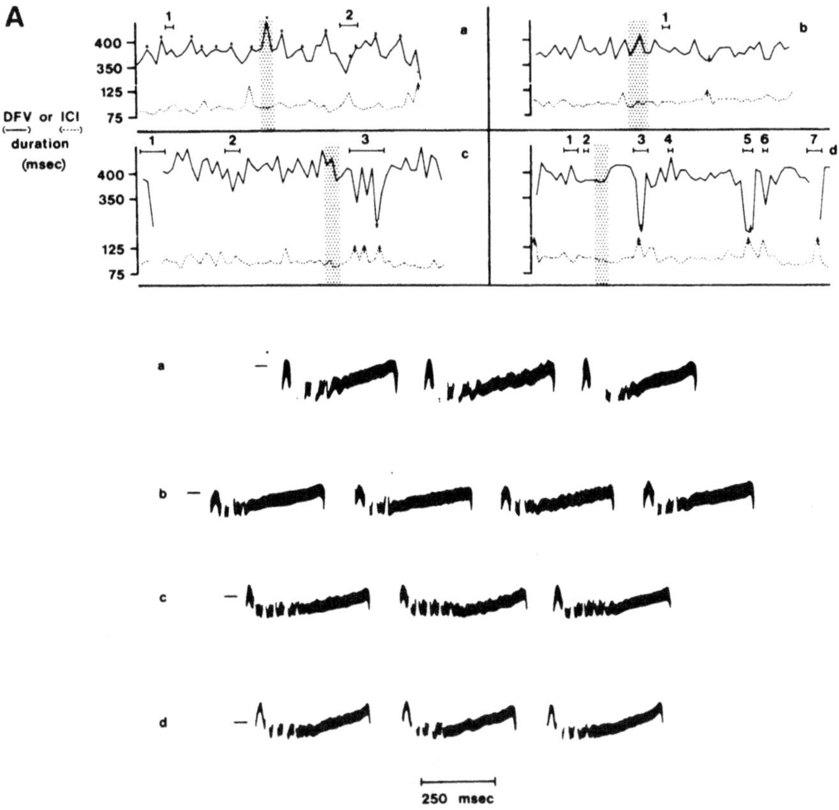

Fig. 9. Sequence plots of portions of call sequences in aerial displays of male Least Sandpipers (*Calidris minutilla*), illustrating variations in form and duration of components. DFV, display flight vocalization; ICI, intercall interval.

(A) Sequence plots and typical elements. The stippled portion in each of the four upper diagrams covers the call sequences illustrated in the lower figures. The numbers in each plot refer to the segments shown in (B). Downward- and upward-pointing arrows indicate that calls were too brief, or intervals between calls too long, respectively, to plot. In the record for male *a*, each call with conspicuous frequency modulation (FM) of the long terminal portion is signified by a small "x" above the line for DFV durations. Examples of FM are shown in the short sequences below: see *a*—second, *b*—third (weakly, second), and *c*—second.

(B) Segments of calling sequences for the males referred to in (A). The letters refer to those males' records, and the numbers correspond to those in the plots in the upper part of (A).

This figure is of tracings of the first harmonic overtones of the calls. Analyzing filter bandwidth, 300 Hz. Frequency markers, 3 kHz.

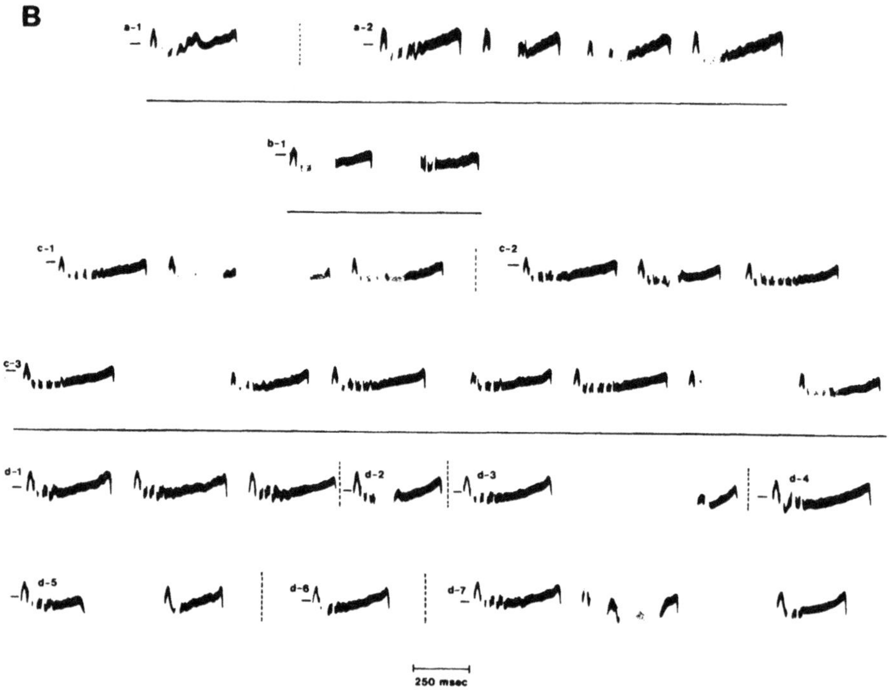

Fig. 9. (*Continued*)

Acoustic morphology also varies among sound classes (by definition), and varies within sound classes because of numerous proximate causes (anatomical, neural, endocrinological, etc.) and correlates (age, gender, individuality, etc.). Extensive structural intergradation of sounds is often referred to as "grading," a term lacking adequate definition or use. Consider the assignment of a sample of sounds, from a population or species, into different sound classes. If classes emerge from the study only with difficulty, often because of the presence of numerous sounds of intermediate form, the system is often considered to be graded. If classes are readily established, the repertoire can be considered as relatively "discrete" (the term generally used as an antonym to "graded"), but variation within classes becomes of interest, and those classes with high levels of variation are relatively graded. Yet within sound classes we must recognize many important sources of variation that vary systematically, including age, gender, individuality, breeding status, etc. We could therefore focus on, say, a particular sound type emitted by breeding males, and assess the extent to which it is graded. Finally, characteristics of sound types will exhibit different levels of variation, so it becomes nec-

Table VI. Variation in "Contact Ernts" of Wilson's Phalarope (*Phalaropus tricolor*) by Gender, Individual Identity, and Acoustic Dimension[a]

		Duration (msec)			Frequency[b] (Hz)		
Sex	Individual	$\bar{Y}$ ± S.E.	(*N*)	C.V.	$\bar{Y}$ ± S.E.	(*N*)	C.V.
Female	A	44 ± 3.2	(8)	20.9	930 ± 17.0	(8)	5.2
	B	36 ± 4.0	(5)	24.8	910 ± 12.5	(5)	3.0
	C	38 ± 2.6	(12)	23.1	1000 ± 6.4	(12)	2.2
	D	46 ± 2.2	(5)	11.9	1040 ± 22.4	(5)	4.8
	E	65 ± 5.0	(4)	15.4	930 ± 12.0	(4)	2.5
	F	36 ± 8.5	(5)	31.2	960 ± 12.5	(5)	2.9
	G	60 ± 2.3	(9)	20.4	1000 ± 1.0	(9)	0.3
Male	H	54 ± 1.8	(8)	9.6	1080 ± 32.0	(4)	5.9
	I	48 ± 2.5	(4)	10.5	1070 ± 22.0	(4)	4.2
	J	38 ± 2.3	(9)	17.6	1110 ± 11.7	(9)	3.2
	K	39 ± 2.0	(12)	17.1	1110 ± 9.5	(12)	3.0
	L	50 ± 3.8	(7)	20.0	1240 ± 9.8	(7)	2.1
	M	30 ± 0.0	(5)	0.0	1230 ± 18.3	(5)	3.3
Female average		46		21.1	967		3.0
Male average		43		12.5	1140		3.6

[a] Data after Table 12 of Howe (1972).
[b] Maximal frequency estimated on first harmonic overtone.

essary to specify the acoustic dimension in which grading is being estimated.

These points can be illustrated with Wilson's Phalarope (*Phalaropus tricolor*), acoustic signaling in which has been studied by Howe (1972). One call type used over short distances, the "Contact Ernt," varies substantially among individuals and between sexes, and varies much more in duration than in frequency (Table VI). Duration of this call type shows greater variation within individual females (average C.V. = 21) than males (average C.V. = 13), and maximal call frequency is substantially higher in males than females (in keeping with the smaller size of males in this species). Clearly it is necessary to ask quite specific questions about components of variation in such data; there is no single "natural" level at which grading can be assessed.

The concept of structural grading is most usefully applied to two distinct levels of syntactic analysis: (1) the entire sound communication system of a species and (2) in reference to emissions of individual animals. The former is of particular interest in ultimate levels of explanation and the latter at proximate levels, especially with regard to the nature of the communication process and perception by receivers (e.g., do receivers

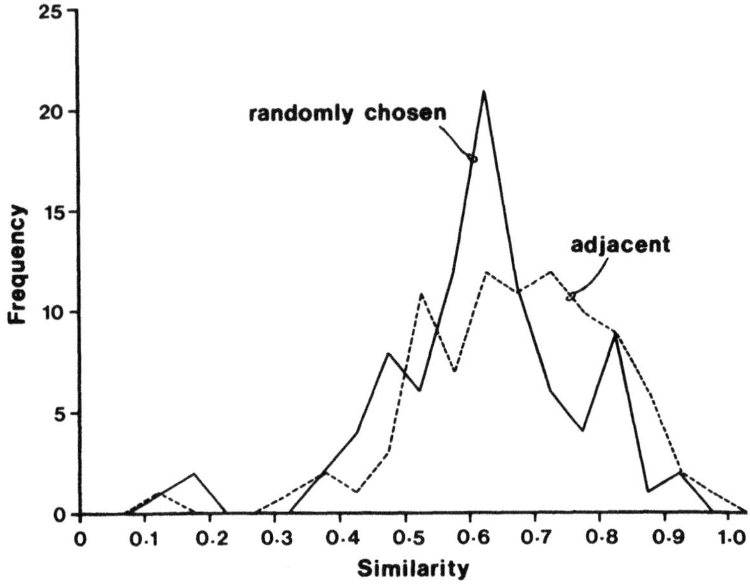

Fig. 10. Frequency polygons for similarity values between adjacent and randomly chosen calls of Blackish Oystercatcher (*Haematopus ater*) disturbed by me near its nest [After Fig. 5 of Miller (1979c)].

perceive and respond to slight variants in a continuously varying fashion?). In terms of individual emissions, there are two logical extremes to structural grading: (1) consecutive sounds are very similar to one another, and less similar to sounds separated by greater intervals; and (2) morphological similarity of sounds occurs at random, in the sense that there is no tendency for consecutive sounds to be more similar to one another than to sounds uttered at other times. These two extremes can be referred to as systems of "adjacent" and "nonadjacent" or "separate" grading, respectively (Marler, 1976; Marler and Tenaza, 1977). Most or all systems fall between the logical extremes. For example, nesting Blackish Oystercatchers (*Haematopus ater*) utter simple "alarm" calls at irregular intervals at the approach of a human (Fig. 8B; Miller, 1979c; Miller and Baker, 1980). Consecutive calls are significantly more similar to one another than are calls sampled at random from the same sequence (Fig. 10; Miller, 1979c). Such a trend is probably widespread, and results from common causation shared by consecutive calls, as well as the likelihood of the stimulus situation being similar for them. Nonadjacent grading is probably uncommon for converse reasons, and also because accurate

Table VII. Summary of Chosen Variables That Differ Significantly between "Repeated-Note Calls" (RNCs) and "Note-Group Calls" (NGCs) of the Northern Jacana (*Jacana spinosa*)[a]

	RNCs	NGCs
Minimal frequency of the dominant frequency band (kHz)	3.92	4.17
Highest detectable frequency (kHz)	7.38	7.18
Bandwidth of the dominant frequency band (kHz)	2.14	1.73
Duration of individual element (msec)	81.9	32.7
Total number of elements	12.3	72.2
Relative loudness (arbitrary scale)	3.1	1.6

[a] Data from Table 22 of Mace (1981).

rapid responses by receivers must be quite difficult in the absence of immediate standards for comparison.

Acoustic signals of the Northern Jacana provide an example of the application of the concept of grading at the levels of syntactics, semantics, and pragmatics. Mace (1981) has described and exhaustively quantified acoustic signals of the species, a polyandrous resident of the neotropics that calls frequently (4–7 calls/hr by breeding birds). Most calls consist of brief broadband elements, and have effective ranges varying from only a few meters to several hundred meters. Classification of calls is difficult, because they consist of from one to many elements each, and each group of elements can be repeated in a regular cadence. Thus, structural grading must be considered at several hierarchical levels. Mace distinguished "repeated-note calls" (RNCs), consisting of from 1 to about 60 (average 12) regularly repeated elements, from "note-group calls" (NGCs), characterized by repetition of *groups* of rapidly repeated elements (average 2.3 elements/group). Elements within RNCs were separated from one another by about 162 msec, on average, as compared with only 46 msec between successive elements *within* groups, and 211 msec *between* groups, for NGCs. Some other important differences between RNCs and NGCs are summarized in Table VII.

Fourteen behavioral contexts were identified by Mace, and RNCs occurred in all but one of them (copulation). In sharp contrast, NGCs occurred in only four: "attentive to mate," "escorting chicks," "copulation," and "nervous/alert." Structural characteristics of both RNCs and NGCs varied significantly across contexts; RNCs can serve as an example. In these calls, numerous variables concerned with frequency and temporal characteristics varied significantly across contexts. For example, the variable "element duration" by itself defined four statistically separable groups of contexts (Fig. 11). Discriminant analysis on these

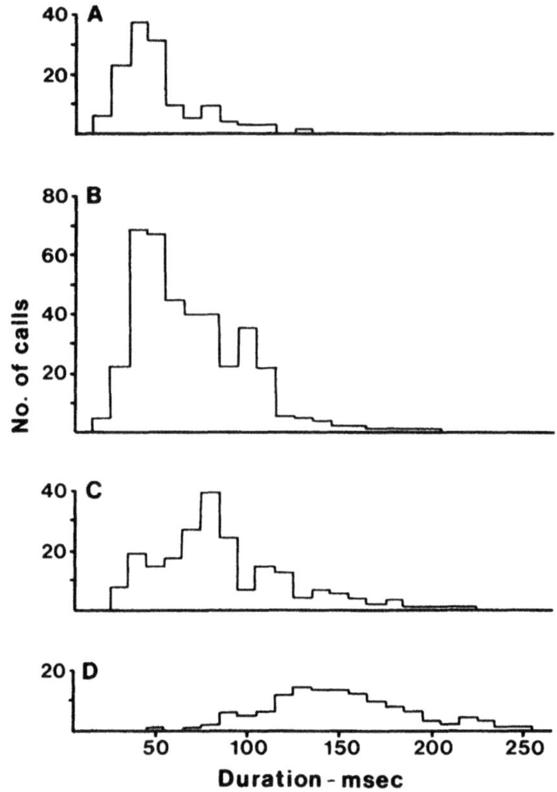

Fig. 11. Frequency histograms for the duration of elements within "repeated note calls" of the Northern Jacana (*Jacana spinosa*), for four groups of contexts: (A) nervous/alert; attentive toward a mate; response to flying conspecific intruder; (B) flight across, to, or from territory; flutter-hop; (C) encounter with predator on the substrate; response to something in the water; male's chase of intruding juvenile; male's encounter with conspecific intruder on substrate; (D) aerial attack on predator; male landing over his mate; female's encounter with conspecific intruder on substrate. [After Fig. 15 of Mace (1981).]

four groups, using an additional four variables, revealed considerable overlap between adjacent groups (Table VIII). These and other analyses affirmed that there was some correspondence between behavioral context and call structure, but it was very general and showed extensive overlap. Thus, structure by itself was not capable of transmitting specific kinds of information.

The final part of Mace's analysis addressed the effects of sound signals upon receivers. He noted substantial variation in the responses of receivers depending upon the context in which particular calls were given.

Table VIII. Classification Results from Discriminant Analysis on Four Groups of Behavioral Contexts in Which Northern Jacanas (*Jacana spinosa*) Call (Five Variables)[a,b,c]

Actual group membership[b]	Predicted group membership (%)[b]			
	1	2	3	4
1	70	17	8	4
2	24	42	28	5
3	12	19	49	21
4	2	1	13	84

[a] Data from Table 41 of Mace (1981).
[b] Contexts in each group are: 1, attentive to mate; response to flying conspecific intruder; nervous/alert; 2, flight across, to, or from territory; flutter-hop; 3, encounter with predator on the substrate; startled; response to something in water; male's encounter with conspecific intruder or substrate; male's chase of intruding juvenile; and 4, aerial attack on predator; female's encounter with conspecific intruder on substrate; male landing over his mate.
[c] Variables are: bandwidth of dominant frequency; element duration; interval between successive elements; relative loudness; and total number of elements.

However, in converse analyses, by examining receivers' responses to signal variants within contexts, much less variation was detected. One situation with clear results was in the context of aerial attack upon a predator by female jacanas; here, calls resulting in the departure of the predator had significantly greater intervals between elements, and significantly more elements.

In summary, Northern Jacanas use an acoustic signaling system built around two structurally defined types of calls. Structural and contextual variation considered together provide a greater understanding of communication and its effects than either does when considered in isolation. Even this detailed study did not reveal whether structurally similar call types evoke a continuous range of responses, and it is possible that uncontrolled field studies cannot answer that question. However, the study documents a clear example of structural grading at the level of the repertoire. It also demonstrates the necessity of including contextual information, and semantic and pragmatic analysis, in such research.

D. Ecological and Social Effects

Communicative behavior must meet both ecological and social demands, so many characteristics of signaling reflect the resolution of complementary or opposing selection pressures from the two spheres. Nevertheless, some ecological and social factors act independently of one another, so certain features of communication can be best examined if their selective bases are discussed separately.

1. Transmission Distance

The distance over which signals are regularly transmitted, plus sources of degradation along the route of transmission, are two crucial forces in the evolution of signaling behavior and of the physical characteristics of signals (Konishi, 1970; Wiley and Richards, 1982). If the average transmission distance is great, there is selection against high-frequency acoustic signals because they attenuate rapidly with distance. Similarly, spectrally complex sounds undergo differential frequency attenuation with distance, and are likely to be selected against as long-distance signals. A long transmission path also exposes sounds to various kinds of degradation from background noise, wind, etc., which should place a premium on signals with simple patterns of amplitude modulation, and that contrast with the background. In general, then, long-distance sound signals are predicted to be high in energy, low in frequency, spectrally pure, and with simple or no amplitude modulation. To compete with background noise they should also be stereotyped and repetitive; the latter feature also improves locatability [see Green and Marler (1979) for other comments on characteristics that enhance locatability]. Do long-distance sound signals of shorebirds have characteristics that agree with these predictions?

Many species of shorebirds in open-country environments emit loud, repeated sounds to attract mates and repel other conspecifics, in displays given in flight, which further increases transmission distance (Fig. 12; see Miller, 1983a,b). All meet with general predictions about optimal physical characteristics for faithful transmission and accurate locatability over long distances, though there are many differences among species (see Section II.E.2). The calls used by parental shorebirds in the presence of potential predators may be designed for long-distance transmission, to recruit other birds (even heterospecifics) in "mobbing"; these calls' features are also in generally good agreement with those predicted.

Wiley and Richards (1982) discuss many factors additional to those summarized above that must be considered for shorebird species that live in physically complex environments like forests (e.g., Subantarctic Snipe, *Coenocorypha aucklandica*; Solitary Sandpiper; etc.), or reside near sources of loud frequent noise, such as seacoasts or rivers. Certainly, the latter must have helped effect the remarkably loud piercing quality of many call types of oystercatchers, Wandering Tattler [*Heteroscelus incanus* (E.H. Miller, unpublished observations)] and Diademed Sandpiper-

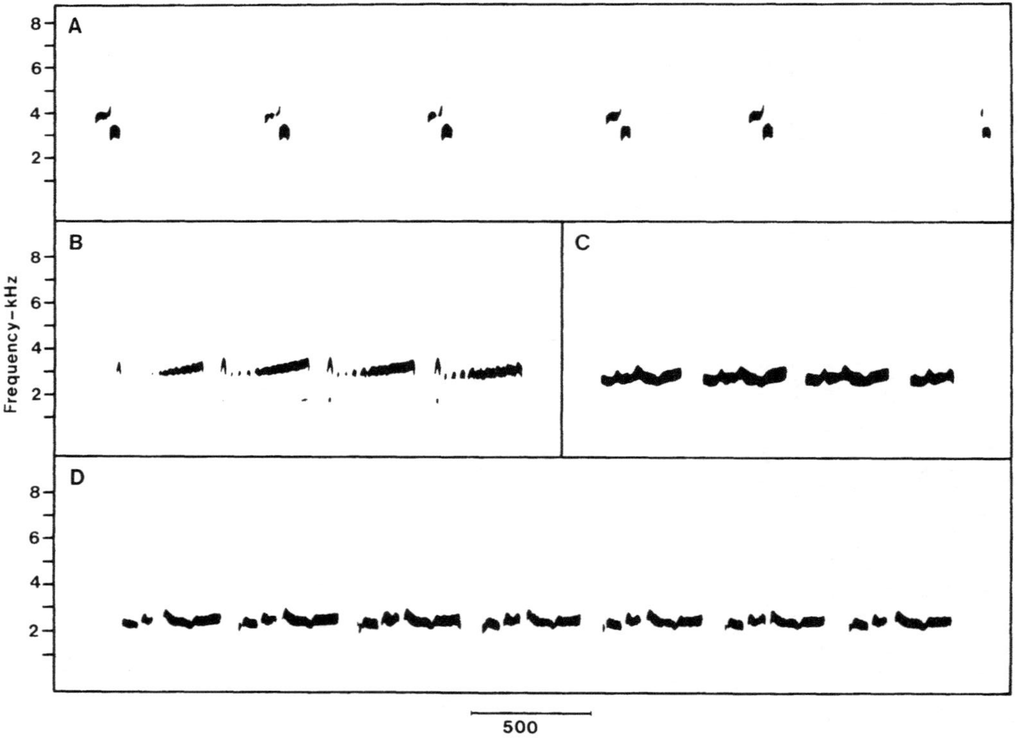

Fig. 12. Sound spectrograms of call sequences during aerial displays. (A) American Golden Plover (*Pluvialis dominica*; near Churchill, Manitoba). (B) Least Sandpiper (*Calidris minutilla*; near Churchill, Manitoba). (C) Wandering Tattler (*Heteroscelus incanus*; Blackstone River Valley, Yukon Territory). (D) Lesser Yellowlegs (*Tringa flavipes*; near Fort St. John, British Columbia). Analyzing filter bandwidth throughout, 300 Hz. Time marker is in msec.

plover [*Phegornis mitchellii* (Johnson and Goodall, 1965)]. Short-distance sounds in general are not subject to any of the same considerations as long-distance ones, for there is so little degradation or frequency-dependent attenuation of their features. Finally, many features are not predictable based just on considerations like those above, and certainly the differences among species are not. Phylogeny does explain many such differences, however (see Section II.E).

Visual signals are far less effective than acoustic ones for long-distance communication. The aerial maneuvers and strong markings of large birds like lapwings are visible from great distances though, and acoustic and visual components of such displays are mutually complementary in drawing attention to a displaying bird. Most visually inconspicuous species that use aerial displays have ritualized motor components that are visible from afar, including wing action, dives, undulations, ascents, rolls, etc. (see Cramp, 1983; Glutz von Blotzheim *et al.*, 1975, 1977; Miller, 1983a,b; Nowicki, 1973). Such components certainly increase a displayer's transmission distance, but the relative value of different components can only be speculated on at present. As for acoustic signals, some features of motor components can be explained most parsimoniously with reference to phylogeny (Section II.E.2).

Because many shorebird species do not signal very frequently from elevations, they are limited in long-distance visual communication except when in flight. Numerous species accentuate landing or takeoff by exposure of wings or tail, or both, however (Hale, 1980). The Upland Sandpiper exemplifies this trend, with both sexes (but particularly males) going through elaborate extension then slow folding of the wings and lowering of the tail after landing on fenceposts (Fig. 5). Males of two lekking species, the Ruff and the Buff-breasted Sandpiper (*Tryngites subruficollis*), greatly increase their conspicuousness by leaping up from the ground surface in a ritualized form of flutter-fighting, so common in their relatives (Fig. 16a; see also Fig. 3 of Holmes, 1973). The Buff-breasted Sandpiper has strikingly white underwing surfaces, which allow even birds on the ground that raise and "flash" their wings to be seen for long distances (Myers, 1979; Oring, 1964; Parmelee *et al.*, 1967).

2. Activity Budgets

The preceding discussion addressed the adaptiveness of sound structure, but omitted consideration of the facultative responses of shorebirds to ecological factors. For example, do individuals call more frequently, or engage in higher wider-ranging aerial displays, in years or habitats of low density? Here, some studies on activity budgets are discussed. These should be extremely labile according to local ecological conditions, and need detailed work to evaluate the ecological significance of how time spent communicating is apportioned. It is difficult and somewhat artificial to estimate the amount of time spent by shorebirds in display, because communication can coincide with other activities and because static-optical signaling may be essentially continuous, especially in well-marked open-country species.

Gibson (1978) used general functional categories in his study of the American Avocet, and quantified the fraction of time spent in "aggressive behavior," "mating behavior," and "diversionary behavior." Certain other of his categories probably include much communication, as he points out, including nest building, parental care, and vigilance behavior. For comparative purposes, it is best to specify functional categories like his, or to focus on discrete and conspicuous display types. An alternative approach could be built around different classes of signaling modes, distinguishing different forms of active and passive communication, but this has not been attempted, to my knowledge. In Gibson's study, prenesting avocets spent 4.6% (males) and 3.5% (females) of their time in predator distraction, mating, and aggression. The latter two classes accounted for 4.4 and 3.4% of male and female time then, respectively, and declined sharply in incubation (to 0.8 and 1.0%). Mating did not appear in samples later in the nesting cycle, but aggression returned to a high level: about 3.2% in males and 1.6% in females, during the period of brood attendance. Aggression subsequently decreased again. This study was the first to provide rough estimates of the time budget of a breeding shorebird, and established a lower bound to time spent in active display. Inclusion of appropriate contexts of nest building, nest relief, and parental care could easily double the estimate of 4.6% for males in the prenesting period, when most time is spent in obvious display. Estimates of time spent in passive forms of display would further increase this figure, but would be of greatest significance to understanding social dynamics, for the energetic cost is low (though an elevated risk of predation may figure into this component).

A detailed study of time and activity budgets of the Semipalmated Sandpiper (*Calidris pusilla*) was carried out by Ashkenazie and Safriel (1979). This species is much more difficult to observe than the colonial avocet, so behavioral categories necessarily omitted or lumped some important displays including ground courtship and scraping. Unpaired Semipalmated Sandpiper males engage in a conspicuous aerial display, and time spent in this is easily quantified, however. Ashkenazie and Safriel estimated that unpaired males spent 18% of their time in flight, most of which was in aerial display. Flight required around 56 kJ/day, nearly 7%

Table IX. Daily Time and Energy Allocations to Behavior That Included Much Communication, for Lesser Sheathbills (*Chionis minor*) Rearing Chicks[a]

		Pair displays	Brooding	Territorial defense	Antipredator aggression
Males	%[b]	0.1	18	1.3	1.4
	kJ/day	0.4	33	20	7
Females	%	0.1	22	1.8	0.8
	kJ/day	0.4	37	25	4

[a] Data from Tables 2 and 3 of Part Six of Burger (1980c).
[b] Percent of diurnal time budget.

of total requirements for this phase of the breeding cycle. Clearly, this one display form by itself figures significantly in both time and energy budgets (see Erckmann, 1981).

The time and energy budgets of Lesser Sheathbills were estimated by Burger (1980c), at Marion Island in the subantarctic. His study began at hatching, and continued until the chicks left the nest to follow their parents (at around 7 weeks). Of the eight behavioral categories recognized by Burger, three were dominated by communicative acts ("territorial defense," "antipredator aggression," "pair displays"), and one presumably included much communication ("brooding"). Displays between paired birds occupied only a few minutes each day, and were energetically cheap (Table IX). Aggressive acts toward potential predators and territorial defense occupied more time and were more energy-demanding, together accounting for 5% of the daily energy budget. Brooding was the most significant behavior in terms of both time and energy (Table IX). It was particularly important when chicks were young, occupying over 40% of the parents' time then. Territorial adults advertised conspicuously but cheaply at other times, simply by resting or preening on prominences, which could be done with only normal maintenance costs. The pairs studied by Burger had noncontiguous territories centered on Rockhopper Penguins (*Eudyptes chrysocome*), which they kleptoparasitized; those pairs must have had much lower costs in active territorial defense than pairs occupying smaller, contiguous territories in colonies of King Penguins (*Aptenodytes patagonicus*).

The most detailed study on activity budgets of breeding shorebirds is by Maxson and Oring (1980), who made extensive observations on Spotted Sandpipers over several years. This species is polyandrous throughout much of its range, with males taking major or sole responsibility for incubation and brood rearing. The main investments made by females are in egg production, and in agonistic activity related to defense

Table X. Relationship of the Percent of Time Spent in Agonistic Behavior, to Gender and to Stage of Breeding Cycle in the Spotted Sandpiper (*Actitis macularia*)[a]

	Percent of time in agonistic behavior[b,c]					
Stage of breeding	BAE	CH	U	A	FT	Overall percent in agonistic behavior
Prelaying	52/53	29/31	7.4/2.6	9.1/10.6	3.1/2.5	8.4/5.7
Laying	51/68	13/18	8.6/2.9	23/10.1	3.9/0.7	4.1/3.0
Incubation	58/59	28/28	4.7/3.3	8.0/6.9	1.3/2.1	4.7/1.7
Brood-rearing	—/61	—/29	—/1.9	—/5.2	—/2.4	—/5.5
Averages	53/60	23/27	6.9/2.7	13.5/8.2	2.8/1.9	

[a] Data after Table 4 of Maxson and Oring (1980).
[b] BAE, balanced aggressive encounters; CH, chasing; U, upright displays; A, appeasement; FT, fighting.
[c] Each cell entry is shown as: male/female.

of mate or territory. Overall, time spent in agonistic behavior differed insignificantly between the sexes, but forms of behavior varied according to sex and stage of breeding cycle (Table X). Females spent about twice as much time in agonistic behavior in the prelaying stage as subsequently, and components varied substantially. In particular, appeasement was very high during the laying stage, perhaps to minimize energy expenditure and potential damage to developing eggs. Males showed different trends, with less time in agonistic behavior at all stages (partly because of incubation duties), and much less use of upright displays. The resurgence in agonistic behavior by males after hatching was due to the need for defense of chicks, who sometimes suffered severe injury or death from attacks by foreign adults. Activity budgets were strongly influenced by food availability. For example, females spent 70% of their time foraging and 5% in agonistic activity in one season with little food (terrestrial arthropods), compared with 48% and 9%, respectively, for two seasons with more abundant food. Comparable trends also occurred within years, corresponding to seasonal trends in food availability.

Other species for which time budgets are available are described in Erckmann (1981) and Walters (1980; this volume). For the polyandrous Red Phalarope, D. Schamel estimated that females spend about 8% of their time in ground display (cited by Ashkenazie and Safriel, 1979). As for the other species, this figure must be considered an underestimate of time spent in active display. Though there are large differences among species, it is clear that active forms of signaling constitute a significant fraction of the time budget, especially in the prenesting phase, and for that sex subject to strongest sexual selection. Mundahl (1977) summarizes some interesting trends for communicative behavior throughout the breeding cycle of the Killdeer (*Charadrius vociferus*), which support this generalization. Detailed comparative work on this topic would surely repay the investment.

3. Diel Rhythms and Reactions to Predators

Display behavior in shorebirds is subject to numerous influences from weather, temperature, topography, time of day, etc., as in most bird species. One of the most clearly documented cases concerns evening and early morning "peent" display by male American Woodcock. The onset of display is closely tied to light level, and begins about 20 min before civil twilight from late March (when civil twilight occurs around 0640) through early June [when civil twilight is around 2010 (Leopold and Eynon, 1961)]. A similar relationship prevails for early morning display (Leopold and Eynon, 1961; Fig. 13). When there is reduced light because of clouds, fog, etc., evening display begins earlier and ends later throughout the breeding season (Fig. 13; Weeks, 1969). Such pronounced but modifiable display rhythms are a mixture of flexible and inflexible responses and patterns, the inflexible ones in this case relating to responsiveness to light intensity.

Favorable physical conditions for transmission presumably explain why strong display rhythms occur in at least some shorebird species. Many others must be compromises, set particularly by the nature and intensity of predation. Crepuscular or nocturnal display by taxa like woodcock (*Scolopax*), Great Snipe (*Gallinago media*), Burhinidae, and Rostratulidae seems likely to have evolved in response to predation pressures of some sort. However, the only form of display behavior in which species differences can be confidently linked to predation is in diversion or distraction. Sordahl (1981) reviewed the behavior of 51 species of breeding North American shorebirds in response to potential predators, and noted a strong tendency to mob by large species. None of the *Charadrius* plovers exhibits mobbing (by Sordahl's definition), but both large *Pluvialis* species do, as well as the Jacanidae, Haematopodidae, and Recurvirostridae. The most telling evidence is from Tringini, for all mob except the two smallest species (Solitary and Spotted Sandpipers) plus the Wandering Tattler. Sordahl suggests that small species expose themselves to danger when they mob. This factor may help explain the trends, though other ones must be involved. For example, why do turnstones mob, but not dowitchers? The Killdeer does not mob according to Sordahl, yet Nol (1980) documented outright attacks by this species on model gulls and raccoons, and aerial chases of live gulls. It seems likely that increased attention to details of reactions by nesting or parental birds will improve our under-

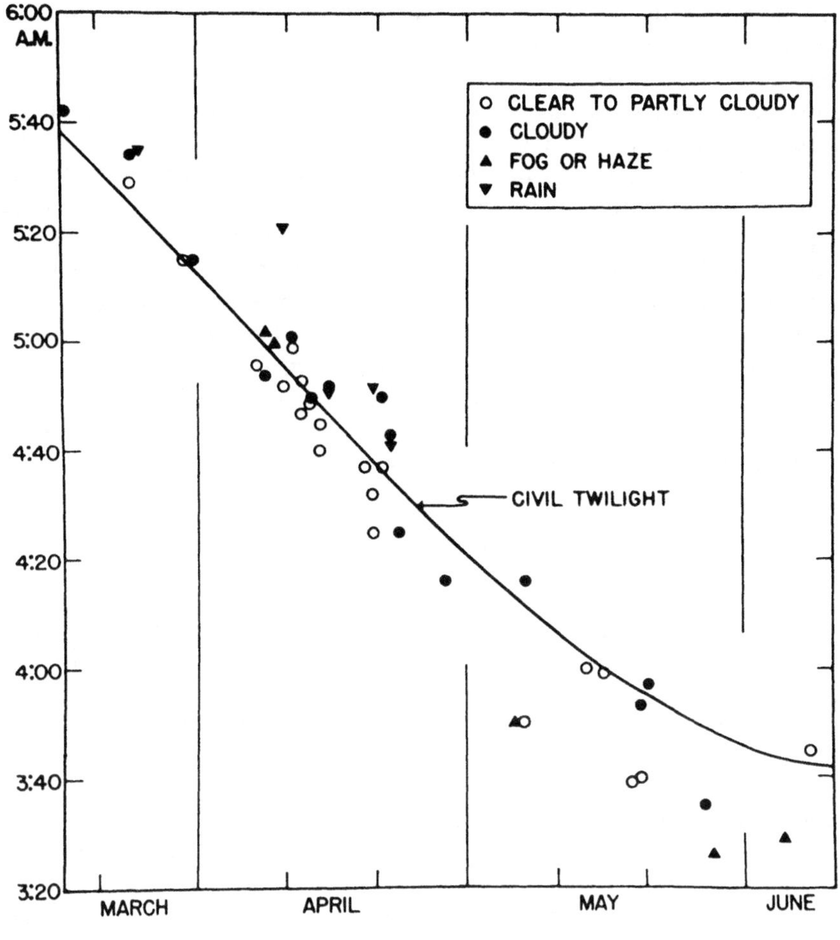

Fig. 13. Relationship of the cessation of "peent" calls by male American Woodcock (*Scolopax minor*) to civil twilight. Each circle or triangle represents a single observation for the last call heard in a morning. [From Leopold and Eynon (1961, Fig. 8).] A similar relationship holds for the onset of calling (Leopold and Eynon, 1961, Fig. 7).

standing of their adaptive responses to predators (see Chapter 8, this volume)

Responses to predators by nesting shorebirds vary geographically, but are not easily explained just by reference to predation pressures. The best documented cases of such variation in various forms of display behavior evoked by predators are by Williamson (1943, 1946, 1948, 1950, 1952). He described mobbing and diversionary behavior by several species of shorebirds on the Faeroes, and noted that the expression of such behavior there was quite different from that recorded elsewhere. For example, for the Eurasian Oystercatcher (*Haematopus ostralegus*) "in Holland and Cumberland displacement butterfly-flight with its accompanying 'kleu, kleu' call-note, is the normal reaction to human intrusion at the critical hatching time and when the young are small. At Fair Isle butterfly-flight, modified by the absence of the call-note and the low altitude of the display, is the usual reaction, sometimes culminating in an incipient lure display. At Unst, nearly 100 miles to the north, the butterfly-flight has become modified to the extent that a terrestrial pattern embracing its main elements is developing; whilst in the Faeroes, 200 miles farther N. W., the evolution of the new and striking lure pattern and its ritualization into mobile and stationary figures is virtually complete" (Williamson, 1952, p. 93). The geographic variants he noted for this and other species do not appear to be related to predation pressure in any simple way.

4. Social System and Sexual Selection

Social characteristics evolve within broad sets of ecological constraints, and many characteristics of signaling can be related most easily just to social pressures. For example, in colonial species like stilts and avocets, much communication reflects adaptation to coloniality, and not to the important factors related to (say) food and predation which have led to coloniality. In this section, I discuss a few features of social systems that have had important influences on some characteristics of signals: parental feeding; monogamy; and intrasexual selection.

Parents feed their young in Dromadidae, Glareolidae, Burhinidae, Chionididae, Pluvianellidae, Haematopodidae, and Scolopacidae (*Philomachus, Scolopax,* and *Gallinago* only). Communication between parents and young in the first two of these is facilitated by the young remaining in one place, for crab plovers are burrow-dwellers, while glareolids return to their young to feed them. In the Australian Dotterel (*Peltohyas australis*) and Australian Pratincole (*Stiltia isabella*), such sedentariness is related to the need for chicks to locate and remain in relatively cool microhabitats, to avoid hyperthermia and dehydration (Maclean, 1976a,b). Species in which parents feed their chicks must have particular signals used in soliciting feeding, or attracting young, a matter that has been studied only in the Eurasian Oystercatcher, by Lind (1965) and Norton-Griffiths (1969). These two workers have provided the most detailed studies on shorebird parent–young behavioral relationships to date.

Adult oystercatchers initially bring food to the newly hatched young just around the time of nest relief, but the precision of this timing weakens quickly. Within a day or so of hatching, adults lure the young from the nest by calling and probing near the nest. Parents also lure the chicks "by presenting food 'just out of reach' . . . so that the young has to move away from the nest in order to be fed" (Norton-Griffiths, 1969, p. 79). During this phase, adults may briefly hold the bill in a posture resembling the feeding posture (see further), and the chicks closely follow the adult's bill. During the chicks' early life away from the nest, parents use several distinctive calls and visual signals in attracting them to feed and in feeding them. Upon obtaining food, a parent runs toward the chick quickly with neck withdrawn and bill pointing slightly down (at greater distances the adult flies). During the adult's approach, it uses one of two different call types, which depend on the mode of locomotion: loud Weep calls are used in 87% of flights, just before landing, and quite Chuck calls in the remainder; and Chuck calls occur in 99% of runs (generally when within about 10 m of the chicks) (Norton-Girffiths, 1969). Most commonly, parents stop some distance from the chicks, and stand while pointing the bill roughly vertically to within 2–3 cm of the ground [see Fig. 2 of Lind (1965) and Fig. 10 of Glutz von Blotzheim *et al.*, (1975)]. The food may be mandibulated or moved up and down the bill, dropped on the ground and then picked up again, or left on the ground while the parent "points" to it. In addition, parents sometimes hold the food while moving the bill around near the chick's face, or may just deposit the food on the ground and then leave. Static holding plus mandibulating occur in nearly all (90% or more) food presentations; other components are less common (Norton-Griffiths, 1969). The orientation and color of the adult's bill are important in eliciting feeding responses from chicks (Hørlyk and Lind, 1978; Lind, 1965), though initial responsiveness to it is in the context of brooding (Norton-Griffiths, 1969).

Developmental changes in how parents attract and feed their young, and in how chicks solicit food; interrelations between chick and adult behavior; and variations in all facets of the parent–young feeding relationship all speak for a complex, labile communication system. This is well illustrated by the tendency of adults to be less attentive to older chicks and less solicitous of them, which in turn affects the behavioral responses of older chicks: "When begging, the [old] chick assumes a posture similar to the submissive posture of the adult; the neck is withdrawn, the head is held at or below the level of the back, and the body is puffed out. . . . The chick approaches . . . usually rubs its head and neck against the flanks and, especially, the breast of the adult. It moves from one side of the parent to the other . . ." sometimes jerking its head upwards, or pecking at the parent's head, bill, legs, and toes, and calling (Lind, 1965, pp. 15–16). In contrast, young chicks simply respond to the sight or calls of an approaching parent by direct approach with calls, and directly taking food from the adult (sometimes preceded by pecking at the food at the tip of the adult's bill).

Monogamous species that feed their young would be good subjects for research on intersexual selection, though none has yet been carried out on this topic, to my knowledge (see further). Monogamous mating systems have other signaling characteristics about which scattered information is available, though. Many authors have noted that display activity in experienced males is reduced or absent compared with inexperienced males, and this is particularly true of males that re-pair with the previous year's mate (Hale and Ashcroft, 1982; Jehl, 1970). Paired birds communicate with one another in numerous contexts, as in changing over at the nest, or in maintaining contact with one another while foraging. Response-calling or antiphonal calling is common in paired shorebirds, such as the Upland Sandpiper (Fig. 14). Antiphonal calling occurs in Jacanidae and Glareolidae, and is particularly common in all three species of phalaropes, though the calls are soft and short-range, often while the birds are close to one another while swimming/feeding (Howe, 1972; Maclean, 1967; Mathew, 1964; E. H. Miller, unpublished observations). Perhaps the loudest and most conspicuous form of simultaneous display by paired birds is piping in oystercatchers (Cramp, 1983; Glutz von Blotzheim *et al.*, 1975; Heppleston, 1970; Huxley and Montague, 1925; and references in Table V). This short-range display occurs in all Haematopodidae, and is characterized by striking postures and movements, and loud lengthy calling. It is not just restricted to communication within a pair though, but occurs commonly as a communal display.

Some of the most elaborate and intricate displays by paired birds are shown by *Charadrius* and Recurvirostridae. In many species of the former, there is a close relationship between nest-scrape displays and copulation, as nicely described by Clark (1982, pp. 120–121) for Kittlitz's Plover (*Charadrius pecuarius*): "the male sat in the hollow, scraping, [etc.] . . . he was approached by the female . . . and she replaced him; almost immediately the female got out and did a slow walk, with head level with body . . . . This was the invitation walk . . . As he approached the female the male started 'high-stepping.' His head was lifted and his whole body raised on almost straight legs. The female stopped a few centimetres to 1 m away from the nest site, and lowered her body slightly. The male continued 'high-stepping' . . . lifting his tarsi and toes above the horizontal . . . and finally made a sudden jump. With tarsi resting on the female's back the male . . . adjusted his position [for 80–150 sec then]

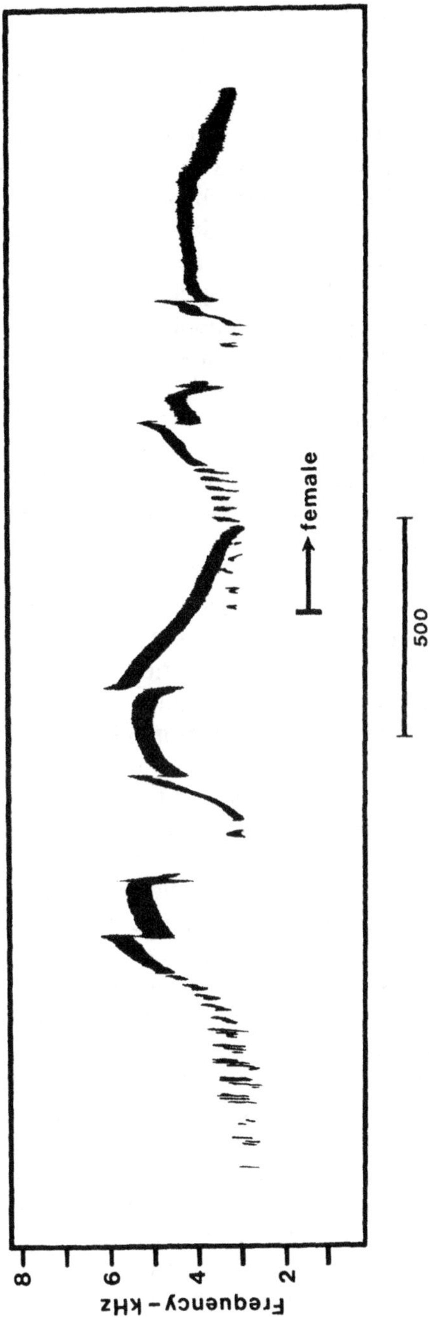

Fig. 14. Response-singing of female to male Upland Sandpiper (*Bartramia longicauda*; near Oak Lake, Manitoba). The male was standing on a knoll, and the female was feeding nearby. [From E. H. Miller and D. R. M. Hatch (unpublished observations).] Analyzing filter bandwidth, 300 Hz. Time marker is in msec.

. . . took the female's nape feathers in his bill and, falling backwards, made cloacal contact. The male finished up on his back with wings spread and the female on top of him with legs in the air" [parts of a homologous sequence for the Snowy Plover, *Charadrius alexandrinus*, are shown as Fig. 39 of Glutz von Blotzheim *et al.* (1975)]. Similar remarkable performances occur in other *Charadrius* species (Clark, 1982; review by Cairns, 1977, 1982). The elaborate display sequences may be given throughout the egg-laying period, as in the Piping Plover, *Charadrius melodus* (Cairns, 1977, 1982), and may also begin well before it starts. In Kittlitz's Plover, Clark (1982) noted copulations up to 11 days before egg laying, and these were repeated frequently (at intervals of 10–45 min). Thus, it seems likely that display behavior involved with scraping and copulation in *Charadrius* species is widespread and is at least partly ritualized, serving functions associated with pairbond maintenance in these generally monogamous species.

In Recurvirostridae and Vanellinae, complex ritualized mutual displays occur between mated birds, and in contexts suggestive of a function in pairbond maintenance. In *Vanellus*, a postcopulatory wing-raised run (and calls) is widespread, for example (Maclean, 1972). Pre- and postcopulatory displays in the colonial, monogamous Black-winged Stilt have been described in detail by Hamilton (1975) and Goriup (1982, pp. 20–21); the latter's description is excerpted here: "As soon as the female assumed the copulation posture [Fig. 15a] . . . the male became very excited, puffing out his feathers to appear much larger than his mate, and striding in semicircles from one side of her to the other, always passing behind . . . . Each time that he came up to her shoulder, he paused to bill-dip and preen his breast or underwing [Figs. 15b–d]. This cycle was repeated two to five times, ending when the male adopted an erect posture prior to mounting [Fig. 15e]. To achieve cloacal contact, the male flexed his legs so that . . . the tarsi rested on the female's back . . . [Fig. 15f]. After dismounting, both male and female adopted upright postures and performed the bills-crossed ceremony: standing close to the female, the male crossed his bill over hers and extended his wing over her back; in this pose, both walked about a metre . . . [Figs. 15g,h]" [see also Fig. 17 of Hamilton (1975) and Fig. 129 of Glutz von Blotzheim *et al.* (1977)].

The elaborate ritualized nature of displays such as those outlined above suggests that intersexual selection is an important factor in monogamous mating systems of shorebirds. Very little on this topic has been published, though Jehl (1975) notes that courtship feeding occurs in the Magellanic Plover, a species that feeds its chicks; such behavior could be very important in assessing potential mates. McGilp and Morgan (1931) observed Banded Stilts (*Cladorhynchus leucocephalus*), a poorly known

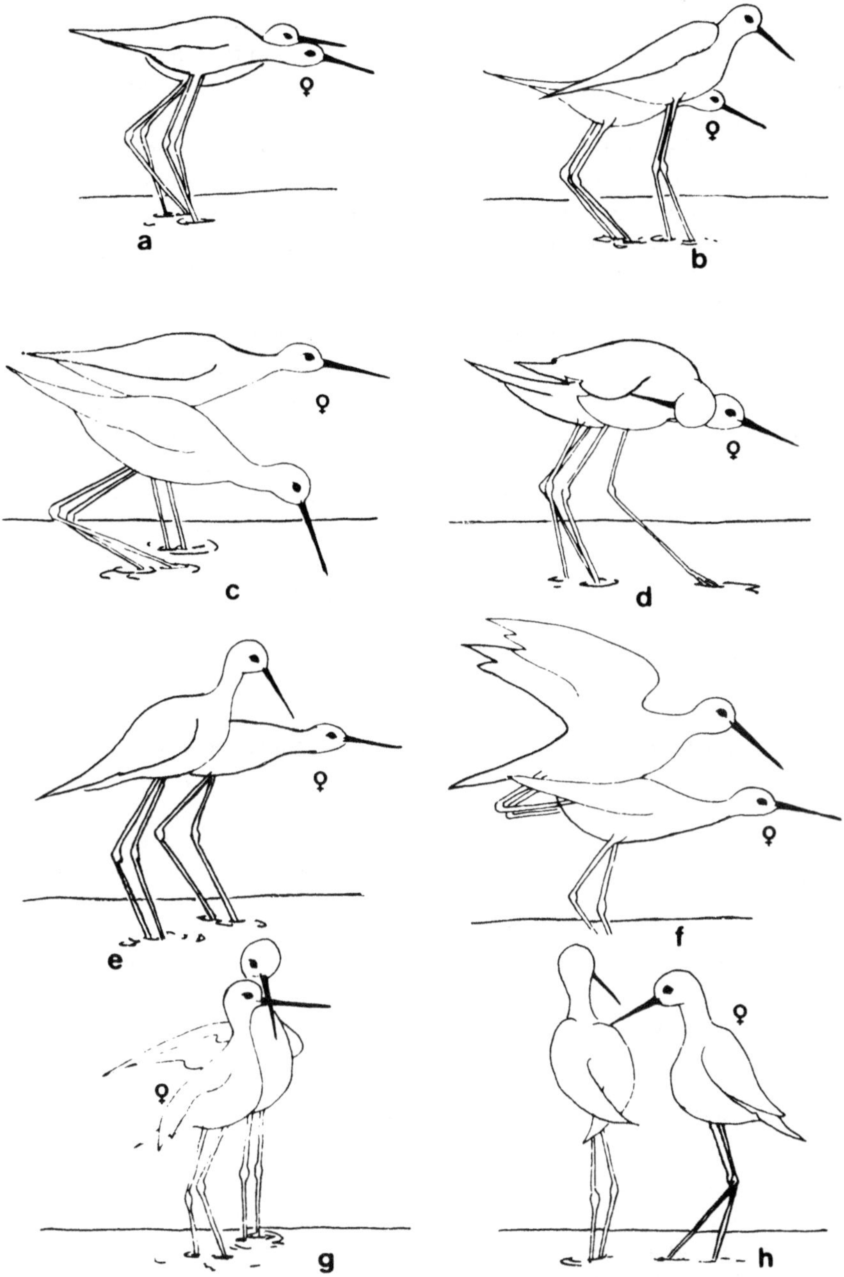

Fig. 15. Sequence of pre- and postcopulatory displays by Black-winged Stilts (*Himantopus himantopus*). See text for discussion. [From Goriup (1982, Fig. 6).]

Australian species that nests in large colonies depending upon the unpredictable occurrence of rainfall: "every now and then a bird would run up from the water and feed its sitting mate" (p. 41). I know of no comparable observations on other species.

A few lekking species of shorebirds have been well studied, and contributions of male behavior to mate choice by females have been examined. The Ruff is best known of such species, but despite intensive research by various workers, very little evidence has been found for display characteristics that seem to influence female choice. Some compelling data are discussed by Hogan-Warburg (1966) and van Rhijn (1973), who determined that female responsiveness was related to quantitative variations in certain forms of male behavior. However, no major conclusions could be formed. Display behavior of lekking Great Snipe has also been described in detail, but no simple relationships between signaling and mate choice have been detected (Avery and Sherwood, 1982; Ferdinand, 1966; Ferdinand and Gensbøl, 1966; Swanberg, 1965). The lack of simple correlations in social species is perhaps unsurprising, for competition among males for positions on leks has already winnowed the number of males from which females must choose: much of the choice has already been made for them (Miller, 1975).

Contest behavior among males has been studied little. In the Least Sandpiper, males early in the breeding cycle (especially unpaired males) engage in several kinds of vigorous hostile display (Fig. 16). One of these involves lengthy sprinting without reference to any territorial boundaries (which do not exist as such in this species). If one male assumes too great a lead, the other bird often makes a short low flight to catch up, or just to move ahead; several quick exchanges like this give the appearance of leapfrogging. Such running may continue between two males, with pauses of various durations, for hours. It seems plausible that this demanding form of contest exerts strong intrasexual selection upon males. In general, in this and in other shorebird species, it is difficult to attribute variations in displays to the action of sexual selection, though it is often straightforward to assign epigamic or hostile functions to displays.

E. Evolutionary and Phylogenetic Considerations

Several lines of evidence suggest that shorebird displays may provide good material for inferring evolutionary trends, and may be sensitive indicators of phylogenetic affinities. In this section, several examples that support this impression are outlined. For example, numerous visual displays and display sequences in Recurvirostridae are nearly invariant even

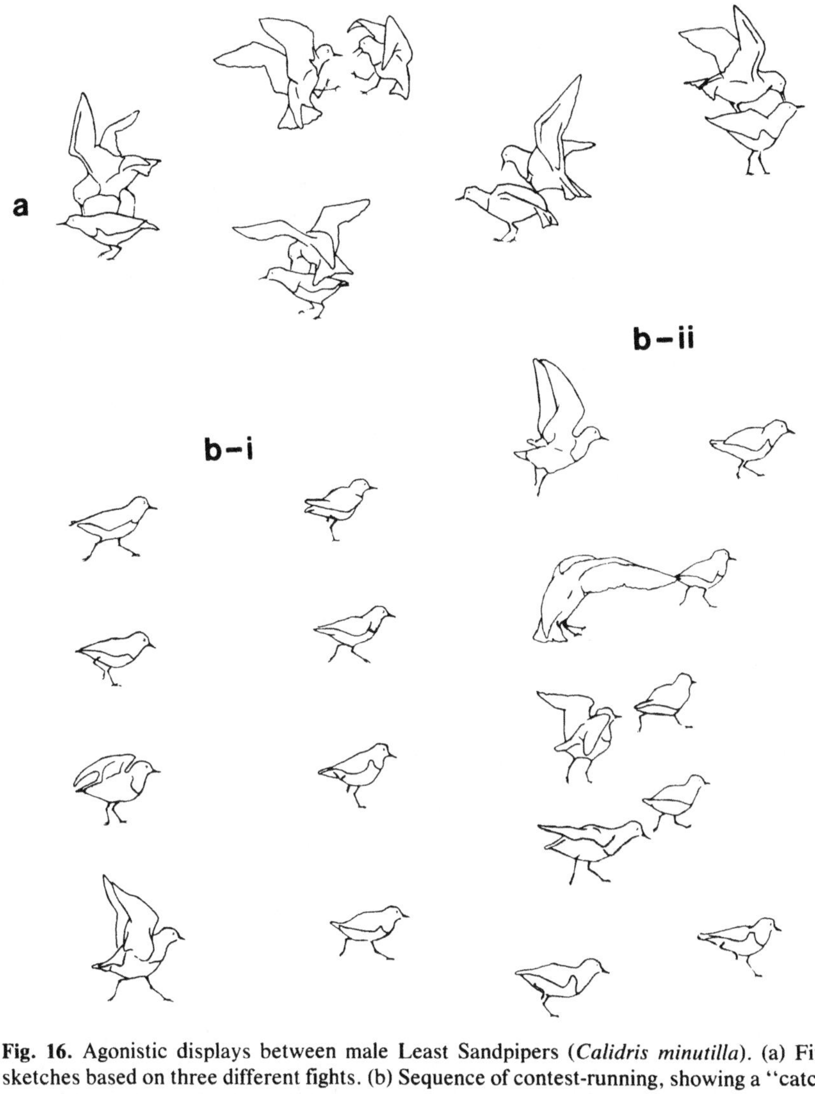

Fig. 16. Agonistic displays between male Least Sandpipers (*Calidris minutilla*). (a) Five sketches based on three different fights. (b) Sequence of contest-running, showing a "catch-up" flight by the trailing bird. Columns b-i and b-ii are viewed from the top down; the top sketch in b-ii follows the bottom sketch in b-i. [Sketches are based on super-8-mm movie films; drawings by A. McClung.]

between continents, and are easily homologized across genera (Cramp, 1983; Goriup, 1982; Glutz von Blotzheim *et al.*, 1977; Hamilton, 1975); certain expressions of threat are strikingly similar in groups as distant as Burhinidae, Haematopodidae, and Jacanidae (Maclean, 1972); etc.

1. Calls by Adult Birds Tending Their Young

Species distinctiveness is not generally selected for in short-distance "alarm" signals, and these can be examined to determine how conservative their structure is across species. Most, perhaps all, shorebirds call softly to their young during and after hatching, but these calls show an unexpectedly high amount of species distinctiveness (Fig. 17). For example, the Common Greenshank commonly emits a broadband "gruff" sound, while Mountain Plovers (*Charadrius montanus*) and Least Sandpipers use many soft low-frequency trills (Graul, 1974; Nethersole-Thompson and Nethersole-Thompson, 1979; Wallis and Wershler, 1981). The latter species also "clucks" frequently, while the Spotted Sandpiper gives one to several repetitions of a much higher-frequency call (Fig. 17). Close study of related species may reveal more obvious similarities, a prediction upheld by examination of parental calls given in response to human intrusion (Fig. 8). Oystercatchers have two quite different call types, with intermediates: single, brief calls; and fairly long, pure-frequency calls introduced by several brief calls (Figs. 8A,B; Miller and Baker, 1980). Two *Pluvialis* species emit long, pure-frequency calls with sudden frequency shifts (Figs. 8E,F), and all *Calidris* species I am familiar with (except the White-rumped Sandpiper) have two apparently homologous calls: a single type characterized by conspicuous frequency modulation; and a trill (Figs. 8I–L). No analyses exist to permit discussion of other species illustrated in Fig. 8 (*Vanellus, Charadrius, Phalaropus*), but the utility of "alarm" calls in systematic research certainly seems to be limited to lower categories, to judge from these examples.

2. Aerial Display

Most species of shorebirds inhabit fairly open environments during the breeding period and, like many open-country bird species, have well-developed aerial displays (Armstrong, 1963). The ultimate explanation of this broad correlation presumably lies in the increased transmission distance that results from displaying at some height above the ground surface (Wiley and Richards, 1982; see Section II.D.1). The accomplished flying skills of shorebirds have resulted in great elaboration and complexity in their aerial displays, which have been accompanied in some cases by the

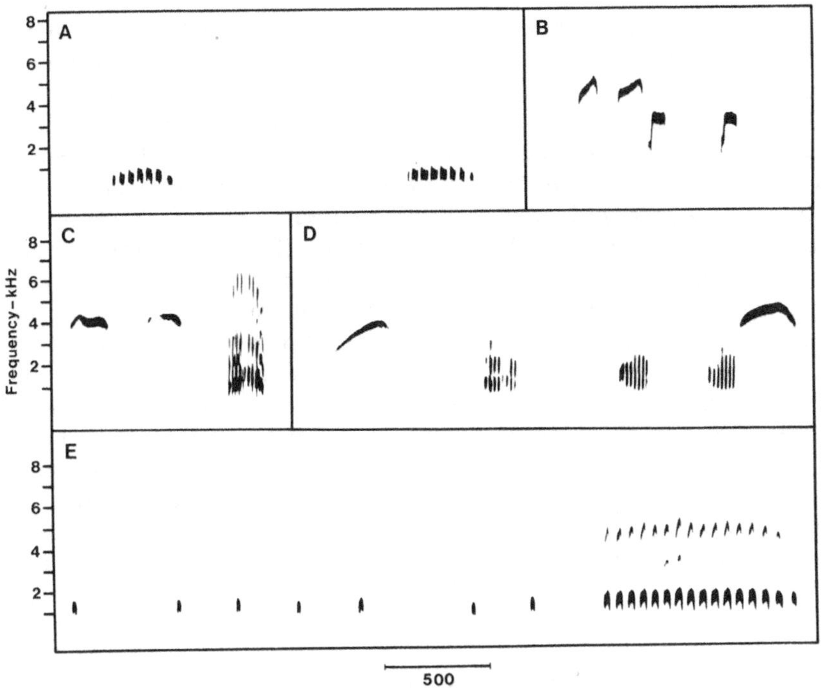

Fig. 17. Sound spectrograms of calls by undisturbed adults attending chicks at the nest, around the time of hatching (all are shown as natural sequences). (A) Mountain Plover (*Charadrius montanus*; made from recording by W. D. Graul). (B) Spotted Sandpiper (*Actitis macularia*; Sable Island, Nova Scotia). Two chick calls precede the parent's two calls. (C, D) Common Greenshank (*Tringa nebularia*). Two chick calls precede the parent's call in (C), and one precedes and follows the parent's calls in (D). [From Nethersole-Thompson and Nethersole-Thompson (1979, Figs. 42 and 45, respectively).] (E) Sequence of clucks followed by trill of Least Sandpiper (*Calidris minutilla*; Sable Island, Nova Scotia).

Analyzing filter bandwidth throughout, 300 Hz. Time marker is in msec.

evolution of morphological features increasing display conspicuousness. The evolutionary impetus behind such elaborations comes from sexual and other forms of social selection, plus adaptations of display features to the physical environment (see Section II.D.1). Considering the diversity of these selective pressures, it is clearly not realistic to consider aerial display as a simple, unitary phenomenon even just within shorebirds. Nevertheless, the structure of this display form reveals evolutionary trends and patterns of adaptive radiation, and may also shed much light on issues surrounding variation, information content, and social significance of a stereotyped, complex display.

Aerial display includes elaborate forms of descent, which may be terminal and return the bird to the ground, or may occur throughout display. A simple form occurs in the Least Seedsnipe, in which males fly silently to a height of about 10 m, begin to sing as they flutter, then continue to sing while they glide slowly to the ground with tail spread; similar display occurs in the gray-breasted Seedsnipe, *Thinocorus orbignyianus* (Johnson and Goodall, 1965; Maclean, 1969; E. H. Miller, unpublished observations). Most or all species of Tringini engage in some form of undulating aerial flight, in which they beat the wings as they rise, then glide down with spread tail, flutter and rise again, and so on in alternating rhythm (Cramp, 1983; Glutz von Blotzheim *et al.*, 1977; Grosskopf, 1958; Hale and Ashcroft, 1982; Rowan, 1929; von Frisch, 1956). Much more complex flight patterns occur, as in Red Knot (Nettleship, 1974), Jacksnipe [*Lymnocryptes minimus* (Kliebe, 1974)], and Northern Lapwing (Dabelsteen, 1978; see Section II.B.2). Most of the striking display attributes are related to some kind of descent, often coupled with distinctive vocalizations. In *Gallinago*, embedded dives during protracted high display flights are typical, and have been studied in detail in the Common Snipe (reviews by Reddig, 1978; Tuck, 1972). A somewhat stylized description of aerial display in this species follows (after Reddig, 1978; see Fig. 18).

During ascent, the bird keeps the tail closed. After reaching maximal altitude, he leans to one side (reaching a sideways angle of 70° or more), spreads the tail, then dives at an angle of about 41° relative to the horizontal. The two outermost rectrices are widely spread, to about 70–80° relative to the long body axis, and the extended wings are held so that the tenth primary is about 55° relative to the long body axis. The wings make fluttering motions during the dive, and this is the only time that sounds are emitted, due to vibration of the outer rectrices in the air current modified by the wing action. The sound's frequency increases slightly during each dive, whereas amplitude increases markedly, being loudest at the dive's end, when velocity is greatest. A glide follows, then the bird rises again; the bird drops about 15 m in a typical dive (Fig. 18A). Display flights of snipe may cover an area of 250-m diameter or even more, and often show a roughly circular flight path. Birds almost always orient their dive to the outside, then recover subsequently (Fig. 18B).

Aerial display in Common Snipe differs from many species in that nonvocal sounds predominate. These are coincident with the repeated phases of ritualized descent. The tail of snipe has been modified as a display structure and the mechanics of sound production by the specialized outer rectrices have been studied by numerous workers (reviews by Reddig, 1978; Rutschke, 1976; Tuck, 1972). There is only a rough cor-

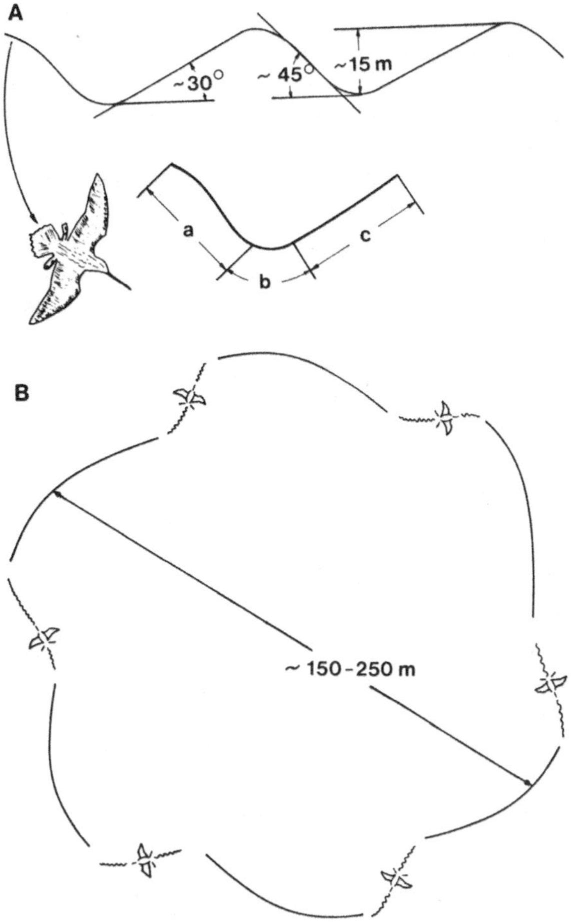

Fig. 18. Schematic diagrams of aerial display in the Common Snipe (*Gallinago gallinago*). (A) Phases of aerial display: a, dive with wing flutter; b, glide; c, steered flight during ascent. (B) Flight path during aerial display, showing the direction during drumming descent (wavy lines) and compensation during ascent and between dives (curved lines). [After Reddig (1978, Figs. 4 and 5, respectively).]

respondence between specialization of the tail and elaboration of the flight display, however (Sutton, 1981; see also Stettenheim, 1976). The Pin-tailed Snipe (*Gallinago stenura*) possesses up to 42 rectrices compared with about a dozen in Common Snipe, and the outermost of these are extremely stiff and narrow, "being about 1 mm wide from base to tip" (Sutton, 1981, p. 461). Males of this species apparently display communally in toks (a rough equivalent of an aerial lek) (Sutton, 1981; Tuck, 1972).

It is easy to envision the ritualization of descents from flight displays, because they are conspicuous and encode much information about the signaler's change in behavior. Further, because some nonvocal sound is invariably produced during rapid descent, the raw material for evolutionary elaboration of displays like those of snipe is present. A clear example of convergence on snipe aerial display is described by Labutin *et al.* (1982) and Veprintsev and Zablotskaya (1982) for the Little Curlew (*Numenius minutus*). This species shows embedded dives during its aerial displays, with remarkably snipe-like nonvocal sounds during the dives.

The taxonomic value of aerial displays has been discussed for vocal characteristics in Calidridini, most monogamous species of which use stereotyped, simple call sequences (Fig. 123; Miller, 1983a,b). The adaptiveness of such structure has presumably promoted convergence by other species, including American Golden Plover, *Pluvialis dominica* (Fig. 12A), and many or most *Charadrius* species (Cairns, 1977, 1982; Cramp, 1983; Glutz von Blotzheim *et al.*, 1975). Such convergence will make some taxonomic judgements difficult, but even in closely related groups like Tringini there are distinctive characteristics that set aerial displays apart. In many Tringini, calls uttered in aerial displays are roughly pure-tonal with stereotyped patterns of frequency modulation or occasional very short breaks, and in many contexts calls are given in bouts separated by longer pauses (Figs. 12C,D; Weeden, 1965). In general, the long-distance aerial displays of shorebirds do seem markedly conservative—at least as conservative as certain calls discussed in the previous section.

3. Agonistic and Sexual Ground Display

Visual components of agonistic and sexual ground display in shorebirds are very conservative across species. Three *Charadrius* species studied by Simmons (1953a,b) differ in their ground threat displays in ways that correspond to display intensity within species. Thus, the Ringed Plover (*Charadrius hiaticula*) normally leans slightly forward with the tail fanned and depressed (or elevated, at highest intensity), with the dorsal plumage fluffed, and with the black collar conspicuously displayed (Fig. 3 of Simmons, 1953a; Fig. A of Cramp, 1983). One male briefly assumed an upright hunched posture like that typical of the Kentish Plover. In addition, Ringed Plovers in low-intensity threat do not fluff the dorsal plumage, and initial phases of threat in the three species are highly similar. At very high intensity, threatening Snowy and Little Ringed Plovers (*Charadrius dubius*) may fluff the dorsal plumage, and the former may also fan the tail, thus resembling the Ringed Plover (see also Cairns, 1977, 1982; Edwards *et al.*, 1947; Glutz von Blotzheim *et al.*, 1975; Graul, 1973b;

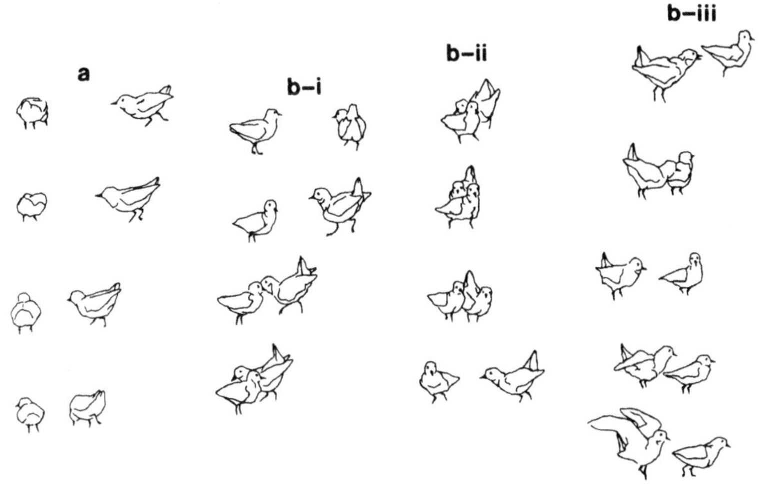

Fig. 19. "Courtship" approaches of female by male Least Sandpipers (*Calidris minutilla*) (two sequences). (a) The male approaches from the right with tail slightly erected, but begins feeding in the last sketch. (b) The male obliquely approaches the female from the right, with tail strongly erected and with body tipped up. The male sings in the top sketch of b-iii, and begins precopulatory flutter in the bottom sketch.

Columns are viewed from the top down; the top sketch in b-ii follows the bottom sketch in b-i, and similarly for b-iii. [Sketches are based on super-8-mm movie films; drawings by A. McClung.]

Phillips, 1972; Smith and Hosking, 1955). In the Little Ringed Plover, the typical threat posture is with a horizontal carriage, the flank feathers spreading over the closed wings, and the head retracted, resulting in a strikingly dorsoventrally flattened appearance in frontal view (Fig. 1 of Simmons, 1953a; Fig. B of Cramp, 1983). An almost identical posture is assumed in the Mountain Plover (Fig. 2 of Graul, 1973a) and the Semipalmated Plover [*Charadrius semipalmatus* (E. H. Miller, unpublished observations)]. Other widespread displays are precopulatory, including stereotyped high-stepping by the male preparatory to mounting (Boyd, 1972; Cairns, 1977, 1982; Clark, 1982; Cramp, 1983; Glutz von Blotzheim *et al.*, 1975).

In Scolopacidae, wing exposure and tail elevation are common in ground displays. Male Least Sandpipers approaching females or other males often just tilt forward and elevate the tail, both components becoming more extreme at close quarters (Fig. 19). Sexual approaches can often only be distinguished from aggressive ones by their slower tempo and more stereotyped actions. At higher intensity, one or both wings are lifted partially or fully, and (in the one-wing attitude) the wing may be

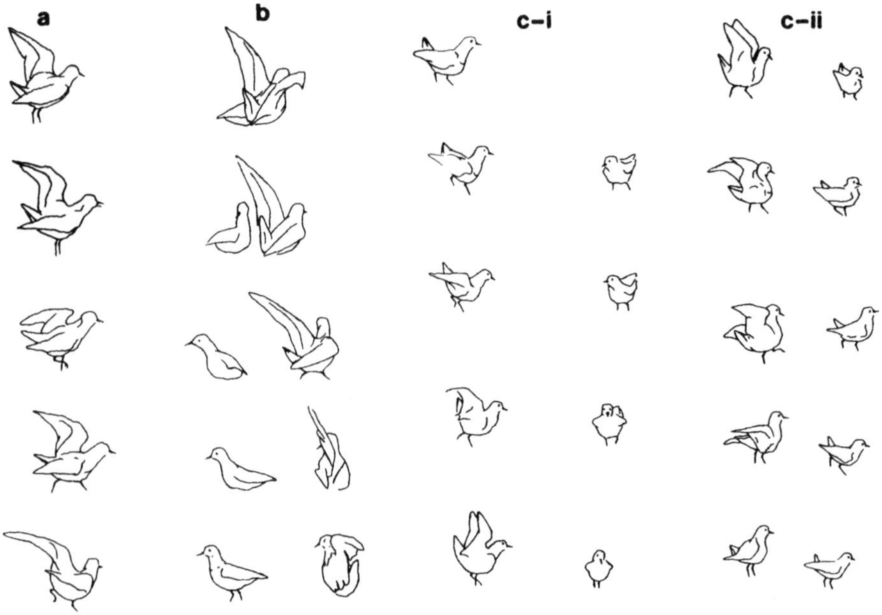

Fig. 20. "Courtship" approaches of female by male Least Sandpipers (*Calidris minutilla*) (three sequences). (a) The male stands (upper sketch), sings (second sketch), then rapidly approaches female, who is not illustrated. (b) The male fully erects his inside wing toward the female, adjusting his orientation to the female as she moves. (c) The male approaches the female from the left, with tail slightly erected, assumes a two-wing-up/tail-up display, then resumes a weak tail-up display (last sketch).

Columns are viewed from the top down; the top sketch in c-ii follows the bottom sketch in c-i. [Sketches are based on super-8-mm movie films; drawings by A. McClung.]

slowly and rhythmically pumped up and down (this is usually the wing closest to the other bird; Figs. 20, 21). Wing-up displays are so common in shorebirds that inferring their homology is difficult, though in the Calidridini alone strikingly similar displays to those mentioned have been described in many species (Bengtson, 1970; Cramp, 1983; Drury, 1961; Glutz von Blotzheim *et al.*, 1975). By using both structural and functional features (e.g., using contextual information), such inference can be strengthened, and it is thereby reasonable to point to tail-up postures of courting Upland Sandpipers and other curlew species as being homologous to those of calidridines (Fig. 22; Allen, 1980; Trimble, 1940).

4. Geographic Variation

No comprehensive studies of geographic variation in display behavior have been carried out, though Williamson made many significant obser-

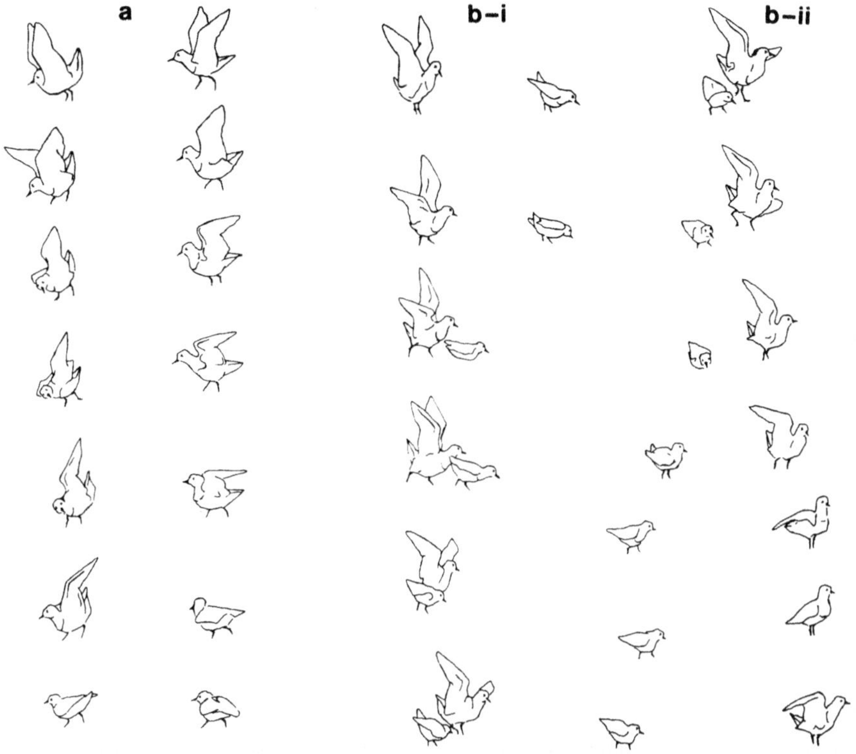

Fig. 21. Agonistic displays between male Least Sandpipers (*Calidris minutilla*) (three sequences). (a) Sequence during contest-running, beginning after the birds landed following a brief low flight together. The left-most bird has just landed in the top sketch, and twists to display broadside to the other (following four sketches). Meanwhile, the right-most bird assumes one-wing-up display and begins to run again; the left-most bird joins him (bottom sketch). (b) Approach by one male to another one feeding. The feeding male pays very little attention to the instigator, except to elevate the tail slightly and (bottom sketch in b-i) to tip.

Columns are viewed from the top down; the top sketch in b-ii follows the bottom sketch in b-i. [Sketches are based on super-8-mm movie films; drawings by A. McClung.]

vations in this regard for behavior in response to potential predators (see Section II.D.1). Thönen (1968) noted that drumming of Common Snipe in northern Alaska is different from that of European birds, and I have made a few comments on aerial displays of Least Sandpipers, Semipalmated Sandpipers, and Dunlins from different parts of North America (Miller, 1983a,b). Aerial song of Short-billed Dowitchers (*Limnodromus griseus*) is very similar in structure and organization throughout the breed-

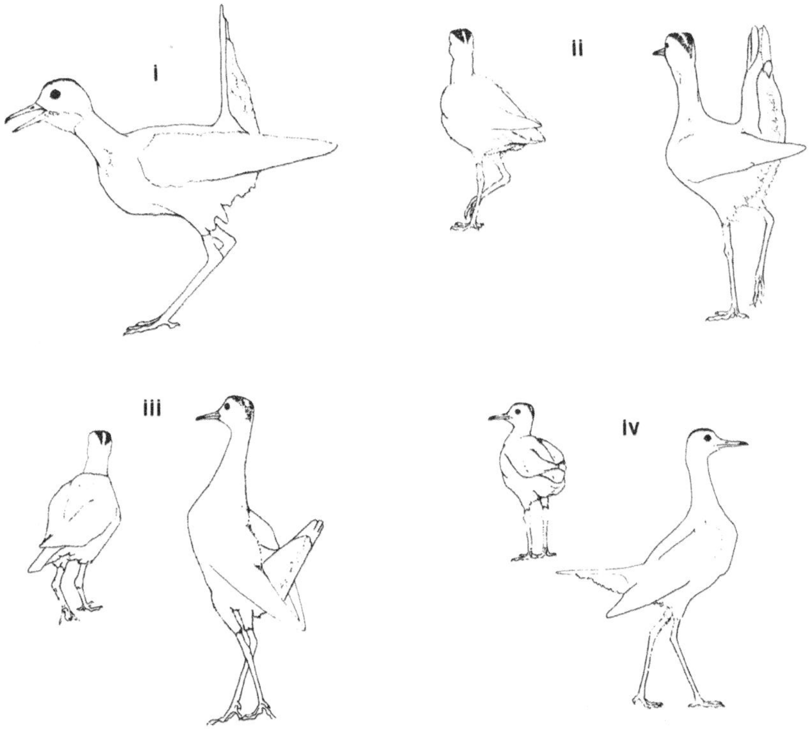

Fig. 22. "Courtship" approach of female by male Upland Sandpipers (*Bartramia longicauda*). The male quickly approaches the female (not shown) while calling softly (i), then chases her (ii); when the female slows down, the male stops and stretches vertically (iii), then both birds relax (iv). [Sketches are based on photographs by E. H. Miller and D. R. M. Hatch; drawings by A. McClung.]

ing range, based on recordings from northwestern British Columbia, northern Manitoba, and Labrador (Miller *et al.*, 1983). Though significant quantitative differences among these localities were detected, the same song components were present and they occurred in the same sequence in nearly all samples.

Fuller understanding of the significance of geographic variation can come through study of species with narrow or wide ranges, of closely related species, and of species on migration routes or wintering grounds. Ontogeny of display behavior is another crucial arena of investigation in this regard; preliminary work has been reported by Tikhonov and Fokin (1979,1980).

III. CONCLUDING REMARKS

Our knowledge of communication in breeding shorebirds is based upon detailed studies of a few species, plus great numbers of general descriptions and anecdotes. In view of the large geographic ranges, ecological diversity, and variety of social systems exhibited even within many species of shorebirds, this is an unfortunate situation for it is not likely to be representative or fully revealing. Consider species like Common Snipe, which breed from Patagonia to Alaska, and in much of Europe and Asia; Killdeers, which breed in North and South America; Snowy Plovers of North America and Europe; and Eurasian Oystercatchers, which breed in Europe, Asia, and Australasia. Such widespread species are particularly attractive for research on geographic variation and species-specificity. Another important area for research is ontogeny, which could be studied for certain taxa able to be reared in captivity (e.g., see Schwinn, 1964; von Frisch, 1956). Species that are poorly known because of their inaccessibility, like Dromadidae, or their crypticity, like Rostratulidae and *Scolopax*, also await study. There is essentially no knowledge of tactual communication in shorebirds, though intriguing descriptions are available for a number of species: in Willets and many *Charadrius* species, males grasp the female during copulation (Cairns, 1982; Hansen, 1979); in the Piping Plover and Long-billed Plover (*Charadrius placidus*), females place their heads underneath the tail of the male in precopulatory display, and in the former species at least the female actually seems to place the tip of her bill against the male's cloaca (Cairns, 1977, 1982; Panov, 1963); and male Long-billed Curlews ruffle the shoulder feathers of the female in the stereotyped precopulatory "shaking" display (Allen, 1980). The nature, extent, and significance of tactual displays and display components merit detailed study. Finally, mechanisms of vocalization, and the relationship of sound characteristics to anatomy of syringeal and extra-syringeal sound sources are unexplored (see Gaunt, 1983b; Niethammer, 1966; Warner, 1969).

The burgeoning interest in shorebird biology has resulted in increased attention to all details of shorebird natural history, including social behavior. Detailed descriptions and basic documentation of behavioral biology in most shorebird species and populations are needed, but there are many ways to describe and quantify behavior, and these reveal different sorts of information and permit us to ask different kinds of questions. Further, our view of animal communication has been altered radically over the past few years, and any research into social behavior must take this into account. For purely documentary purposes, one might hope that behavioral descriptions are of the most general but explicit form so as to make possible the posing of many different questions, and this is especially true for rare species and for those difficult to study. In general studies, or in studies with other purposes, workers should be particularly cautious about using interpretive or anthropomorphic terms as shorthand, and should try to base their descriptions upon observable motor patterns as much as possible.

ACKNOWLEDGMENTS

Many people have generously given me encouragement, advice, experience, support, assistance, reprints, translations, references, pictures, or data, and I express my deepest appreciation to them all: Allan Baker, Daryl Boness, Alan Burger, Adolf Ceska, Pete Connors, Peter Ferns, Bob Frisch, Erica Gillard, Walter Graul, Bill Gunn, Gay Hansen, Rod Hay, David Hatch, Harold Hosford, Don Jenni, Paul Johnsgard, Julie Kehoe, Nicole Kelly, Nancy Lilburn, Ed Longprés, Angie McClung, Terry Mace, Ian McLaren, Steve MacLean, Pete Myers, Desmond Nethersole-Thompson, Dennis Paulson, Frank Pitelka, Dave Samuel, Marty Stephens, Joe Strauch, Jeff Walters, and Maria Wollemann. My research has been supported by Individual Operating Grants from the Natural Sciences and Engineering Research Council of Canada, and by the Friends of the British Columbia Provincial Museum.

REFERENCES

Adret, P., 1982, The sound signals of the adult avocet *Recurvirostra avosetta* during the perinatal phase, *Ibis* **124**:275–287.

Ailes, I. W., 1976, Ecology of the Upland Sandpiper in central Wisconsin, M.S. thesis, University of Wisconsin, Stevens Point.

Allen, J. N., 1980, The ecology and behavior of the Long-billed Curlew in southeastern Washington, *Wildl. Monogr.* **73**.

Andersen, F. S., 1951, Contributions to the biology of the Ruff (*Philomachus pugnax* (L.)), III, *Dan. Ornithol. Foren. Tidsskr.* **45**:145–173.

Armstrong, E. A., 1963, *A Study of Bird Song*, Oxford University Press, London.

Ashkenazie, S., and Safriel, U. N., 1979, Time–energy budget of the Semipalmated Sandpiper *Calidris pusilla* at Barrow, Alaska, *Ecology* **60**:783–799.

Avery, M., and Sherwood, G., 1982, The lekking behaviour of Great Snipe, *Ornis Scand.* **13**:72–78.

Baker, A. J., 1974, Ecological and behavioural evidence for the systematic status of New Zealand oystercatchers (Charadriiformes: Haematopodidae), *R. Ont. Mus. Life Sci. Contrib.* **96**:1–34.

Baker, M. C., 1982, Individuality of vocalizations in Dunlin: A possible acoustic basis for recognition of parent by offspring, *Auk* **99**:771–774.

Beer, C. G., 1975, Multiple functions and gull displays, in: *Function and Evolution in Behaviour—Essays in Honour of Professor Niko Tinbergen, F.R.S.* (G. P. Baerends, C. G. Beer, and A. Manning, eds.), pp. 16–54, Oxford University Press, London.

Beightol, D. R., 1972, Sonagraphic analysis of the American Woodcock's 'peent' call as a research and management tool, M.S. thesis, West Virginia University, Morgantown.

Beightol, D. R., and Samuel, D. E., 1973, Sonagraphic analysis of the American Woodcock's *peent* call, *J. Wildl. Manage.* **37**:470–475.

Bengtson, S.-A., 1970, Breeding behaviour of the Purple Sandpiper *Calidris maritima* in west Spitsbergen, *Ornis Scand.* **1**:17–25.

Bock, W. J., 1958, A generic review of the plovers (Charadriinae, Aves), *Bull. Mus. Comp. Zool. Harv. Univ.* **118**:27–97.

Bourgeois, J.-C., 1977, Contribution a l'étude des problèmes relatifs à l'interpretation des recensements de populations de bécasses d'Amérique (*Philohela minor*), Thèse M.Sc., Université Québec à Trois-Rivières, Trois-Rivières.

Bourgeois, J.-C., and Couture, R., 1977, A method for identifying American Woodcock males based on peent call sonagraphic analysis, in: *Proceedings of the Sixth Woodcock Symposium* (D. A. Keppie and R. B. Owen, Jr., eds.), pp. 171–184, Fredericton.

Boyd, R. L., 1972, Breeding biology of the Snowy Plover at Cheyenne Bottoms Waterfowl Management Area, Barton County, Kansas, M.S. thesis, Kansas State Teachers College, Emporia.

Burger, A. E., 1979, Breeding biology, moult and survival of Lesser Sheathbills *Chionis minor* at Marion Island, *Ardea* **67**:1–14.

Burger, A. E., 1980a, Sexual size dimorphism and aging characters in the Lesser Sheathbill at Marion Island, *Ostrich* **51**:39–43.

Burger, A. E., 1980b, An analysis of the displays of Lesser Sheathbills *Chionis minor*, *Z. Tierpsychol.* **52**:381–396.

Burger, A. E., 1980c, Behavioural ecology of Lesser Sheathbills *Chionis minor* at Marion Island, Ph.D. thesis, University of Cape Town, Cape Town.

Bursian, S. J., 1971, The structure and function of Killdeer vocalizations, M.S. thesis, University of Minnesota, Minneapolis.

Cadman, M. D., 1980, Age-related foraging efficiency of the American Oystercatcher (*Haematopus palliatus*), M.Sc. thesis, University of Toronto, Toronto.

Cairns, W. E., 1977, Breeding biology and behaviour of the Piping Plover (*Charadrius melodus*) in southern Nova Scotia, M.Sc. thesis, Dalhousie University, Halifax.

Cairns, W. E., 1982, Biology and behavior of breeding Piping Plovers, *Wilson Bull.* **94**:531–545.

Clark, A., 1982, Some observations on the breeding behaviour of Kittlitz's Sandplover, *Ostrich* **53**:120–122.

Connors, P. G., 1983, Taxonomy, distribution, and evolution of Golden Plovers (*Pluvialis dominica* and *Pluvialis fulva*), *Auk* **100**:607–620.

Cramp, S. (ed.), 1983, *Handbook of the Birds of Europe, the Middle East and North Africa: The Birds of the Western Palearctic*, Vol. 3, *Waders to Gulls*, Oxford University Press, London.

Cunningham, J. M., 1973, The Banded Dotterel, *Charadrius bicinctus*: Pohowera or Tuturiwhatu?—Call notes and behavior, *Notornis* **20**:21–27.

Dabelsteen, T., 1978, An analysis of the song-flight of the lapwing (*Vanellus vanellus* L.) with respect to causation, evolution and adaptations to signal function, *Behaviour* **66**:136–178.
Dowsett, R. J., Lemaire, F., and Stjernstedt, R., 1977, The voice of the courser *Rhinoptilus cinctus*, *Bull. Br. Ornithol. Club* **97**:73–75.
Drury, W. H., Jr., 1961, The breeding biology of shorebirds on Bylot Island, Northwest Territories, Canada, *Auk* **78**:176–219.
Edwards, G., Hosking, E., and Smith, S., 1947, Aggressive display of the Ringed Plover (studied by means of a stuffed specimen and a mirror), *Br. Birds* **40**:12–19.
Edwards, P. J., 1982, Plumage variation, territoriality and breeding displays of the Golden Plover *Pluvialis apricaria* in southwest Scotland, *Ibis* **124**:88–95.
Erckmann, W. J., Jr., 1981, The evolution of sex-role reversal and monogamy in shorebirds, Ph.D. thesis, University of Washington, Seattle.
Ferdinand, L., 1966, Display of the Great Snipe (*Gallinago media* Latham), *Dan. Ornithol. Foren. Tidsskr.* **60**:14–34.
Ferdinand L., and Gensbøl, B., 1966, Maintenance of territory in the Great Snipe (*Gallinago media* Latham) on the display ground, *Dan. Ornithol. Foren. Tidsskr.* **60**:35–43.
Ferns, P., 1978, Individual differences in the head and neck plumage of Ruddy Turnstones (*Arenaria interpres*) during the breeding season, *Auk* **95**:753–755.
Ferns, P. N., 1981, Identification, subspecific variation, ageing and sexing in European Dunlins, *Dutch-Birding* **3**:85–98.
Ferns, P. N., and Green, G. H., 1979, Observations on the breeding plumage and prenuptial moult of Dunlins, *Calidris alpina*, captured in Britain, *Gerfaut Rev. Belge Ornithol.* **69**:286–303.
Forsythe, D. M., 1967, Vocalizations of the Long-billed Curlew, M.S. thesis, Utah State University, Logan.
Forsythe, D. M., 1970, Vocalizations of the Long-billed Curlew, *Condor* **72**:213–224.
Gaunt, A. S., 1983a, On sonograms, harmonics, and assumptions, *Condor* **85**:259–261.
Gaunt, A. S., 1983b, An hypothesis concerning the relationship of syringeal structure to vocal abilities, *Auk* **100**:853–862.
Gibson, F., 1978, Ecological aspects of the time budget of the American Avocet, *Am. Midl. Nat.* **99**:65–82.
Glutz von Blotzheim, U. N., Bauer, K. M., and Bezzel, E. (eds.), 1975, *Handbuch der Vögel Mitteleuropas*, Vol. 6, Charadriiformes, Part 1, Akademie Verlagsgesellschaft, Wiesbaden.
Glutz von Blotzheim, U. N., Bauer, K. M., and Bezzel, E. (eds.), 1977, *Handbuch der Vögel Mitteleuropas*, Vol. 7, Charadriiformes, Part 2, Akademie Verlagsgesellschaft, Wiesbaden.
Goriup, P. D., 1982, Behaviour of Black-winged Stilts, *Br. Birds* **75**:12–24.
Gratto, C. L., Cooke, F., and Morrison, R. I. G., 1981, Hatching success of yearling and older breeders in the Semipalmated Sandpiper *Calidris pusilla*, *Bull. Wader Study Group* **33**:37–38.
Graul, W. D., 1973a, Possible functions of head and breast markings in Charadriinae, *Wilson Bull.* **85**:60–70.
Graul, W. D., 1973b, Adaptive aspects of the Mountain Plover social system, *Living Bird* **12**:69–94.
Graul, W. D., 1974, Vocalizations of the Mountain Plover, *Wilson Bull.* **86**:221–229.
Green, S., and Marler, P., 1979, The analysis of animal communication, in: *Handbook of Behavioral Neurobiology* (P. Marler and J. G. Vandenbergh, eds.), Vol. 3, pp. 73–158, Plenum Press, New York.

Greenewalt, C. H., 1968, *Bird Song: Acoustics and Physiology*, Smithsonian Institution Press, Washington, D.C.
Grosskopf, G., 1958, Zur Biologie des Rotschenkels (*Tringa t. totanus*), I, *J. Ornithol.* **99**:1–17.
Groves, S., 1978, Age-related differences in Ruddy Turnstone foraging and aggressive behavior, *Auk* **95**:95–103.
Grudzien, T. A., 1976, The breeding behavior of the Common Snipe (*Capella gallinago delicata* Ord) in central Michigan, M.S. thesis, Central Michigan University, Mount Pleasant.
Hailman, J. P., 1977, *Optical Signals: Animal Communication and Light*, Indiana University Press, Bloomington.
Hailman, J. P., 1982 [Review of] Ritualization and the evolution of movement signals, *J. Field Ornithol.* **53**:186–187.
Hale, W. G., 1980, *Waders*, Collins, Glasgow.
Hale, W. G., and Ashcroft, R. P., 1982, Pair formation and pair maintenance in the Redshank *Tringa totanus*, *Ibis* **124**:471–490.
Hamilton, R. B., 1975, Comparative behavior of the American Avocet and the Black-necked Stilt (Recurvirostridae), *A.O.U. Ornithol. Monogr.* **17.**
Hansen, G. L., 1979, Territorial and foraging behaviour of the eastern Willet *Catoptrophorus semipalmatus semipalmatus* (Gmelin), M.Sc. thesis, Acadia University, Wolfville.
Hay, R., 1983, The Wrybill, *Anarhynchus frontalis*, in: *Reader's Digest Book of New Zealand Birds* (in press).
Heckenlively, D. B., 1972, Responses of adult Killdeers to a downy young distress call, *Condor* **74**:107–108.
Heidemann, M. K., and Oring, L. W., 1976, Functional analysis of Spotted Sandpiper (*Actitis macularia*) song, *Behaviour* **56**:181–193.
Heppleston, P. B., 1970, The function of oystercatcher piping behaviour, *Br. Birds* **63**:133–135.
Heppleston, P. B., 1971, The feeding ecology of oystercatchers (*Haematopus ostralegus* L.) in winter in northern Scotland, *J. Anim. Ecol.* **40**:651–672.
Hindwood, K. A., 1934, Observations on the Lotus-bird, *Emu* **34**:51–56.
Hirons, G., 1977a, The roding behaviour of the European Woodcock, in: *Supplement to the Proceedings of the Sixth Woodcock Symposium* (D. A. Keppie and R. B. Owen, Jr., eds.), no pagination, Fredericton.
Hirons, G., 1977b, The Game Conservancy woodcock project, *Game Conservancy Annu. Rev.* **9**:x–xx.
Hogan-Warburg, A. J., 1966, Social behavior of the Ruff, *Philomachus pugnax* (L.), *Ardea* **54**:109–225.
Holmes, R. T., 1973, Social behaviour of breeding Western Sandpipers, *Ibis* **115**:107–123.
Holmes, R. T., and Pitelka, F. A., 1961, Behavior and taxonomic position of the White-rumped Sandpiper, *Proc. Alaska Sci. Conf.* **12**:19–20.
Hørlyk, N.-O., and Lind, H., 1978, Pecking response of artificially hatched oystercatcher *Haematopus ostralegus* young, *Ornis Scand.* **9**:138–145.
Howe, M. A., 1972, Pair bond formation and maintenance in Wilson's Phalarope, *Phalaropus tricolor*, Ph.D. thesis, University of Minnesota, Minneapolis.
Howell, T. R., 1979, Breeding biology of the Egyptian Plover, *Pluvianus aegyptius*, *Univ. Calif. Berkeley Publ. Zool.* **113.**
Huxley, J. S., and Montague, F. A., 1925, Studies on the courtship and sexual life of birds. V. The oyster-catcher (*Haematopus ostralegus* L.), *Ibis* **1925**:868–897.

Jehl, J. R., Jr., 1970, Sexual selection for size differences in two species of sandpipers, *Evolution* **24**:311–319.

Jehl, J. R., Jr., 1973, Breeding biology and systematic relationships of the Stilt Sandpiper, *Wilson Bull.* **85**:115–147.

Jehl, J. R., Jr., 1975 *Pluvianellus socialis*: Biology, ecology, and relationships of an enigmatic Patagonian shorebird, *Trans. San Diego Soc. Nat. Hist.* **18**:25–74.

Jenni, D. A., Gambs, R. D., and Betts, B. J., 1974, Acoustic behavior of the Northern Jacana, *Living Bird* **13**:193–210.

Johnsgard, P. A., 1981, *The Plovers, Sandpipers, and Snipes of the World*, University of Nebraska Press, Lincoln.

Johnson, A. W., and Goodall, J. D., 1965, *The Birds of Chile and Adjacent Regions of Argentina, Bolivia and Peru*, Vol. 1, Platt Establecimientos Graficos S.A., Buenos Aires.

Kliebe, K., 1974, Beobachtungen zur Flugbalz der Zwergschnepfe (*Lymnocryptes minimus*), *Vogelwelt* **95**:30–33.

Klingholz, F., and Meynhardt, H., 1979, Lautinventare der Saugetiere-diskret oder kontinuierlich?, *Z. Tierpsychol.* **50**:250–264.

Knudson, M. L., 1972, Functional analysis of song in the Spotted Sandpiper, *Actitis macularia* (L.), M.S. thesis, University of North Dakota, Grand Forks.

Konishi, M., 1970, Evolution of design features in the coding of species-specificity, *Am. Zool.* **10**:67–72.

Kroodsma, D. E., and Miller, E. H. (eds.), 1982a, *Acoustic Communication in Birds*, Academic Press, New York.

Kroodsma, D. E., and Miller, E. H. (eds.), 1982b, Introduction, in: *Acoustic Communication in Birds*, Vol. 1, pp. xxi–xxxi, Academic Press, New York.

Labutin, Y. V., Leonovitch, V. V., and Veprintsev, B. N., 1982, The Little Curlew *Numenius minutus* in Siberia, *Ibis* **124**:302–319.

Lemnell, P. A., 1978, Social behaviour of the Great Snipe *Capella media* at the arena display, *Ornis Scand.* **9**:146–163.

Leopold, A., and Eynon, A. E., 1961, Avian daybreak and evening song in relation to time and light intensity, *Condor* **63**:269–293.

Lind, H., 1961, *Studies on the Behaviour of the Black-tailed Godwit* (*Limosa limosa*(*L.*)), Munksgaard, Copenhagen.

Lind, H., 1965, Parental feeding in the oystercatcher (*Haematopus o. ostralegus* (L.)), *Dan. Ornithol. Foren. Tidsskr.* **59**:1–31.

Mace, T. R., 1981, Causation, function, and variation of the vocalizations of the Northern Jacana, *Jacana spinosa*, Ph.D. thesis, University of Montana, Missoula.

McGilp, J. N., and Morgan, A. M., 1931, The nesting of the Banded Stilt (*Cladorhynchus leucocephalus*), *South Aust. Ornithol.* **11**:37–53.

Maclean, G. L., 1967, The breeding biology and behaviour of the Double-banded Courser *Rhinoptilus africanus* (Temminck), *Ibis* **109**:556–569.

Maclean, G. L., 1969, A study of seedsnipe in southern South America, *Living Bird* **8**:33–80.

Maclean, G. L., 1972, Problems of display postures in the Charadrii (Aves, Charadriiformes), *Zool. Afr.* **7**:57–74.

Maclean, G. L., 1976a, A field study of the Australian Dotterel, *Emu* **76**:207–215.

Maclean, G. L., 1976b, A field study of the Australian Pratincole, *Emu* **76**:171–182.

Maclean, G. L., 1977, Comparative notes on Black-fronted and Red-kneed Dotterels, *Emu* **77**:199–207.

Maier, V., 1982, Acoustic communication in the Guinea Fowl (*Numida meleagris*): structure and use of vocalizations, and the principles of message coding, *Z. Tierpsychol.* **59**:29–83.

Mal'chevskii, A. S., 1981, New data on sound communication in birds, *Vestn. Leningr. Univ. Biol.* **4**:58–64 (in Russian).

Mal'chevskii, A. S., 1982, Biological foundations of sound communication in birds, *Zool. Zh.* **61**:1000–1008 (in Russian).

Marcström, V., 1974, On the courtship flight of the woodcock (*Scolopax rusticola* L.) in Sweden, in: *Proceedings, Fifth American Woodcock Workshop*, no pagination, Athens.

Marler, P., 1976, Social organization, communication and graded signals: The chimpanzee and the gorilla, in: *Growing Points in Ethology* (P.P.G. Bateson and R. A. Hinde, eds.), pp. 239–280, Cambridge University Press, London.

Marler, P. R., and Tenaza, R., 1977, Signaling behavior of apes with special reference to vocalization, in: *How Animals Communicate* (T. A. Sebeok, ed.), pp. 965–1033, Indiana University Press, Bloomington.

Mathew, D. N., 1964, Observations on the breeding habits of the Bronzewinged Jacana [*Metopidius indicus* (Latham)], *J. Bombay Nat. Hist. Soc.* **61**:295–301.

Maxson, S. J., and Oring, L. W., 1980, Breeding season time and energy budgets of the polyandrous Spotted Sandpiper, *Behaviour* **74**:200–263.

Miller, E. H., 1975, Social and evolutionary implications of territoriality in adult male New Zealand fur seals, *Arctocephalus forsteri* (Lesson, 1828), during the breeding season, *Rapp. P.-V. Réun. Cons. Int. Explor. Mer* **169**:170–187.

Miller, E. H., 1977, Breeding biology of the Least Sandpiper, *Calidris minutilla* (Vieill.) on Sable Island, Nova Scotia, Ph.D. thesis, Dalhousie University, Halifax.

Miller, E. H., 1979a, Functions of display flights by males of the Least Sandpiper, *Calidris minutilla* (Vieill.), on Sable Island, Nova Scotia, *Can. J. Zool.* **57**:876–893.

Miller, E. H., 1979b, Flight displays of Least Sandpipers, *Bull. Wader Study Group* **26**:44–45.

Miller, E. H., 1979c, An approach to the analysis of graded vocalizations of birds, *Behav. Neural. Biol.* **27**:25–38.

Miller, E. H., 1982, Character and variance shift in acoustic signals of birds, in: *Acoustic Communication in Birds* (D. E. Kroodsma and E. H. Miller, eds.), Vol. 1, pp. 253–295, Academic Press, New York.

Miller, E. H., 1983a, Structure of display flights in the Least Sandpiper, *Condor* **85**:220–242.

Miller, E. H., 1983b, The structure of aerial displays in three species of Calidridinae (Scolopacidae), *Auk* **100**:440–451.

Miller, E. H., 1983c, Sounds of shorebirds. 1: Introduction and methods of analysis, *Bull. Wader Study Group* **38**:35–37.

Miller, E. H., and Baker, A. J., 1980, Displays of the Magellanic Oystercatcher (*Haematopus leucopodus*), *Wilson Bull.* **92**:149–168.

Miller, E. H., Gunn, W. W. H., and Harris, R., 1983, Geographic variation in aerial song of the Short-billed Dowitcher (Aves, Scolopacidae), *Can. J. Zool.* **61**:2191–2198.

Morton, E. S., 1977, On the occurrence and significance of motivation-structural rules in some bird and mammal sounds, *Am. Nat.* **111**:855–869.

Morton, E. S., 1982, Grading discreteness, redundancy, and motivation-structural rules, in: *Acoustic Communication in Birds* (D. E. Kroodsma and E. H. Miller, eds.), Vol. 1, pp. 183–212, Academic Press, New York.

Mundahl, J. T., 1977, Role specialization in the parental and territorial behavior of the Killdeer (*Charadrius vociferus*), M.S. thesis, Utah State University, Logan.

Myers, J. P., 1979, Leks, sex, and Buff-breasted Sandpipers, *Am. Birds* **33**:823–825.
Myers, J. P., 1981, Cross-seasonal interactions in the evolution of sandpiper social systems, *Behav. Ecol. Sociobiol.* **8**:195–202.
Myers, J. P., 1982, The promiscuous Pectoral Sandpiper, *Am. Birds* **36**:119–122.
Nemetschek, G. 1977, Beobachtungen zur Flugbalz der Waldschnepfe (*Scolopax rusticola*), *J. Ornithol.* **118**:68–86.
Nethersole-Thompson, D., and Nethersole-Thompson, M., 1979, *Greenshanks*, Poyser, Berkhamsted.
Nettleship, D. N., 1974, The breeding of the Knot *Calidris canutus* at Hazen Camp, Ellesmere Island, N.W.T., *Polarforschung* **44**:8–26.
Niethammer, G., 1966, Sexualdimorphismus am Ösophagus von *Rostratula, J. Ornithol.* **107**:201–204.
Nol, E., 1980, Factors affecting the nesting success of the Killdeer (*Charadrius vociferus*) on Long Point, Ontario, M.Sc. thesis, University of Guelph, Guelph.
Norton-Griffiths, M., 1969, The organization, control and development of parental feeding in the oystercatcher (*Haematopus ostralegus*), *Behaviour* **34**:55–114.
Nowicki, T., 1973, A behavioral study of the Marbled Godwit in North Dakota, M.S. thesis, Central Michigan University, Mount Pleasant.
Oring, L. W., 1964, Displays of the Buff-breasted Sandpiper at Norman, Oklahoma, *Auk* **81**:83–86.
Oring, L. W., 1968, Vocalizations of the Green and Solitary Sandpipers, *Wilson Bull.* **80**:395–420.
Oring, L. W., 1973, Solitary Sandpiper early reproductive behavior, *Auk* **90**:653–663.
Oring, L. W., and Lank, D. B., 1982, Sexual selection, arrival times, philopatry and site fidelity in the polyandrous Spotted Sandpiper, *Behav. Ecol. Sociobiol.* **10**:185–191.
Panov, E., 1963, Taxonomic position of the Ussuri Plover *Charadrius hiaticula placidus* Grey and Gray (on the basis of ethological data), *Zool. Zh.* **42**:1546–1553 (in Russian. English translation in Josselyn van Tyne Memorial Library, Bird Division, Museum of Zoology, University of Michigan, Ann Arbor).
Parmelee, D. F., Stephens, H. A., and Schmidt, R. H., 1967, The birds of southeastern Victoria Island and adjacent small islands, *Bull. Natl. Mus. Can.* **222**.
Parr, R., 1980, Population study of Golden Plover *Pluvialis apricaria*, using marked birds, *Ornis Scand.* **11**:179–189.
Phillips, R. E., 1972, Sexual and agonistic behavior in the Killdeer (*Charadrius vociferus*), *Anim. Behav.* **20**:1–9.
Phillips, R. E., 1977, Notes on the behaviour of the New Zealand Shore Plover, *Emu* **77**:23–27.
Phillips, R. E., 1980, Behaviour and systematics of New Zealand plovers, *Emu* **80**:177–197.
Pitelka, F. A., 1959, Numbers, breeding schedule, and territoriality in Pectoral Sandpipers of northern Alaska, *Condor* **61**:233–264.
Pitelka, F. A., Holmes, R. T., and MacLean, S. F., Jr., 1974, Ecology and evolution of social organization in Arctic sandpipers, *Am. Zool.* **14**:185–204.
Potter, J., 1934, Lotus-birds found breeding on Hawkesbury River, N.S.W., *Emu* **33**:298–305.
Prater, A. J., Marchant, J. H., and Vuorinen, J., 1977, Guide to the Identification and Ageing of Holarctic Waders, British Trust for Ornithology, Guide No. 17.
Reddig, E., 1978, Der Ausdrucksflug der Bekassine (*Capella gallinago gallinago*), *J. Ornithol.* **119**:357–387.
Reed, S. M., 1981, New Zealand dotterel (*Charadrius obscurus*)—An endangered species?, *Notornis* **28**:129–132.

Richards, D. G., and Wiley, R. H., 1980, Reverberations and amplitude fluctuations in the propagation of sound in a forest: Implications for animal communication, *Am. Nat.* **115:**381–399.

Rowan, W., 1929, Notes on Alberta waders included on the British list, VII, *Br. Birds* **23:**2–17.

Rutschke, E., 1976, Zur struktur der Schallfedern von Bekassinen (*Capella*), *Beitr. Vogelkd.* **22:**12–25.

Samuel, D. E., and Beightol, D. R., 1972, Monitoring woodcock singing sites through sonagrams, *Proc. Annu. Conf. Southeast. Assoc. Game Fish Comm.* **26:**301–305.

Samuel, D. E., and Beightol, D. R., 1973, The vocal repertoire of male American Woodcock, *Auk* **90:**906–909.

Schwinn, M., 1964, Raising shorebirds in captivity, M.S. thesis, Kansas State Teachers College, Emporia.

Shepard, J. M., 1975, Factors influencing female choice in the lek mating system of the Ruff, *Living Bird* **14:**87–111.

Shissler, B. P., Samuel, D. E., and Burkhart, D. L., 1982, An aging technique for American Woodcock on summer fields, *Wildl. Soc. Bull.* **9:**302–305.

Simmons, K. E. L., 1953a, Some aspects of the aggressive behaviour of three closely related plovers (*Charadrius*), *Ibis* **95:**115–127.

Simmons, K. E. L., 1953b, Some studies on the Little Ringed Plover, *Avic. Mag.* **59:**191–207.

Skeel, M. A., 1976, Nesting strategies and other aspects of the breeding biology of the Whimbrel (*Numenius phaeopus*) at Churchill, Manitoba, M.Sc. thesis, University of Toronto, Toronto.

Skeel, M. A., 1978, Vocalizations of the Whimbrel on its breeding grounds, *Condor* **80:**194–202.

Smith, S., and Hosking, E., 1955, *Birds Fighting: Experimental Studies of the Aggressive Displays of Some Birds*, Faber & Faber, London.

Smith, W. J., 1977, *The Behavior of Communicating: An Ethological Approach*, Harvard University Press, Cambridge, Mass.

Sordahl, T. A., 1979, Vocalizations and behavior of the Willet, *Wilson Bull.* **91:**551–574.

Sordahl, T. A., 1980, Antipredator behavior and parental care in the American Avocet and Black-necked Stilt (Aves: Recurvirostridae), Ph.D. thesis, Utah State University, Logan.

Sordahl, T. A., 1981, Predator-mobbing behaviour in the shorebirds of North America, *Bull. Wader Study Group* **31:**41–44.

Stead, E. F., 1932, *The Life Histories of New Zealand Birds*, Search, London.

Stettenheim, P. R., 1976, Structural adaptations in feathers, in: *Proceedings of the 16th International Ornithological Congress* (H. J. Frith and J. H. Calaby, eds.), pp. 385–401, Australian Academy of Science, Canberra.

Sutton, G. M., 1932, The birds of Southhampton Island, *Mem. Carnegie Mus.* **12**(Part 2, Sect. 2).

Sutton, G. M., 1981, On aerial and ground displays of the world's snipes, *Wilson Bull.* **93:**457–477.

Swanberg, P. O., 1965, Studies of less familiar birds. 138. The Great Snipe, *Br. Birds* **63:**504–508.

Thomas, D. W., and Dilworth, T. G., 1980, Variation in *peent* calls of American Woodcock, *Condor* **82:**345–347.

Thönen, W., 1968, Auffallender Unterschied zwischen den instrumentalen Balzlauten der europaischen und nordamerikanischen Bekassine *Gallinago gallinago*, *Ornithol. Beob.* **65**:6–13.

Tikhonov, A. V., and Fokin, S. Y., 1979, Acoustic signalling and behavior of shorebirds in early ontogenesis. I. Prenatal stages of development, *Biol. Nauki* **10**:33–40 (in Russian).

Tikhonov, A. V., and Fokin, S. Y., 1980, Acoustic signalization and the behavior of shorebirds in early ontogenesis. II. Signalization and the behavior of nestlings, *Biol. Nauki* **10**:45–54 (in Russian).

Tikhonov, A. V., and Fokin, S. Y., 1981a, Lek and calls of shorebirds, *Ornitologia* **16**:64–72 (in Russian).

Tikhonov, A. V., and Fokin, S. Y., 1981b, Acoustic signaling and behavior during the nesting period in shorebirds, *Byull Mosk. Ova Ispyt. Prir. Otd. Biol.* **86**:31–42 (in Russian).

Tikhonov, A. V., and Fokin, S. Y., 1982, Acoustic signaling and behaviour of waders in early ontogenesis, in: *Ornithological Studies in the U.S.S.R.* (V. M. Gavrilov and R. L. Potapov, eds.), vol. 2, pp. 262–273, Nauka, Moscow.

Trimble, R., 1940, Habits, in: *Birds of Western Pennsylvania* (by W. E.C. Todd), pp. 216–218, University of Pittsburgh Press, Pittsburgh.

Tuck, L. M., 1972, The snipes: A study of the genus *Capella*, *Can. Wildl. Serv. Monogr.* No. 5.

van Rhijn, J. G., 1973, Behavioural dimorphism in male Ruffs, *Philomachus pugnax* (L.), *Behaviour* **47**:153–229.

van Rhijn, J. G., 1983, On the maintenance and origin of alternate strategies in the Ruff *Philomachus pugnax*, *Ibis* **125**:482–498.

van Tets, G. F., D'Andria, A. H., and Slater, E., 1967, Nesting distribution and nomenclature of Australasian vanelline plovers, *Emu* **67**:85–93.

Veprintsev, B. N., and Zablotskaya, M. M., 1982, Acoustic signals of the Little Curlew *Numenius minutus*, Pushchino (in Russian).

von Frisch, O., 1956, Zur Brutbiologie und Jugendentwicklung des Brachvogels (*Numenius arquata* L.), *Z. Tierpsychol.* **13**:50–81.

Wallis, C. A., and Wershler, C. R., 1981, Status and breeding of Mountain Plovers (*Charadrius montanus*) in Canada, *Can. Field-Nat.* **95**:133–136.

Walters, J. R., 1980, The evolution of parental care in lapwings, Ph.D. thesis, University of Chicago, Chicago.

Walters, J. R., 1982, Parental behavior in lapwings (Charadriidae) and its relationships with clutch sizes and mating systems, *Evolution* **36**:1030–1040.

Warham, J., and Bell, B. D., 1979, The birds of Antipodes Island, New Zealand, *Notornis* **26**:121–169.

Warner, R. W., 1969, The anatomy of the avian syrinx, Ph.D. thesis, University of London, London.

Weeden, R. B., 1965, Further notes on Wandering Tattlers in central Alaska, *Condor* **67**:87–89.

Weeks, H. P., Jr., 1969, Courtship and territorial behavior of some Indiana woodcocks, *Proc. Indiana Acad. Sci.* **79**:162–171.

Weir, N. L., 1979, The use of sonagraphic analysis of the American Woodcock (*Philohela minor*) peent call as a method for identifying individual males, M.S. thesis, Pennsylvania State University, University Park.

Weir, N. L., and Graves, H. B., 1982, Discriminant analysis of the peent call for identification of individual male American Woodcock, in: *Woodcock Ecology and Management* (T. J. Dwyer and G. L. Storm, eds.), pp. 34–39, *U.S. Fish Wildl. Serv. Wildl. Res. Rep.* 14.

Wiley, R. H., and Richards, D. G., 1982, Adaptations for acoustic communication in birds: Sound transmission and signal detection, in: *Acoustic Communication in Birds* (D. E. Kroodsma and E. H. Miller, eds.), Vol. 1, pp. 131–181, Academic Press, New York.

Williamson, K., 1943, The behaviour pattern of the Western Oyster-catcher (*Haematopus ostralegus occidentalis* Neumann) in defence of nests and young, *Ibis* **85**:486–490.

Williamson, K., 1946, Field-notes on the breeding-biology of the Whimbrel, *North West. Nat.* **21**:167–184.

Williamson, K., 1948, Field-notes on nidification and distraction-display in the Golden Plover, *Ibis* **90**:90–98.

Williamson, K., 1950, The distraction behaviour of the Faeroe snipe, *Ibis* **92**:66–74.

Williamson, K., 1952, Regional variation in the distraction displays of the oyster-catcher, *Ibis* **94**:85–96.

Witherby, H. F., Jourdain, F. C. R., Ticehurst, N. F., and Tucker, B. W., 1941, *The Handbook of British Birds*, Vol. 4, Witherby, London.

Wollemann, M., 1978, Two kinds of the alarm sound of the Black-winged Stilt (*Himantopus himantopus*), *Puszta* **7**:22–23 (in Hungarian).

Wollemann, M., and Olaszy, G., 1977, Spectrogram analysis of different alarm calls in gulls and waders, *Agressologie* **18**:97–102.

Zvonov, B. M., and Tikhonov, A. V., 1981, Denervation of the syrinx and the structure of radiated sounds in non-Passeriformes, *Izv. Akad. Nauk SSSR Ser. Biol.* **6**:924–930 (in Russian).

Chapter 7

THE EVOLUTION OF PARENTAL BEHAVIOR AND CLUTCH SIZE IN SHOREBIRDS

Jeffrey R. Walters*

Allee Laboratory of Animal Behavior
The University of Chicago
Chicago, Illinois 60637

I. INTRODUCTION

In this chapter I consider two aspects of shorebird reproduction that are intimately related as components of life-history strategies: clutch size and parental behavior. That shorebirds with few exceptions do not feed their young, and that young are precocial, nidifugous, and relatively self-reliant, has led to the assumption that adults have little difficulty caring for their young. That is, it is generally assumed that parental care imposes few constraints on time and energy budgets of adult shorebirds (Kendeigh, 1952; Parmelee and Payne, 1973; Graul, 1973; Emlen and Oring, 1977; Welty, 1982, p. 293). This assumption has no empirical base, and data I collected from lapwings (Charadriidae, Vanellinae) indicate that parental care in fact imposes considerable demands on adults in at least some species. This justifies a reexamination of the parental care of shorebirds and of the assumption that clutch sizes are not limited by the ability of adults to care for young. This is further prompted by the lack of a satisfactory explanation of the limitation of clutch size in shorebirds.

In this context I present data on the parental behavior of lapwings, review parental behavior and clutch size in shorebirds generally, and consider the evolution of these components of life history within this group.

* Present address: Zoology Department, North Carolina State University, Raleigh, North Carolina 27695.

II. PARENTAL BEHAVIOR

Parental care is defined as behavior directed by adults toward young that promotes the welfare of the young. Considering the volume of quantitative data on parental behavior collected from altricial birds, the absence of comparable data for precocial species like shorebirds is striking. These data are lacking because the things precocial birds do for their young (e.g., protect them from predators) are generally more difficult to quantify than is feeding young, the parental behavior most often measured in altricial birds. The difficulties of quantifying the cost of parental behavior to adults are magnified by the elusive behavior of adults with young, their habitation of environments in which observation is difficult, and their mobility. The lack of data is evident in Ricklefs' comparison of the energetic costs of reproduction in precocial and altricial birds over the various stages of the breeding cycle: only for the nestling stage of the precocial species was no estimate of energetic cost possible (Table 31 in Ricklefs, 1974).

A. Studies of Lapwings

Lapwings comprise a distinct subfamily of plovers containing 24 species (Bock, 1958; Morony *et al.*, 1975). Most live in open habitats with short vegetation and are conspicuous, bold, and sedentary when breeding. Hence, they are more readily observed than most precocial species. The legendary alertness, aggressiveness, and incessant alarm-calling of these birds when breeding (Crawshay, 1907; Hudson, 1920; Wetmore, 1926; Archer and Godman, 1937; Jackson, 1938) indicate that parental care is well developed and natural history studies substantiate this (Bent, 1929; Laven, 1941; Skead, 1955; Hall, 1959, 1964; Jayakar and Spurway, 1965a,b, 1968; Little, 1967; Thomas, 1969; Barlow *et al.*, 1972; Maclean, 1972a). Lapwings are thus ideal subjects for a quantitative analysis of parental behavior.

I studied four species of lapwings in order to (1) quantitatively document the forms that parental behavior takes and (2) obtain an estimate of the cost to adults of caring for young.

1. Methods

Study Areas and Populations. I studied lapwings at two sites. Long-toed Lapwings (*Vanellus crassirostris*), Blacksmith Plovers (*V. armatus*), and Crowned Plovers (*V. coronatus*) were studied in Amboseli National Park, Kenya (3°S, 37°E) during 14 months spanning 3 years (July–August 1975; August 1976–July 1977; December 1978). Southern Lapwings (*V. chilensis*) were studied for 4 months (April–July 1978) near Mantecal, Venezuela (7°N, 69°W).

Amboseli is comprised of flat open savannah, with extensive short-grass plains, open *Acacia xanthophloea* woodlands, and denser *A. tortilis* woodlands. Although rainfall is seasonal and sparse (250–300 mm/year), permanent water is available at several waterholes and two large marshes (see Altmann and Altmann, 1970; Western, 1973). Mean daily temperature varies little from month to month in Amboseli (20°C to 25°C), but because of its elevation (1100 m), diurnal variation is large, typically about 20°C.

The Venezuelan site is a flat, open, seasonally flooded llanos, with scattered patches of forest and many small ponds. Open bajio and estero areas are especially common (see Ramia, 1967; Troth, 1979). The rainy season extends from April to October, and rainfall averages about 1500 mm/year. Mean daily temperatures are high, and vary relatively little monthly (24°C to 30°C). Diurnal variation is also slight, typically 10°C, due to the low elevation of the site (see also Eisenberg, 1979; Troth, 1979).

Demographic data and qualitative data on parental behavior were obtained by monitoring a local population of each species. For all species except Crowned Plovers, additional quantitative behavioral data were obtained by systematic observational sampling of a small segment (10–20 pairs) of the population (Walters, 1980).

Behavioral Sampling. No birds were banded, but adult Long-toed Lapwings and Southern Lapwigs were individually identifiable from slight variations in plumage. Neither species is sexually dimorphic, but sex could be determined from copulation position, which was consistent within pairs and has been shown to be a reliable indicator in a marked population of lapwings (*V. miles*) in Australia (Barlow *et al.*, 1972).

Time budgets and spatial relationships within families were estimated from scan samples (Altmann, 1974) of position and behavior taken 30 min apart. Each such sample included one or two pairs of lapwings and their young, and every few hours sampling was rotated among the 10–20 pairs observed. The behavioral categories used are defined in the Appendix. Approximately equal numbers of samples were collected during all daylight hours. Observations were made at close range from a vehicle that functioned as a blind, using binoculars.

The intersample interval is sufficient to ensure statistical independence of successive samples (Walters, 1980). Furthermore, because lapwings are highly visible and sedentary, individuals designated to be sampled were rarely out of sight. Estimates of time budgets and spatial relationships are therefore relatively free of the biases that usually plague such estimates, and their accuracy is a simple function of sample size. Furthermore, breeding was asynchronous, so that all stages of the breeding cycle could be sampled simultaneously within a population. This greatly reduces confounding effects of seasonality on changes in behavior over the breeding cycle, simplifying interpretation of results.

Interspecific aggression and antipredator behavior were examined by focal sampling (Altmann, 1974) performed during the interval between scan samples. Samples were 10–20 min in duration, depending on species, and included one or two pairs. For all instances of these behaviors, the following were recorded: species of predator or intruder; behavioral response of lapwing adults and young to predator or intruder; effect of the lapwings' response on the predator or intruder; distance between lapwing and predator or intruder when the response occurred.

Spatial relationships were examined by analysis of variance of interindividual distances using a square-root transformation (SPSS, Nie *et al.*, 1975), and time budgets were compared by chi-square analyses of raw frequencies.

2. Parental Behaviors

I first briefly describe the parental behaviors observed and then in the following sections assess their benefits to young and costs to adults (see also Walters, 1980).

Brooding. Chicks were brooded frequently during the first week after hatching, and during inclement weather for several weeks thereafter. The frequency of brooding varied greatly among species according to weather conditions encountered during breeding. Brooding was most common in the Southern Lapwing, the only species that bred during the wet season.

Antipredator Behavior. Lapwings protected their young from predators by distraction displays, alarm calling, and aggression. Distraction consisted primarily of false incubation or running in a crouched posture while calling loudly; under estreme duress, Southern Lapwings performed an injury-feigning display. Swooping was the primary form of aggression, whether predators were on the ground or in the air.

The general antipredator behavior of lapwings was to detect dangerous predators, induce a response from their young through communicatory behavior, and provide additional protection of young through predator-directed behaviors. Within this framework there was complex variability. Alarm calls were complex, perhaps forming a graded series. Furthermore, adult response varied in consistent ways with species, location, and behavior of the predator, and with stage of lapwing breeding cycle. The patterns evident in the more than 400 encounters with potential predators sampled indicate that adults have sophisticated discriminatory abilities, and may even be able to evaluate the danger to young inherent in each encounter with a potential predator and respond appropriately (Walters, 1980). For example, adults responded to falcons and harriers (*Circus* sp.) when these predators were still far away (typically > 100 m), but ignored vultures, snail-eating kites, and fish-eating raptors unless they were within a few meters of the young. Southern Lapwings emitted alarm calls in response to avian and mammalian predators that caused young to crouch and remain motionless, but emitted a much different alarm call in all 11 observed encounters with reptilian predators, and it caused the young to flee and remain upright. Certain species of potential predator evoked responses only during particular stages of the breeding cycle and were ignored during others. What is known of the behavior and diets of the various predators and the consequences of the various antipredator behaviors suggests that the variation in response to predators by lapwings is functional, i.e., it reflects the degree and kind of danger represented by the predator (Walters, 1980).

Interspecific Aggression. The selectivity, location, and timing of aggression by lapwings toward other species of birds intruding on their territories indicated that such aggression was a form of parental behavior (Walters, 1979, 1980). Such aggression was most frequent when young were present, and was associated spatially with the young. The mean frequency of interspecific aggression by lapwings tending downy chicks ranged from 3 to 20 attacks per pair per hour, depending on species.

Vigilance. Vigilant behavior, i.e., alert and rest-alert behavior (Appendix), was strongly associated with the presence of young, although it occurred infrequently in other circumstances (see below). Vigilance was directed both toward the young and toward the environment.

Foraging Away from Young. Adults moved away from their young to forage. This avoidance of young was specific to foraging, but was neither a necessary consequence of the nature of foraging nor a reflection of differences in microhabitat preference. I therefore interpret avoiding foraging near young as a parental behavior.

The specificity of the avoidance is evidenced by data on interindividual distances. Adult Long-toed Lapwings foraged far from downy young, but nonforaging adults stayed close to young regardless of chick behavior (Fig. 1). Adults ceased foraging away from young as the young matured, and eventually were as near to them when foraging as when not foraging. (Other parental behaviors waned with chick age similarly; see below.) Neither did adults forage away from their mates when young were not present (Fig. 1).

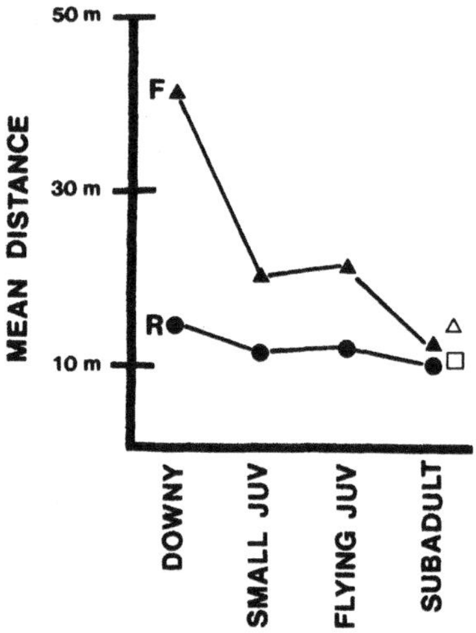

Fig. 1. Distance between adult Long-toed Lapwings and their young as a function of adult behavior and stage of the breeding cycle. Values are given for both foraging (F) and loafing (R; resting, sitting, or preening) adults. Stages of the breeding cycle are identified according to the development of the young. Downy plumage persisted for 3 to 4 weeks after hatching (DOWNY stage) until juvenile plumage was acquired (SMALL JUV stage). Attainment of flight at 6 to 8 weeks marked the beginning of a second juvenile phase (FLYING JUV stage), which ended with the appearance of first basic plumage at about 4 months (SUBADULT stage). The distances between a foraging adult and its mate (△) and between a loafing adult and its mate (□) during the period of territory defense prior to breeding are given for comparison. Sample sizes per point ranged from 9 to 118, with a mode of 32, and standard deviations of raw data were of similar magnitudes as mean values. ANOVA performed on transformed data revealed significant effects of stage and adult behavior, and a significant interaction.

The other species exhibited a similar tendency to avoid foraging near small young. Adult Southern Lapwings continued to forage away from young even after the young were able to fly (Fig. 2).

There was no indication that adults and young foraged in different habitats. Adults and young often foraged in the same locations, but at different times (Fig. 3). In cases in which the young foraged on only a

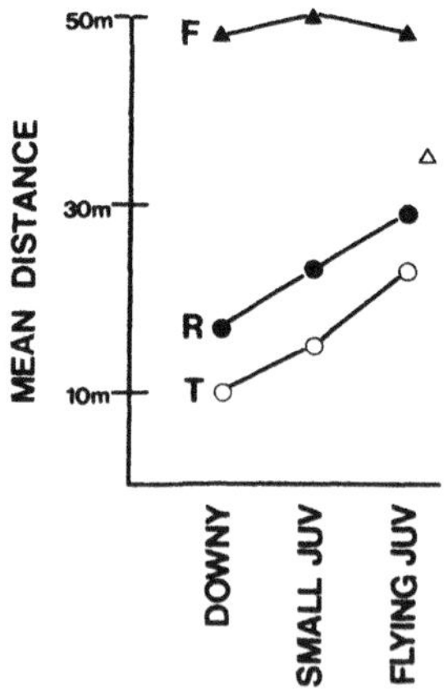

Fig. 2. Distances between adult Southern Lapwings and their young as a function of adult behavior and stage of the breeding cycle. Distances are given for foraging adults (F), loafing adults (R), and adults engaged in parental behavior (T). Stages of the breeding cycle are defined in Fig. 1. The distance between a foraging adult and its mate during the period of territory defense prior to breeding is given for comparison (△). Sample sizes ranged from 24 to 96, with a mode of 39, and standard deviations of raw data were of roughly equivalent magnitude as mean values. ANOVA of transformed data revealed significant effects of stage and adult behavior, and a significant interaction.

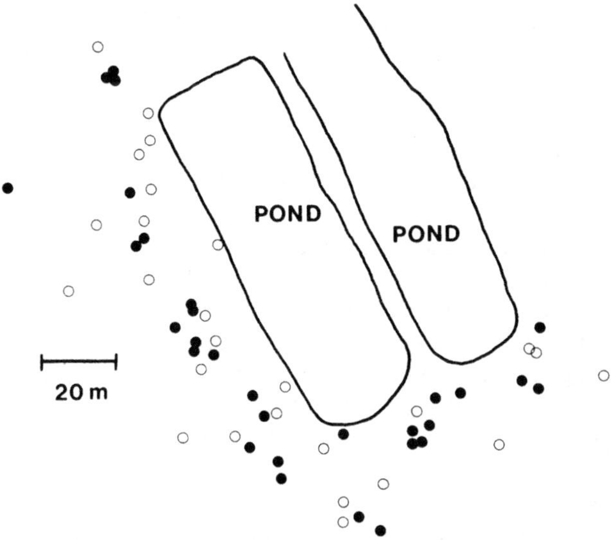

Fig. 3. Distribution of members of a Southern Lapwing family when foraging during the downy stage (defined in Fig. 1). Adults (○) and young (●) foraged in the same areas, although at different times. Each point represents one individual observed foraging during a scan sample. The family was comprised of two adults and two young.

restricted portion of a territory, adults had used that portion before young were present (Walters, 1980).

The possibility of differential microhabitat requirements was further investigated in Long-toed Lapwings. The habitat was divided into grid units of known vegetation type. For all vegetation types, a grid unit was used less by adults if it was also used by the young than if the young never foraged in it. Furthermore, relationships between vegetation preference and vegetation-specific rates of prey capture attempts were nearly identical in adults and young (Walters, 1980).

These data suggest that avoidance of young by foraging adults was deliberate. Behaviorally it was accomplished by flying or rapidly walking away from young when switching from tending to foraging behavior (see below). The behavior of adults when they did encounter prey in the vicinity of the young may also be revealing. In this situation, adults gave an impression of ambivalence, perhaps between desire to feed and inhibition to feed: strikes at prey were often aborted, and vocalizations indicative of mild alarm were emitted, events that were rarely observed in other foraging contexts.

Showing Food. Only Crowned Plovers showed food to their young. This distinctive behavior involved a unique posture and call that attracted chicks to an adult that had discovered a food source, typically a dung pat containing insect larvae. Once the young arrived and began feeding, the adult moved away. Except for the call, the adult's behavior was reminiscent of the inhibited foraging exhibited by other species when near young.

Leading, Following, and Gathering Young. These activities were characterized by alert postures and calling that attracted the young. They served to keep chicks near their parents. In leading, adults preceded chicks and often they flew back and forth between the young and a destination, calling, ceasing only when the young finally reached the area. Such episodes were clearly goal-directed, and functioned to move young to new foraging areas. Leading was also employed in short-distance movements.

Adults sometimes maintained proximity with chicks by following them, employing the same calls and postures as when leading. Gathering refers to similar behavior that lacked clear orientation to a goal or a particular chick. The adult moved about, calling, until all the chicks were assembled. Loss of visual contact with chicks and increasing separation between adults and young seemed to trigger this behavior.

3. Benefits of Parental Behaviors to Young

Benefits of parental behaviors to young generally were not documented. I state the following possible benefits as hypotheses, based on the research of others and speculation.

Brooding helps chicks thermoregulate, and is particularly important in shorebirds because thermoregulatory ability develops later in their chicks than in those of other precocial birds (Norton, 1973).

The obvious benefit to young of adult antipredator behavior is that it reduces predation on them. There is some evidence that antipredator behavior is effective: Goransson *et al.* (1975) found that untended eggs placed near nests of Northern Lapwings (*V. vanellus*) were less likely to be taken by predators, in this case chiefly gulls, than those placed farther from lapwing nests. Perhaps young also benefit from adults' monitoring the environment and making decisions about appropriate responses to potential predators for the young, because this enables chicks to concentrate on foraging. Finally, the young may learn to discriminate dangerous predators from other species by observing the responses of their parents.

Interspecific aggression may reduce competition for food experienced by young, and thus enable them to avoid starvation, grow faster, or spend less time foraging and hence exposed to predators (Orians and Willson, 1964; Walters, 1979). Interspecific aggression may also reduce predation on the young if the species lapwings attack attract predators, which may then discover the lapwing chicks (Myers, 1980).

Benefits of vigilance are likely manifested through other parental behaviors: it presumably functions in detecting predators and intruders on the territory, and in monitoring the young.

Foraging away from young may reduce predation on young if moving adults attract predators (Pitelka *et al.*, 1974). This behavior may also reduce competition for food between adults and young, which may benefit young as outlined above.

Showing food to young presumably increases their foraging efficiency. Leading behavior, and perhaps even following and gathering may also increase the foraging efficiency of chicks by guiding them to suitable foraging areas. Leading, following, and gathering behavior also keep chicks in proximity to a tending adult, which may benefit the young in several ways. The probability that a predator will approach a chick closely enough to discover it before being detected by the adults may increase with adult–young distance (Safriel, 1975). So too may the probability that a chick will become lost. Becoming lost is apparently a realistic and fatal possibility in a variety of precocial birds (Ridpath, 1972; Beaver, 1978), including shorebirds (Parmelee *et al.*, 1968; Lenington, 1980).

4. Costs of Parental Behaviors to Adults

Most costs to adults of parental behaviors, like benefits to young, were not documented. However, I did quantify one cost, time adults spent performing parental behaviors. This time could be partitioned into two components: (1) time spent being vigilant and (2) time spent performing other more active parental behaviors (antipredator behavior, interspecific aggression, brooding, showing food, leading, following, and gathering young) (Walters, 1982). The increased time devoted to these activities when young were present was accompanied by a sizable reduction in time spent foraging relative to the nonbreeding (territorial) stage in both Southern Lapwings and Long-toed Lapwings, but not in Blacksmith Plovers (Fig. 4). This reduction in foraging cannot be attributed to seasonal changes in food availability, because the various stages of the breeding cycle were sampled simultaneously in each population. Instead, it suggests that the demands of parenting constrain adults, raising the possibility of energetic stress. In all species, time spent loafing (i.e., preening, rest-

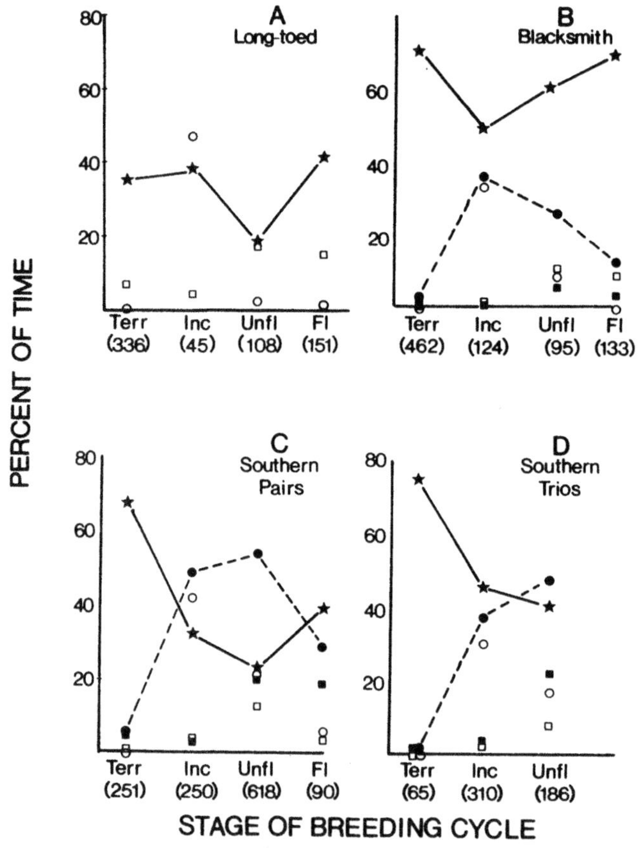

Fig. 4. Percent of time devoted to parental behavior (●) and foraging (★) as a function of the breeding cycle in (A) Long-toed Lapwings, (B) Blacksmith Plovers, (C) Southern Lapwings breeding in pairs, and (D) Southern Lapwings breeding in cooperative trios. The downy and small juvenile stages (defined in Fig. 1) are combined as the unfledged young stage (Unfl), and the flying juvenile and subadult stages are combined as the fledged young stage (Fl). The other stages are incubation (Inc) and the territorial stage (Terr), the period of territory defense prior to breeding and between breeding attempts. Parental behavior is subdivided into active parental behaviors (○), extreme vigilance (alert of Appendix) (□), and moderate vigilance (rest-alert of Appendix) (■). No estimate of moderate vigilance and hence total parental behavior was obtained for the Long-toed Lapwing. Sample sizes are given in parentheses; 95% confidence intervals of the estimates ranged from ± 1% to ± 15%, with most being ± 4–5%. In all four cases, the time budget is not independent of stage of breeding cycle (X^2, $p < 0.05$). [From Walters (1982).]

ing, or sitting) decreased to an apparent minimum level with the onset of incubation, and thereafter changes in time devoted to parental behavior were directly reflected in foraging time (see Fig. 8).

Antipredator behavior and interspecific aggression probably require considerable energy, but most other parental behaviors probably do not. Therefore, the time required for parental behavior likely is more constraining than the energy required.

There are other possible costs of parental behavior. First, adults risk their lives when attacking predators. Myers (1978) observed the capture and subsequent eating of a Southern Lapwing by a Crested Caracara (*Polyborus plancus*) during an attack by the lapwing upon the caracara (see also Furrer, 1976).

Another possible cost is that avoiding foraging near young may reduce adult foraging efficiency. Also, adult behavior followed a diurnal cycle prior to breeding, but not when young were present. That the diurnal structure of time budgets reverted to the nonbreeding pattern once the young could fly suggests that restructuring activity when young are present is costly to adults. For example, being active at midday, which was characteristic only of adults tending young, might increase costs associated with thermoregulation (Walters, 1980; see also Gibson, 1978).

Finally, vigilance may interfere with other activities, especially foraging. These activities are not compatible because vigilance requires attentiveness toward the young and the surrounding environment, whereas foraging requires attentiveness toward the ground. Therefore, a bird alternating rapidly between these two activities (see below) might be an inefficient forager.

5. Tending Strategies

One striking aspect of parental care in lapwings, besides the behaviors employed, was the coordination between mates. I have described this coordination as a division of labor via two distinct modes of behavior, termed the "off-duty" and "tending" roles (Walters, 1982). An off-duty adult stays away from the young and forages, or occasionally loafs or is vigilant, whereas the tending adult stays near the young, is vigilant, performs active parental behaviors, or loafs, depending on the activity of the young, and only rarely forages. This division of labor alleviated the obvious conflict between foraging away from young and tending young.

This coordination between mates may be discerned from patterns in how an adult's behavior was conditional on the behavior of its mate. Simultaneous responses by mates to important stimuli such as predators and intruders on the territory should make behavior slightly more syn-

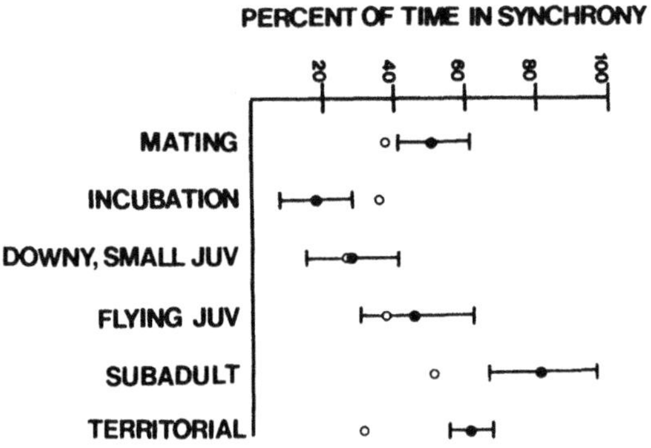

Fig. 5. The percent of time that adult Blacksmith Plovers performed the same activity as their mate as a function of the breeding cycle. Expected values (○), observed values (●), and binomial 95% confidence intervals around the latter are presented. The expected value is the sum, over the behaviors included in the analysis, of the squares of the unconditional probabilities that an adult was in each behavioral state, i.e., the proportion of time spent in that state. The expected value is thus the amount of synchrony expected by chance, i.e., without coordination between mates. If the expected value is significantly different from the observed value (i.e., outside the 95% confidence interval), coordination of activity between mates is indicated. Behavioral states included in the analysis are foraging, loafing (resting, sitting, or preening), alert, away from territory, and other. Similar results were obtained for each of the other species.

chronous than expected by chance in the absence of active coordination between mates. Mates were quite synchronous in their activity when not breeding (Fig. 5, territorial stage), more so than could be accounted for by simultaneous responses to common stimuli. Incubation, of course, introduced an asychronous coordination between mates as they (both incubate) alternated between incubating and foraging. Mates failed to return to synchronous activity once the eggs hatched; the source of asynchrony during this stage (Fig. 5, downy–flying juvenile) was an alternation between off-duty (foraging) and tending behavior. Both adults sometimes tended simultaneously, but almost never were both foraging away from the young simultaneously. Mates were again synchronous in their activity only when parental behavior waned (Fig. 5, subadult stage).

Spatial relationships provide similar evidence of an alternation of roles. One can compute, based on the proportion of time each adult was within a certain distance of the young, the proportion of time both adults were expected to be within that distance, or beyond that distance, simultaneously if the locations of adults with respect to young were independent of one another. I used 10 or 25 m as the critical distance, depending on how closely young were tended in a particular species. When small young were present, adults were both near or both away less often than expected in all species, but no such pattern was evident once the young were able to fly (e.g., Table I).

Table I. Percent of Time Both Adults Were Near or Both Were Away from Young in Southern Lapwings

Stage[a]	Type value	Both < 10 m	Both < 25 m	Both > 25 m	Sample size
Downy	Observed	17[b]	43[b]	2[b]	248
	Expected	22	49	9	
Small juv	Observed	5[b]	38	8	130
	Expected	11	45	11	
Fledged juv	Observed	5	14	32	78
	Expected	10	19	32	

[a] Stages are described in Figs. 1 and 4.
[b] Indicates values significantly different from expected values (binomial test, $p = 0.05$).

Although the species I studied were generally similar in their parental behavior, there were important differences between them, especially in the behavior of the tending (as opposed to off-duty) adult. Southern Lapwings and Crowned Plovers were active tenders: they followed, led, and gathered young constantly, and thus kept near them. In contrast, Blacksmith Plovers and Long-toed Lapwings were inactive tenders: the tending adult was essentially stationary, as it only moved to a new position nearer its young occasionally.

Differences in the tending strategies of these species were manifested in spatial relationships between adults and young, and time devoted to tending behaviors. Distance between nearest (tending) adult and young was greater in inactive tenders than active tenders, and the difference was intensified when young were active, as when foraging (Walters, 1982). Inactive tenders, of course, spent less time performing active parental behaviors, but devoted as much time to vigilance as active tenders. Total time expended on parental behavior was higher in active tenders (Fig. 4).

The Blacksmith Plover differed from all other species in that the tending adult often foraged. The off-duty adult still moved far from the young and foraged, but the adult that remained with the young often foraged too, unlike in Long-toed Lapwings and Southern Lapwings. The tending adult performed some foraging in Crowned Plovers, and this species was unique in showing food to young and in failing to perform interspecific aggression.

These differences between species are mostly quantitative. That is, species differed primarily in the frequency with which various behaviors from a common repertoire were employed. For example, Long-toed Lapwings were inactive tenders, and thus normally did not lead young or call young, but they possessed a call and posture for leading like those of active tenders, and employed them in exceptional circumstances. Among the parental behaviors exhibited by lapwings, only food-showing was completely absent from the repertoires of some species. Hence, inactive and active tending represent two ends of a continuum rather than a dichotomy.

6. Evolution of Parental Behavior

I now develop an interpretive model of the evolution of parental behavior based on hypothesized effects of several environmental features on the proposed benefits to young of these behaviors. The model is speculative, and is intended as a framework for future research, not a description of results.

Before presenting the model, it is necessary to summarize information about the habitat of each species, emphasizing those aspects included in the model. These are (1) abundance of food, (2) openness of habitat, and (3) abundance of terrestrial predators. Relative abundance of terrestrial predators was estimated from rate of encounters with such species in focal samples. The other two features were not measured. Relative abundance of food was judged from the density of lapwings and other insectivores (which was measured repeatedly) in the habitats of each lapwing species. Relative openness of habitat was judged from my assessment of the amount of obstructing vegetation present (see also Hall, 1959, 1964; Skead, 1955). The model thus considers only very gross differences in habitat features.

Differences between species in these habitat features and in parental behavior are summarized in Table II. Crowned Plovers inhabited dry, open plains far from water. This was the most open habitat, and it contained few terrestrial predators and sparse food. Blacksmith Plovers inhabited open areas of various kinds adjacent to water. This habitat was nearly as open as that of Crowned Plovers, and contained more food and more terrestrial predators. Southern Lapwings occupied a habitat in Venezuela roughly equivalent to that of Blacksmith Plovers in Kenya. However, Southern Lapwing habitat was not as open as that of Blacksmith Plovers (but see below). Long-toed Lapwings inhabited surface vegetation within marshes, an unusual, distinct habitat that was open, rich in food, and of course devoid of terrestrial predators.

Activeness of Tending. The model focuses on adult–young distance because of the hypothetical influence of this variable on the effectiveness of antipredator behavior and the probability of young becoming lost (see above). Adults control adult–young distance through activeness of tending: the more actively they tend, the smaller the average (tending) adult–

Table II. Features of Parental Behavior and Ecology of Lapwing Species

Lapwing species	Parental behaviors[a]				Environmental factors		
	Showing food	Active tending	Interspecific aggression	Feeding by tending adult	Abundant food	Open habitat	Abundant terrestrial predators
Long-toed	−	−	+	−	+	+	−
Blacksmith	−	−	+	+	0	+	+
Southern	−	+	+	−	0	0	+
Crowned	+	+	−	−	−	+	0

[a] Parental behavior and environmental factors are described in text. For parental behaviors, a + indicates that the behavior is present in the species, a − that it is absent. For environmental factors, the development of a factor is rated as +, 0, or − for each species, + indicating highest development, − lowest development, and 0 intermediate.

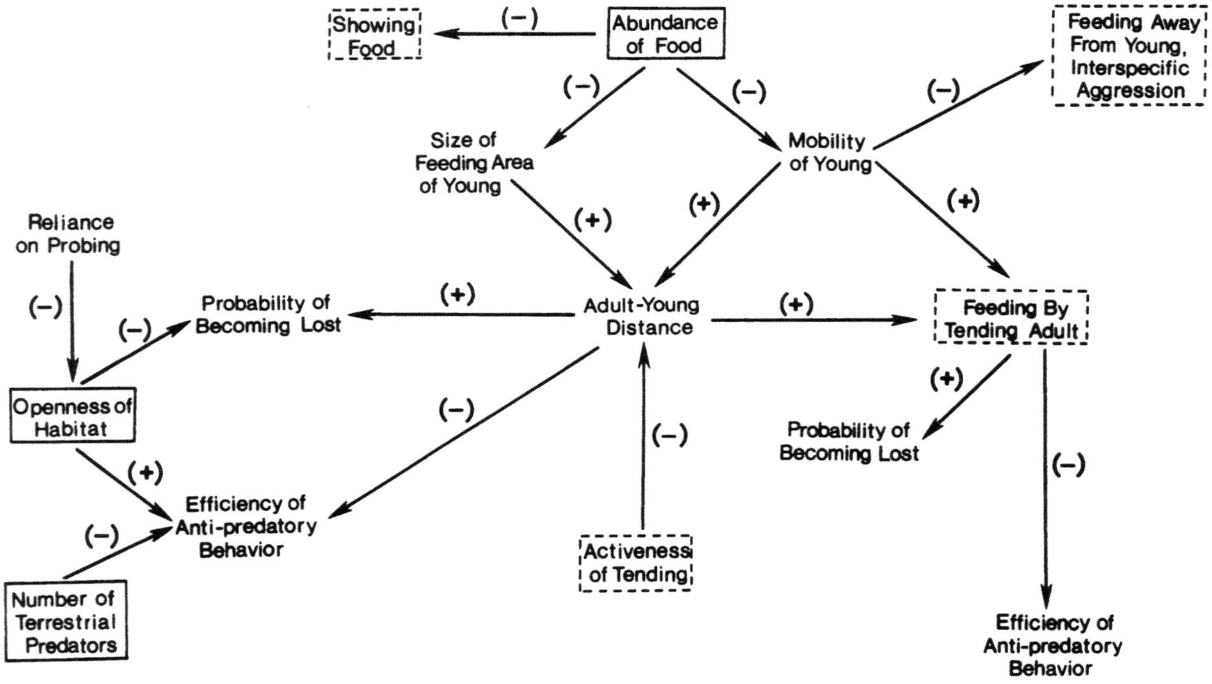

Fig. 6. Model of the evolution of lapwing parental behavior. Parental behaviors are outlined by broken lines, environmental factors by solid lines. A (+) denotes a direct relationship between two variables, a (−) an inverse relationship. For example, A $\xrightarrow{(+)}$ B indicates that an increase in A causes an increase in B, whereas A $\xrightarrow{(-)}$ B indicates that an increase in A causes a decrease in B.

young distance (Fig. 6). Several ecological factors are expected to increase adult–young distance, independent of adult behavior. Chief among these factors is sparse food, because it increases mobility of young by forcing them to forage over broader areas. These factors are expected to lead to more active tending, because this is the adults' means to balance environmental effects on adult–young distance. If two species exhibit similar adult–young distances, that with the sparsest food supply must tend more actively to maintain it.

Other ecological factors may reduce the need to stay near young, and hence allow adults to tend less actively. These factors include high visibility (openness of habitat) and low density of terrestrial predators. [Aerial predators are not considered because they were detected at such great distances that a small change in adult–young distance likely has a much smaller effect on exposure to these predators than to terrestrial predators (Walters, 1980).] The balance here is on probability of becoming lost and efficiency of antipredator behavior (Fig. 6). Parents can allow their young to stray without risk of predation if there are few terrestrial predators or if high visibility allows them to detect predators at great distances. Also, parents can more easily monitor straying chicks in open habitats, and hence can prevent them from becoming lost without staying as near to them.

This model can account for differences in activeness of tending among the species studied. The Long-toed Lapwing differed from the Southern Lapwing in being an inactive, rather than active, tender. There were four differences between these two species that might affect activeness of tending (Table II). (1) Long-toed Lapwings were not vulnerable to terrestrial predators, but Southern Lapwings were. (2) Long-toed Lapwing habitat was more open than Southern Lapwing Habitat (but see below). The ability of Southern Lapwings to occupy more closed habitat may be attributed to their reliance on probing in the substrate when foraging (Fig. 6). Other species relied more on the pecking methods characteristic of charadriids, which are ineffective in tall or dense vegetation (Klomp, 1954; Burton, 1974). (3) Young Long-toed Lapwings were less mobile than young Southern Lapwings owing to a richer food supply and a less negotiable habitat. (4) Long-toed Lapwing territories were much smaller than those of Southern Lapwings, again presumably due to their richer food supply (Walters, 1980). All of these factors reduce the need for active tending in Long-toed Lapwings compared to Southern Lapwings.

Active tending in Crowned Plovers is attributed to the great mobility of their young, necessitated by a sparse food supply. Inactive tending in Blacksmith Plovers is attributed to their open habitat. This influence must be strong enough to outweigh the effects of high predator density and having young of intermediate mobility, both of which select for active tending.

Other Parental Behaviors. In addition to activeness of tending, the model also treats (1) interspecific aggression, (2) showing food, (3) foraging away from young, and (4) foraging while tending (Fig. 6). Interspecific aggression is hypothesized to benefit chicks by reducing the number of other birds in their vicinity, because intruders either compete with young or attract predators (see above). If young are highly mobile, the benefits of defending the area around them may be greatly reduced. In the case of defending against competitors, reducing the density of intruders in the area surrounding the young will alter the amount of food removed by competitors from a particular location only during the brief period the young are there. The intruders could forage there prior to and subsequent to the chicks' passing through. Unless resources renewed very rapidly, aggression would provide little increase in availability of food.

Benefits derived from defending against birds that might attract predators may also be reduced by mobility of young, depending on the distance and time interval over which the relevant predators respond to prey density. If predators react to changes in prey density over periods of days or hours, or over great distances, defending the area around the chicks will have little effect on attraction of predators by intruders. Victims of aggression, by fleeing only a short distance, would remain sufficiently near the young that they might still attract predators to them.

The model therefore asserts that the presence of mobile young selects against interspecific aggression (Fig. 6). This accounts for the lack of interspecific aggression in the species with the most mobile young, Crowned Plovers (Table II).

The benefits of parents foraging away from young are the same as those of interspecific aggression (see above), and therefore the former should be affected by mobility of young in the same way as the latter (Fig. 6). This accounts for foraging by tending adults in Crowned Plovers. Blacksmith Plovers foraged near young to an even greater degree. The model attributes this to the combination of fairly mobile young and large tending adult–young distances resulting from inactive tending (Table II). Both should reduce the effect of foraging by tending adults on competition between adults and young, and possibly that on attraction of predators by adults.

The model relates showing food directly to food abundance (Fig. 6). The species with the sparsest food supply (Crowned Plover), and hence the greatest need to improve the foraging efficiency of its young, is the only one that exhibited food-showing.

B. Parental Behaviors of Other Shorebirds

1. Types of Behavior

Quantitative data on parental behavior comparable to these lapwing data have not been collected for other shorebirds. There are, however, numerous qualitative descriptions of parental behavior, plus a few quantitative time budget studies from which information about parental behavior can be inferred. These data are not sufficient to evaluate the model of the evolution of parental behaviors presented above, but they enable a general description of shorebird behavior and a limited review of tending strategies.

The parental behaviors of lapwings described above typify those of shorebirds in general. Brooding, vigilance, and crouching by the young in response to adult alarm calls are apparently universal. Leading seems to occur in all species except those few whose young are nidicolous (Table III). It is most often employed in moving chicks from a nesting area to a foraging area after hatching (Miller, 1979a). A rare behavior with perhaps the same function as leading is carrying of young, which occurs in woodcocks (*Scolopax* sp.) (Ingram, 1978) and jacanas (Jacanidae) (Hopcraft, 1968). In both groups the behavior has evolved in a context in which leading young is a difficult means of moving them, either because of the distances covered (woodcocks) or the difficulty in negotiating the substrate (jacanas).

Feeding of young is common among shorebirds, but is closely correlated with taxonomy. All members of some families feed young (Rostratulidae, Dromadidae, Haematopodidae, Burhinidae, Glareolidae, Chionididae, Pluvianellidae), whereas no members of the remaining families do (Jacanidae, Ibidorhynchidae, Charadriidae, Recurvirostridae, Thinocoridae) with the exception of the Scolopacidae. Feeding of young is typically associated with unusual foraging morphology (Rostratulidae, Glareolidae) or behavior (Haematopodidae, Burhinidae, Dromadidae, Pluvianellidae). Perhaps developmental constraints on the acquisition of such foraging modes lead to feeding of young. However, the Cursoriinae do not have unusual foraging but feed young, and a number of species with seemingly equally difficult foraging requirements do not feed young (Ibidorhynchidae, Recurvirostridae). Of course, young may feed on different prey than adults in exceptional species, and feeding of young may thus correspond closely with ecology. However, in the absence of data on the foraging of shorebird chicks, an important phylogenetic component in the evolution of feeding of young remains a viable possibility.

Table III. Parental Behaviors of Shorebirds[a]

Taxa	Behavior							Source
	Feed young	Forage away from young	Alternate tending	Lead young	Interspecific aggression	Distract predator	Attack predator	
Dromadidae	++	–[b]	?	–[b]	?	?	?	Fjeldså (1977)
Chionididae	++	–[b]	++	–[b]	?	?	++	Burger (1979, 1981)
Pluvianellidae	++	–[b]	–	?	?	–	?	Jehl (1975)
Thinocoridae	–	–	–[c]	++	–	++	–?	Maclean (1969)
Burhinidae	++	?	?	?	?	?	–?	Bannerman (1962)
Haematopodidae	++	–[b]	++	–[b]	?	++	++	Webster (1941), Miller and Baker (1980)
Ibidorhynchidae	–	?	?	?	?	?	?	Bates and Lowther (1952)
Recurvirostridae	–	++	++	++	++	++	++	Gibson (1971), Hamilton (1975)
Cursoriinae	++	?	?	?	–	++	–?	Maclean (1967, and personal observation)
Glareolinae	++	?	?	++	?	++	++	Bannerman (1962), Maclean (1973)
Vanellinae	–	++	++	++	++	++	++	This study, Little (1967)
Charadriinae	–	++	++	++	++	++	+	Simmons (1956), Wilcox (1959), Mayfield (1973), Graul (1975), Hussell and Page (1976), Maclean (1977), Lenington (1980), Phillips (1980)

Jacanidae	–	+	–[c]	+ +	+ +	+ +	+ +	Jenni and Betts (1978)
Rostratulidae	+ +	–	–	?	?	?	?	Personal observation
Tringini, Numeniini, Limnodromini, and Limosini	–	+ +	+ +	+ +	+ +	+ +	+ +	Bent (1929), Hagar (1966), Skeel (1978), Maxson and Oring (1980), Howe (1982)
Phalaropodini	+	+ +	–[c]	+ +	–	+ +	+ +	Höhn (1967), Hildén and Vuolanto (1972), Kistchinski (1975), Mayfield (1979)
Arenariini	–	+ +	+ +	?	?	+ +	+ +	Bent (1929), Bannerman (1960), Pienkowski and Green (1976)
Calidrini	+ ?	+ +	+ +	+ +	+	+ +	+	Brown (1962), Parmelee (1970), Jehl (1973), Pienkowski and Green (1976)
Scolopacini and Gallinagonini	+	?	–[c]	+ +	?	+ +	+	Bent (1927), Sheldon (1967)

[a] Notation as follows: + + indicates the behavior occurs commonly within the taxa; + indicates that it occurs but infrequently; – indicates that it does not occur.
[b] Young are nidicolous; these behaviors are therefore precluded.
[c] Only one parent tends brood, so this behavior is not possible.

Although some scolopacids feed young (Table III), feeding is more limited than in other taxa that feed young. Most of these scolopacids provide their young with only a little food, and only during the first few days after hatching (e.g., Ruthke, 1977). Feeding is most regular and most substantial in snipes and woodcocks, but even in these species the young procure much of their own food initially and all of it once they are older. The scolopacids may therefore be generally classified as a taxon in which young feed themselves.

It is difficult to evaluate the prevalence of foraging away from young and alternating tending and off-duty roles among shorebirds that do not feed young due to lack of appropriate data. Explicit suggestions of such behavior have been reported for a variety of species in which both adults care for young, and these are the basis for the analysis in Table III. Such behavior may be widespread.

In species in which only one adult cares for young, the second adult may be considered to engage in the off-duty away role permanently. In some of these species, the second adult simply deserts the family, but in others it is active in the off-duty role. That is, it joins in antipredator behavior, is vigilant, joins in interspecific aggression, and may even rarely relieve its mate and tend the young. Such behavior has been reported in jacanas (Jenni and Betts, 1978), polygynous lapwings (Laven, 1941), plovers (Phillips, 1980), godwits (Hagar, 1966), phalaropes (Hildén and Vuolanto, 1972), and seedsnipes (Maclean, 1969).

Predator-directed antipredator behaviors are characteristic of shorebirds (Table III), but the particular behaviors employed, and the relative frequency with which each is employed, vary considerably among species (see Gochfeld, this volume). Larger species employ aggressive responses more frequently and to a wider range of predators than smaller species (Sordahl, 1981). However, even small shorebirds react aggressively to certain predators, especially avian ones that are relatively small and pose little threat to adults, such as crows and gulls. Aggression ranges from close swooping to circling above the predator to threats delivered from the ground.

These aggressive behaviors merge into distraction behaviors. For example, calidridine sandpipers sometimes hover in the face of a human passing near their nest or young (Brown, 1962). Behaviors that are more clearly distracting include loud calling and a variety of elaborate stereotyped displays, specifically false-incubation, crouched running, injury-feigning, and the rodent-run, a bizarre display in which the bird runs away with tail spread emitting a squealing vocalization. Which displays are used varies among species. Smaller shorebirds rely more on distraction (as opposed to aggression) than do larger ones, but even the largest, boldest shorebirds employ distraction regularly.

Interspecific aggression is also widespread, but its prevalence is somewhat unclear due to lack of sufficient study (Table III). Even less is known about some of the other parental behaviors observed in lapwings, precluding their inclusion in Table III. Following and gathering have not been explicitly described as defined above for other shorebirds. Hints of sophisticated discrimination of predators exist for other species, e.g., Sociable Plover *V. gregarius* (Meinertzhagan, 1954; Bannerman, 1961), but documentation is nonexistent for shorebirds, and is rare for birds in general (but see Mueller, 1976; Grubb, 1977; Ficken and Witkin, 1977). Similarly, showing food has only been documented for Crowned Plovers, but there are hints of such behavior in other species, notably calidridines (Parmelee, 1970; Fjeldså, 1977).

2. Tending Strategies

Species for which the data are sufficient to evaluate tending strategies and time costs of tending are very few. Lenington (1973, 1980) provides a clear description of inactive tending in Killdeer (*Charadrius vociferus*). Semipalmated Sandpipers (*Calidris pusilla*) are fairly active tenders immediately after hatching, but then become inactive judging from the descriptions of Ashkenazie and Safriel (1979a,b). Time budget data indicate that foraging is reduced during the period of active tending, but time devoted to parental behavior cannot be determined because, as in most such studies, the behaviors comprising the ethogram do not correspond to parental behaviors. That is, a category may include some behavior that is parental and some that is not. This is especially problematic in determining time devoted to vigilant behavior.

The most complete time budget data, and least problematic in this respect, are those provided by Maxson and Oring (1980) for the Spotted Sandpiper (*Actitis macularia*) and Gibson (1978) for the American Avocet (*Recurvirostra americana*). Because both species are inactive tenders, the time cost of caring for young lies primarily in time devoted to vigilant behavior (Fig. 7). Time devoted to parental behavior is detracted mostly from loafing and maintenance activities (preening, bathing), so that foraging is reduced little or not at all when young are present, as in Blacksmith Plovers (compare Figs. 4 and 7). Tending young apparently constrains time budgets less in these species than in Crowned Plovers, Long-toed Lapwings, or Southern Lapwings.

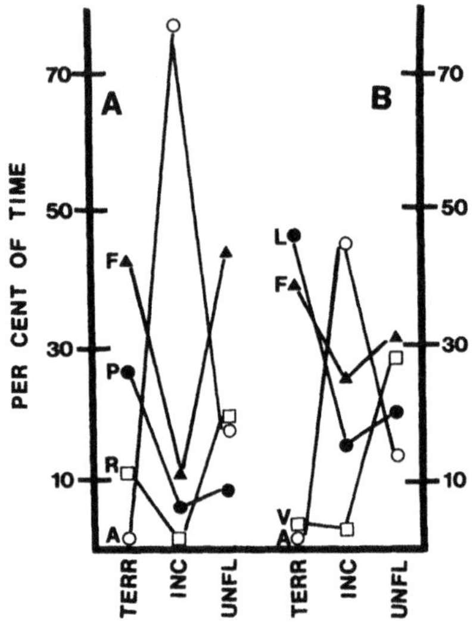

Fig. 7. Time budgets of (A) Spotted Sandpipers and (B) American Avocets as a function of the breeding cycle. Stages of the breeding cycle are defined in Fig. 4. For Spotted Sandpipers, F indicates foraging, P indicates preening, R indicates resting, which includes vigilant behavior, and A indicates walking, brooding, and incubation, which may be considered active parental behaviors. [Derived from Maxson and Oring (1980).] For American Avocets, F indicates foraging, L indicates loafing, which includes preening and resting, V indicates vigilant behavior, and A indicates active parental behaviors. [Derived from Gibson (1978).]

C. Discussion

Study of the parental behavior of shorebirds has barely begun. There is a need for better descriptions of parental behavior, especially with respect to subtleties such as activeness of tending, alternation of tending duties, and predator-discrimination abilities. This area of shorebird biology must then progress beyond the descriptive stage through documentation and quantification of benefits to young and costs to adults of parental behaviors. Specific hypotheses about the evolution of behavior may then be tested. The ideas presented here about costs and benefits and about the evolution of parental behavior are speculative, but provide a heuristic framework for such future research.

The proposed model of the evolution of parental behaviors can easily be tested by comparisons between species. It may also be tested by comparisons of populations of a single species. For example, Blacksmith Plovers breed in some habitats that are much more closed than that of the study population (Walters, 1980); the model predicts they will tend more actively in those habitats. This prediction was verified for Southern Lapwings. Two of the seven successful breeding pairs observed inhabited a field that was much more open than the shrubby areas inhabited by the remaining birds. The mean tending adult–young distance was substantially greater in these two pairs than in any of the other five, presumably because adults tended less actively in the more open area (Walters, 1980).

Although the data are few, it is already clear that parental behavior cannot be regarded as a trivial activity in shorebirds. It imposes considerable demands on adults in some species, but does not in others. Having shown that the usual assumptions about parental behavior in shorebirds are unreasonable, it is necessary to reevaluate the effects of parental behavior on other aspects of shorebird life-history, such as clutch size and mating systems. Time costs of parental behavior and their variation among species have considerable consequences for single-parent care, and thus may affect the evolution of shorebird mating systems (Walters, 1982). The relationship between costs and benefits of parental behavior and brood size has similar implications for the evolution of clutch size (see below). Studies of shorebird parental behavior may therefore be of more general importance than has been previously realized.

III. CLUTCH SIZE

The clutch sizes of shorebirds are described by Maclean (1972b). More recent information has increased the accuracy of our estimates of clutch size for a few species, but these alterations are so few that a new review is unwarranted. Instead, I shall examine the evolution of clutch sizes in shorebirds, using as a basis the data provided by Maclean (1972b) and additional information summarized by Johnsgard (1981).

Variation in shorebird clutch sizes follows a unique pattern. The most striking aspect is a sharp truncation at the maximum clutch size of four. Four is the most common modal clutch size, but no species lays more than four eggs. Many species lay fewer, however. Clutch size theory must therefore answer two questions to explain shorebird clutch sizes: (1) why no species lays more than four eggs and (2) why many species lay clutches that are reduced in size relative to the most common clutch size of four.

Lack's work (1947, 1948, 1954, 1968) laid the foundations of current theories about the evolution of clutch size in birds. He postulated that ability to feed young is the factor that limits clutch size in most birds, and that ability to form eggs is the limiting factor in species that do not feed young. Lack's basic premise was that the number of eggs laid corresponds to the maximum number of offspring that can be raised, which is determined by the proposed limiting factors. This premise has subsequently been qualified to read that the number of eggs laid corresponds to the maximum number of offspring that can be raised with the reproductive effort appropriate for a single clutch (i.e., optimal working capacity; Royama, 1966). That is, individuals maximize their reproductive success over lifetimes, not within a single clutch, and adjust effort in any one year accordingly (Williams, 1966; Charnov and Krebs, 1974; Goodman, 1974). Many birds thus lay a smaller clutch than that which produces the most young with maximum reproductive effort (von Haartman, 1971; Perrins and Moss, 1975; Drent and Daan, 1980).

This does not alter the core of Lack's theory, however. There may still be a single reproductive activity that limits clutch size by reaching the optimal working capacity of adults. The best candidates for this activity are still those proposed by Lack, namely feeding of young for most species and ability to form eggs for species that do not feed young. Clutch size is then determined by the match between the food required for these activities and the food obtained by adults foraging at optimal working capacity.

Clutch size thus varies both between and within species in correspondence with not only limiting factors, but also with differences in how reproductive effort is apportioned, because this affects optimal working capacity (Cody, 1966; Murphy, 1968; Foster, 1974a; Payne, 1974; Brockelman, 1975; Stearns, 1976, 1977; Howe, 1978), and with availability of food during breeding, because this affects production per reproductive effort (Ashmole, 1963; Royama, 1969; Ricklefs, 1970, 1980; Foster, 1974b; Slagsvold, 1975; Owen, 1977).

A. Species That Feed Young

To begin, I divide shorebirds into species that feed their young and those that do not, for the proposed mechanisms of clutch size limitation differ depending on whether or not young are fed.

Species that feed their young generally have clutch sizes less than four (Maclean, 1972b) (Table IV). (The issue of truncation at a clutch size of four is hence moot for this subgroup of shorebirds.) That feeding young leads to reduced clutch sizes can only be tentatively concluded, however, because species that feed young are generally found in geographic regions (tropical, southern) where clutch sizes tend to be reduced even if young are not fed (see below). Furthermore, whether or not young are fed varies only between families, not within, and there are differences at the family level in tendency toward reduction of clutch size among species that do not feed young (see below). One might therefore attribute reduction of clutch sizes in species that feed young to the same factors that cause geographic and taxonomic differences among species that do not feed young (see below), and postulate that clutches are limited by the same factor in both. However, the association between reduction of clutch size

Table IV. Clutch Sizes of Shorebirds That Feed Their Young

Taxa[a]	No. of species	Tropical[b] 4	Tropical[b] <4	Northern 4	Northern <4	Southern 4	Southern <4
Dromadidae	1	0	1	—		—	
Chionididae	2	—		—		0	2
Pluvianellidae	1	—		—		0	1
Burhinidae	9	0	9	—		—	
Haematopodidae	8	—		0	3	0	5
Glareolidae	16	0	13	0	1	0	2
Rostratulidae	2	1	0	—		0	1
Total	39	1	23	0	4	0	11

[a] Scolopacidae are not included for reasons explained in text.
[b] Species are divided into three subgroups according to location of breeding range. Species are considered tropical if most of their breeding range lies between the Tropic of Cancer (23.5°N) and the Tropic of Capricorn (23.5°S), northern if most of their breeding range is north of the Tropic of Cancer, and southern if most is south of the Tropic of Capricorn. Within each subgroup, species are divided according to whether the modal clutch size is four or some number less than four. Blanks indicate null subgroups.

and feeding young is so strong (compare Tables IV and V) that the most parsimonious hypothesis is that feeding of young, not geography or taxonomy, is the key factor. It follows that the factor limiting clutch sizes differs between species that do not feed young and those that do, and feeding of young is implicated in the latter case. According to this hy-

Table V. Clutch Sizes of Shorebirds That Do Not Feed Their Young

Taxa	No. of species	Tropical[a] 4	Tropical[a] <4	Northern 4	Northern <4	Southern 4	Southern <4
Thinocoridae	4	—		—		4	0
Ibidorhynchidae	1	—		1	0	—	
Recurvirostridae	13	3	0	4	0	5	1
Vanellinae	24	11	7	4	0	2	0
Charadriinae	38[b]	0	9	9	6	1	13
Jacanidae	8	8	0	—		—	
Scolopacidae	80[c]	1	7	68	2	0	2
Total	168	23	23	86	8	12	16

[a] Format as in Table IV.
[b] There is one additional species whose clutch size is unknown.
[c] There are six additional species whose clutch sizes are unknown.

pothesis, clutch sizes may be viewed as a match between optimal working capacity, determined by the life-history strategy, and the amount of food provided young per unit of feeding effort, determined by food availability.

There is no evidence that shorebirds are limited in their ability to feed young comparable to that which exists for many altricial birds (e.g., Crossner, 1977). There is also no evidence to the contrary, nor of any other limiting factor. Lack's original hypothesis therefore still stands, but by default rather than due to additional empirical support. Research is needed to distinguish between it and other viable limiting factors, such as all those proposed for species that do not feed young (see below).

B. Species That Do Not Feed Young

The majority of shorebirds do not feed their young. Variation in the clutch sizes of these species may be summarized as follows (Table V).

1. A sharp truncation at the maximum clutch size of four, with over 70% of the species having a modal clutch size of four.
2. A strong tendency toward the maximum clutch size in northern regions, but an approximately equal division between maximum and reduced clutch sizes in tropical and southern regions.
3. A tendency for some taxa (Jacanidae, Vanellinae, Recurvirostridae) to have larger clutches than other taxa (Charadriinae) independent of the effect of geography.

I shall now examine factors proposed to limit the clutch sizes of shorebirds that do not feed their young and determine their adequacy in accounting for these patterns. These are (1) egg-formation ability, (2) incubation ability, (3) nest predation, (4) phylogenetic inertia, (5) and parental behavior.

1. Egg-Formation Ability

I begin with the explanation proposed by Lack, that clutch sizes are limited by the ability of females to form eggs. The critical element in egg-formation ability has not been identified for shorebirds. Most versions seem to assume it is energy, although it could also be some specific nutrient.

The egg-formation hypothesis is less attractive for shorebirds than for other precocial birds that do not feed young because the former lay much smaller clutches than the latter. However, shorebirds produce unusually large eggs (Rahn *et al.*, 1975) so that their clutch weights are in fact not unusually small among precocial birds of comparable body weights. Anatids have much greater clutch weights than shorebirds at comparable body weights (see data in Lack, 1968), but clutch weights of phasianids average only half those of anatids (Rahn *et al.*, 1975), which is even less than those of shorebirds. Shorebirds therefore are not atypical among precocial species in reproductive output. However, they may be atypical in reproductive strategy, investing heavily in each egg at the expense of producing a larger number of eggs. In other birds, egg size is positively correlated with viability of young (Schifferli, 1973; Nisbet, 1978; Al-Murrani, 1978), and shorebird species that are unusual in regularly laying two or more clutches rapidly also have unusually small eggs (Howe, 1975; Osborne and Bourne, 1977; Ross, 1979). General differences between shorebirds and other precocial birds in clutch size may thus reflect a difference in life-history strategy rather than in mechanism of clutch size limitation.

Evidence directly supporting the egg-formation hypothesis is nonexistent for shorebirds, however. That clutch weights in smaller shorebirds often exceed female body weight (Rahn *et al.*, 1975) intuitively reinforces the notion that females have a limited capacity to produce eggs, but this in no way constitutes evidence supporting the hypothesis. There are data indicating that the food supply affects the timing of egg laying (but not clutch size) in Northern Lapwings (Högstedt, 1974), but as in all such experiments, it is not clear whether the effects of food supplements are manifestations of the ultimate or proximate determination of clutch size (Drent and Daan, 1980). Confusion about levels of explanation is a serious problem in clutch size research generally.

Not only are data supporting the hypothesis lacking, but contradictory evidence exists also. Most striking are the results of egg removal experiments with Northern Lapwings. Females of this species normally lay four eggs, but can be induced to lay one or two additional eggs by removing the initial eggs of a clutch as soon as they are laid (Rinkel, 1940; Klomp, 1951). These experiments suggest that clutch size is proximately regulated by feedback from the first egg laid as it is in gulls. Gulls respond to removal of eggs in exactly the same manner as Northern Lapwings, and may be induced to lay up to eight eggs in addition to the usual four. Apparently, stimuli from incubating the first egg halt follicle production, thus determining clutch size (Paluden, 1952; Weidmann, 1956; Parsons, 1976). If these stimuli are removed, eggs continue to be produced until the bird decides to abandon the nest.

One does not expect such a proximate means of limiting clutch size if ability to form eggs is the ultimate limiting factor. Instead, one expects the proximate mechanism to be related to the condition of the female, as

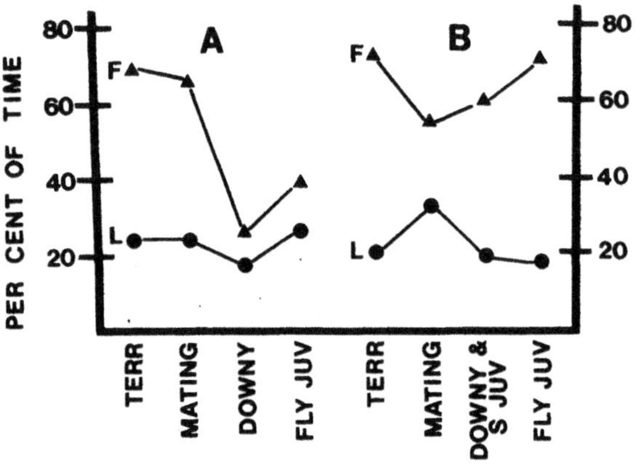

Fig. 8. Percent of time spent foraging (F) and loafing (L; resting, sitting, or preening) as a function of the breeding cycle in (A) Southern Lapwings and (B) Blacksmith Plovers. Stages of the breeding cycle are as defined in Figs. 1 and 4 except that the mating stage is added. This stage lasts from the initiation of nest building until the start of incubation and is the period during which eggs are formed. Time budget is not independent of stage of breeding cycle in either species. Sample sizes per stage range from 43 to 462, with a mode of 158.

it is in Arctic-nesting geese (Klomp, 1970), the only group of nidifugous birds for which there is conclusive evidence that ability to form eggs limits clutch size (Ryder, 1970; Inglis, 1977; Ankney and MacInnes, 1978; Ankney, 1979; Raveling, 1979). In geese, levels of protein reserves appear to proximately control clutch size (Ankney and MacInnes, 1978; Raveling, 1979).

The difference between geese and shorebirds in proximate regulation of clutch size intimates a difference in ultimate regulation. Further evidence contradicting the egg-formation hypothesis comes from lapwing time budget data. If females are limited by their ability to obtain food, foraging time should be at a premium during egg formation (unless foraging profitability varies radically as a function of time). An indication of this would be that loafing is kept to a minimum at this time. Instead, loafing was as frequent during egg laying as during other stages of the breeding cycle, sometimes more frequent (Fig. 8). Decreases in loafing were observed during other periods thought to be demanding, for example when small young were present in Southern Lapwings (Fig. 8A). The limited data collected from birds of known sex indicated that these conclusions held for females specifically as well as adults generally. Time budgets of other shorebirds give similar results (Gibson, 1978; Ashkenazie and Safriel, 1979b; Maxson and Oring, 1980) (Fig. 7). There are no indications that foraging time is at a premium when eggs are being formed in shorebirds.

It is difficult to account for truncation at the maximum clutch size with the egg-formation hypothesis. This requires that differences between species in food availability, life-history strategies, and egg size always balance one another so that the amount of food females are able to procure is sufficient to produce only four eggs, no more. No such truncation exists in any other group of nidifugous birds.

The egg-formation hypothesis can account for reduced clutches only to the extent that species with reduced clutches are less efficient at converting a given amount of available food into eggs. However, one does not expect to account for all variation in egg number by the factor limiting clutch size. This factor will surely account for some of the variation, but its influence must be separated from the suite of other influences on clutch size. In the case of the egg-formation hypothesis, life-history considerations determine the foraging effort, food availability affects the returns on this effort, and ability to convert this food into eggs then determines egg number. Differences between taxa in tendency toward reduced clutch sizes may therefore reflect, at least in part, differences in life-history strategies or food availability related to ecological dissimilarities between these taxa. Similarly, greater tendency toward reduced clutches in southern regions (which house mostly maritime species) and tropical regions compared to northern regions may be due to greater seasonality, and therefore food available per individual, in the latter (Ashmole, 1963; Ricklefs, 1980). There may also be regional differences in life-history attributes if, for example, competition or predation is generally more intense in southern coastal regions and the tropics than in the Arctic regions where most northern shorebirds breed. Until the influences of life-history variables and food availability can be measured, it will be difficult to judge the egg-formation hypothesis on its ability to account for reduced clutch sizes.

In summary, the traditional explanation that food available to the laying female limits shorebird clutch sizes is not well supported by the limited data available. In the absence of more complete data, it remains a viable hypothesis, but the need for alternatives is compelling, and several have been suggested. I shall now review these, and offer yet another.

2. Incubation Ability

The possibility that shorebird clutch sizes are limited by the ability of adults to incubate eggs has received considerable attention recently. Miller (1979b) has proposed that truncation at the maximum clutch size is due to the need to conserve heat during incubation. Due to the shape of shorebird eggs, four eggs can be aligned such that heat lost when the clutch is uncovered is less than that for any possible arrangement of a larger or smaller number of eggs. This hypothesis addresses only the issue of truncation at the maximum; another limiting factor must be postulated where clutch sizes are reduced. Andersson (1978) offers the more general hypothesis that brood patch area limits incubation ability and thus clutch size.

Hills (1980) provides experimental evidence of limitation by incubation ability. Her adding an egg to four-egg clutches resulted in a dramatic decline in hatching success in six species. Similarly, Gibson (1971) reports that American Avocets have difficulty incubating abnormally large clutches that occur naturally. It is not clear which, if either, of the proposed mechanisms accounts for these effects.

Despite these intriguing results, the incubation-ability hypotheses have some serious drawbacks. Most birds have no difficulty incubating abnormally large clutches (e.g., Frederickson, 1969). Some claim that many shorebirds also can incubate abnormally large clutches successfully, although this needs to be documented more conclusively (Shipley, 1980). Also, some species with unusual mating systems may routinely incubate more than four eggs under natural conditions (Walters and Walters, 1980). More importantly, Andersson's hypothesis begs the question of adaptation, i.e., it offers no explanation of variation in brood patch area. Miller's hypothesis does not thus suffer, but has a limited application, as noted above.

Another interpretation of incubation effects is that they reflect secondary adaptations. That is, egg shape may be a secondary adaptation evolved in the context of a clutch size, four, determined by other causes, rather than the factor ultimately limiting clutch size. Similarly, brood patch area may evolve to match the normal clutch size, which is determined by other factors (Klomp, 1970). Andersson (1976) proposes that brood patch number in Jaegers (*Stercorarius longicaudus*) is such a secondary adaptation, and it prevents these birds from successfully incubating experimentally enlarged clutches.

If egg shape and brood patch area are secondary adaptations, they should be evolutionarily malleable, and should change when it becomes otherwise advantageous to have a different clutch size. If these features are instead ultimately limiting clutch size, it should be demonstrable that other hypothesized factors are not limiting. For example, it should be possible to show that shorebirds can form more than four eggs and raise more than four young, but the thermal benefits due to egg shape outweigh the advantages of producing additional offspring.

In conclusion the incubation-ability hypotheses have several theoretical problems, but are better supported by evidence than most alternatives, and certainly are viable.

3. Predation on Nests

It has been suggested that a small clutch may be an adaptation to high nest predation (Oring and Knudson, 1972; Perrins, 1977). Females lay at best one egg per day, so the length of time a nest is exposed to predation increases with clutch size. Because most predators tend to consume entire clutches, a small clutch may on the average produce more young than a larger one due to a difference in the proportion of clutches lost to predators. However, calculations by Perrins (1977) clearly show that shorebird clutch sizes are too small for decreased nest predation to outweigh the advantages of increasing the number of young under realistic predation rates. No other mechanism by which clutch size could be limited by predation has been offered. Predation pressure may of course be one of the factors determining life-history strategy (i.e., apportionment of reproductive effort), but this role should be distinguished from its possible role as a limiting factor. Currently there is no viable predation hypothesis with respect to the latter role.

4. Phylogenetic Inertia

Maclean (1972b) relates shorebird clutch sizes to historical factors, hypothesizing that species have maximum clutch sizes of four in the area in which their taxa originated and reduced clutches in other regions subsequently invaded. He thus relates reduction in clutch size to unknown ecological changes accompanying invasion of new areas, and identifies the maximum of four as the ancestral clutch size. He offers no explanation of the evolution of the ancestral clutch size.

One might allow history an even greater role, although Maclean (1972b) does not do so, and propose that present clutch sizes are not adaptive, but are historical remnants. That is, they result from phylogenetic inertia (Wilson, 1975, p. 32), the heritable variability necessary for a different clutch size to evolve being absent. Known variability is too great for this to be a reasonable general hypothesis, but it may apply to the issue of truncation at maximum clutch size. Clutches of more than four eggs have been reported for many species (Johnsgard, 1981), but some such clutches are known to result from two females laying in a single nest either by egg dumping (Gibson, 1971) or cooperative breeding (Walters and Walters, 1980). Claims that five-egg clutches are laid by a single female (e.g., Bannerman, 1961, p. 97) need to be verified to demonstrate that the variability necessary for a clutch size greater than four to evolve exists, and this variation must be shown to be heritable.

This is the only way in which the phylogenetic inertia hypothesis can be falsified. That four was historically the optimal clutch size cannot be tested. Other hypotheses can be more easily falsified, and this (if I may be permitted an opinion based on discussions with colleagues) seems the primary source of strength for the phylogenetic inertia hypothesis. It is a default option, to which one turns when confronted with the difficulties with the other hypotheses. It has no empirical basis in its own right, and is unlikely to gain one.

5. Parental Behavior

Parental behavior has received little attention as a factor limiting clutch size in shorebirds that do not feed their young because it was assumed that parental behavior is not demanding in such species. This assumption is actually irrelevant (see below), and may be unreasonable in addition (see above). It may therefore be worthwhile to reconsider that some form of parental behavior limits the clutch sizes of shorebirds that do not feed their young.

Safriel and Shipley Models. For parental behavior to limit clutch size, the costs and/or benefits of parental behavior must be sensitive to brood size, regardless of how demanding parental care is. The relationship between costs and/or benefits and brood size is a necessary condition for such a mechanism, whereas a large cost and/or benefit is neither necessary nor sufficient.

Safriel (1975) suggests a mechanism for a relationship between brood size and benefits of parental behavior. He proposes that young must disperse in order to forage effectively, that brood dispersion therefore increases with brood size, and that the distance between adult and young increases with brood dispersion (and thus brood size). Once broods exceed a certain size, chicks are too far from adults to be protected adequately, increasing the likelihood of predation. He conjectures that predators are able to take all or most of a brood once they detect a chick, so that reproductive success decreases as brood size increases above the level at which adults can adequately protect their young. Safriel found that artificially enlarged broods of Semipalmated Sandpipers were more likely to disappear than normal broods, as predicted by the hypothesis, assuming predation is the cause of disappearance.

Shipley (1980) presents an alternative model. He also assumes that the ability of adults to protect their young from predators is a function

Table VI. Mean Distance between Foraging Unfledged Chick and Nearest Adult as a Function of Brood Size in Lapwings[a]

Species	Brood size	
	One	Two
Inactive tenders		
Long-toed Lapwing	9.6 (13)	13.9 (37)
Blacksmith Plover	4.0 (1)	19.6 (21)
Active tenders		
Southern Lapwing	7.4 (74)	6.9 (29)

[a] Distances are in meters. Sample sizes (number of observations during scan samples) are given in parentheses following mean values. None of the differences are statistically significant ($p > 0.05$).

of the dispersion of young, and that dispersion is a function of brood size. He proposes that for geometric reasons, four is the optimal number for minimizing adult–young distance and overlap in chick foraging areas simultaneously. If a fifth chick is added to the brood, either adult–young distance or overlap in chick foraging areas must be greatly increased. Shipley offers no evidence that shorebird broods actually assume the geometric configurations he imagines.

Both models are primarily concerned with explaining truncation at the maximum shorebird clutch size. However, the relationship between predation and foraging efficiency hypothesized can easily be imagined to limit clutch sizes to a reduced number under certain conditions, namely those of extreme predation pressure or foraging dispersion.

Lapwing Model. Although Safriel's (1975) pioneering study provides indirect evidence that parental behavior limits clutch size, the basic assumptions of the above models are speculative. I collected data from lapwings to test these assumptions. The results, described in more detail elsewhere (Walters, 1982), confirmed the assumptions only in part, leading me to propose a more general model describing how parental behavior might limit clutch size in shorebirds. I shall now summarize these results and the model based on them.

The models of Safriel (1975) and Shipley (1980) assume that parent–young distance increases with brood size. This may be true of inactive tenders, but not of active tenders (Table VI). In the actively tending Southern Lapwing, the tending adult stayed as near each chick in a brood of two as it did a single chick. Addition of a third chick did not appear to alter tending adult–young distance either, but the data were insufficient to test this adequately (Walters, 1980). There was no significant increase in tending adult–young distance with increased brood size in inactive

Table VII. Effect of Brood Size on Time Budgets of Adult Southern Lapwings

Stage of breeding cycle[a]	Brood size[b]	Behavioral state				Sample size
		Forage[c]	Parental behavior	Rest, sit, and preen	Other	
Downy	1	27	52	15	5	211
	2 or 3	19	59	14	7	191
Small	1	24	48	22	5	99
juvenile	2 or 3	15	66	17	3	110

[a] Stages are described in Fig. 1.
[b] Time budget is statistically independent of brood size in the downy stage, but not the small juvenile stage χ^2_3, $p = 0.05$).
[c] Values are given as percentages of total time, although statistical analyses were performed on raw frequencies.

tenders either. However, sample sizes were inadequate, and a strong trend toward a positive correlation was evident for both species of inactive tender examined (Table VI). Therefore, that distance increases with brood size in inactive tenders remains a reasonable assumption. Indeed, a brood size effect seems inevitable considering that inactive tenders rarely directed distance-reducing behaviors such as following at young, or gathered broods (see above), and that chicks showed no tendency to group when foraging (Walters, 1980).

The mechanisms for limiting clutch size proposed by Safriel and Shipley are therefore inappropriate for active tenders, and another must be offered. Although Southern Lapwings stayed as near each chick in a large brood as a small one, time budget data suggest that they had to work harder to do so. That is, the amount of time devoted to parental behavior increased with brood size, and time devoted to foraging decreased (Table VII). This suggests a means by which parental behavior might limit clutch size in active tenders: as brood size increases, the demands of parenting increase until the optimal working capacity of adults is reached, which point marks the optimum brood (clutch) size.

The few data on clutch size collected from the study populations suggest that active tenders have reduced clutch sizes, whereas inactive tenders do not (Walters, 1982). The lapwing model of limitation of clutch size by parental behavior therefore incorporates active tending as a factor in reducing clutch sizes, due to the time demands of parental behavior. This clutch size model is thus linked to the model of the evolution of parental behavior presented above (Fig. 6). Those factors that are hypothesized to select for active tending (low visibility, high predation, sparse food) lead to reduced clutch sizes. Conditions that lead to inactive tending favor the maximum clutch size of four.

The lapwing model assumes that adult–young distance is positively correlated with chick mortality, as do those of Shipley and Safriel. The evolution of active tending depends on this relationship (see above), and the model also contends that it is this relationship that limits clutch size to four in inactive tenders. That is, the reduction in benefits of tending as adult–young distance increases with brood size outweighs the advantage of producing additional young above a brood size of four. For this mechanism to be viable, these reductions must be very sharp between broods of five and four.

Safriel's (1975) model may thus be subsumed within mine as the explanation for inactive tenders, if one adds prevention of becoming lost to prevention of predation as a benefit of small adult–young distance, and his experimental results are in accordance with the model if Semipalmated Sandpipers are indeed inactive tenders (see above). The model also includes a second mechanism for active tenders, limitation of clutch size by time available for caring for young, which leads to reduced clutch sizes. Clutch size is then linked to environmental conditions through factors that determine activeness of tending.

In conclusion, the parental behavior hypothesis is also viable. Having received relatively little attention, fewer objections have been raised to it than to other hypotheses, but it also has less supporting evidence than most others.

C. Discussion

The data base for clutch size research is good, although it could be improved by better documentation of differences between populations within species. There is also ample development of theory designed to explain clutch sizes. What is now required is the testing of explicit hypotheses concerning the evolution of clutch size in shorebirds.

To conduct such research, effects of food availability, life-history strategy, and limiting factors must be separated in explaining clutch size variation. Once separated, hypotheses about the effects of each may be developed and tested. Hypotheses about effects of food availability may be tested through measurement of energy intake rates of various species. Using life-history theory, clutch size differences attributable to variation in reproductive effort (e.g., Väisänen, 1977) can be related to factors such as life span (Boyd, 1962). Testing of hypotheses about limiting factors is even more promising. There are three substantial hypotheses: (1) egg-formation ability, (2) incubation ability, and (3) parental behavior (different mechanisms for species that feed young and those that do not), all of which lead to specific predictions. An explicit parental behavior hypothesis was presented above, and the other hypotheses could be expanded to a similar level of specificity. Once specific predictions are formulated, these can be tested by experimental manipulations and further examination of clutch size variation so that hypotheses can be strengthened, modified, or eliminated. It is thus feasible to distinguish between these hypotheses.

Experimental manipulations are an especially promising research vehicle. In interpreting the results of these, one subtlety must always be borne in mind, namely the distinction between levels of explanation. One must specify whether a hypothesis explains proximate or ultimate limitation of clutch size. Often an explanation could function on either level, and it is imperative that levels be explicitly distinguished in the building of theories. This is a major problem in clutch size research.

There has been much speculating and broad generalizing, but it has yet to produce a consensus about the evolution of shorebird clutch sizes. The field awaits those willing to address specifics and to test hypotheses.

ACKNOWLEDGMENTS

The research on lapwings was conducted in partial fulfillment of the requirements of a Ph.D. Biology Department, The University of Chicago. I thank the faculty and students of that institution for intellectual stimulation and helpful suggestions. I am especially grateful to Dr. S. Altmann for his guidance through all stages of the research. I am indebted to C. Casler, Dr. A. R. Kiester, Dr. J. Pacheco, and L. G. Bustos for making the Venezuelan research possible, and to Dr. S. Altmann, Dr. J. Altmann, Dr. G. Hausfater, Warden J. Kioko, and The Kenya Ministry of Tourism and Wildlife for similar support in Kenya. The research was funded by the National Institute of Mental Health through training (MH 13913, MH 15181) and research grants (MH 19617) awarded to Dr. S. Altmann and Dr. J. Altmann, and by the Ridgeway Fund of The University of Chicago.

I thank North Carolina State University for support during manuscript preparation, and A. Hill for helping type the manuscript.

Finally, I thank my wife Beverly for her constant encouragement and aid.

APPENDIX: ETHOGRAM

BEHAVIOR	DEFINITION
Alert	Erect stance with neck extended and nearly perpendicular to the substrate, and trunk tilted downward posteriorly.
Rest-alert	Relaxed posture (see rest), but head held up and away from body.
Rest	All standing postures in which the head is tucked against the trunk; characterized by horizontal orientation of trunk, and often ruffled feathers and one-legged stance. Includes head tucked under wing.
Sit	All postures in which both legs are tucked under the body so that the trunk rests on the substrate, excepting incubation.
Forage	Includes a variety of postures involved in the capture and consumption of prey. These are characterized by an extended neck, and range from a nearly erect posture to one in which the neck extends toward the substrate while the trunk tilts upward posteriorly.
Walk	Distinguished from forage by absence of orientation toward substrate and absence of the usual pace exhibited when seeking food.
Preen	Cleaning and positioning feathers with bill.
Incubate	Sitting on or standing over eggs.
Active parental behaviors	See text.
Other	All behavior not included in above categories. Consists chiefly of drinking, bathing, flying, and social behavior.

REFERENCES

Al-Murrani, W. K., 1978, Maternal effects on embryonic and post-embryonic growth in poultry, *Br. Poult. Sci.* **19**:277.

Altmann, J., 1974, Observational study of behaviour: Sampling methods, *Behaviour* **49**:227.

Altmann, S. A., and Altmann, J., 1970, *Baboon Ecology*, University of Chicago Press, Chicago.

Andersson, M., 1976, Clutch size in the Long-tailed Skua *Stercorarius longicaudus:* Some field experiments, *Ibis* **118**:586.

Andersson, M., 1978, Optimal egg shape in waders, *Ornis Fenn.* **55**:105.

Ankney, C. D., 1979, Does the wing molt cause nutritional stress in Lesser Snow Geese?, *Auk* **96**:68.

Ankney, C. D., and MacInnes, C. D., 1978, Nutrient reserves and reproductive performance of female Lesser Snow Geese, *Auk* **95**:459.

Archer, G., and Godman, E. M., 1937, *The Birds of British Somaliland and the Gulf of Aden,* Vol. II, Gurney & Jackson, London.

Ashkenazie, S., and Safriel, U. N., 1979a, Breeding cycle and behavior of the Semipalmated Sandpiper at Barrow, Alaska, *Auk* **96**:56.

Ashkenazie, S., and Safriel, U. N., 1979b, Time–energy budget of the Semipalmated Sandpiper *Calidris pusilla* at Barrow, Alaska, *Ecology* **60**:783.

Ashmole, N. P., 1963, The regulation of numbers of tropical oceanic birds, *Ibis* **103**:458.

Bannerman, D. A., 1960–1962, *The Birds of the British Isles*, Vols. IX–XI, Oliver & Boyd, Edinburgh.

Barlow, M. L., Miller, P. M., and Sutton, R. R., 1972, Breeding data on the Spur-winged Plover in Southland, New Zealand, *Notornis* **19**:212.

Bates, R. S. P., and Lowther, E. H. N., 1952, *Breeding Birds of Kashmir,* Oxford University Press, London.

Beaver, P. W., 1978, Ontogeny of vocalization in the Greater Rhea, *Auk* **95**:382.

Bent, A. C., 1927, Life histories of North American shorebirds, Part 1, *U.S. Natl. Mus. Bull.* **142**.

Bent, A. C., 1929, Life histories of North American shorebirds, Part 2, *U.S. Natl. Mus. Bull.* **146**.

Bock, W. J., 1958, A generic review of the plovers (Charadriinae, Aves), *Bull. Mus. Comp. Zool. Harv. Univ.* **118**:27.

Boyd, H., 1962, Mortality and fertility of European Charadrii, *Ibis* **104**:368.

Brockelman, W. Y., 1975, Competition, the fitness of offspring, and optimal clutch size, *Am. Nat.* **109**:677.

Brown, R. G. B., 1962, The aggressive and distraction behaviour of the Western Sandpiper, *Ereunetes mauri, Ibis* **104**:1.

Burger, A. E., 1979, Breeding biology, moult, and survival of Lesser Sheathbills *Chionis minor* at Marion Island, *Ardea* **67**:1.

Burger, A. E., 1981, Time budgets, energy needs and kleptoparasitism in breeding Lesser Sheathbills (*Chionis minor*), *Condor* **83**:106.

Burton, P. J. K., 1974, *Feeding and the Feeding Apparatus in Waders*, Trustees of the British Museum, London.

Charnov, E. L., and Krebs, J. R., 1974, On clutch-size and fitness, *Ibis* **116**:217.

Cody, M. L., 1966, A general theory of clutch size, *Evolution* **20**:174.

Crawshay, R., 1907, *The Birds of Tierra del Fuego,* Quaritch, London.

Crossner, K. A., 1977, Natural selection and clutch size in the European Starling, *Ecology* **58**:885.

Drent, R. H., and Daan, S., 1980, The prudent parent: Energetic adjustments in avian breeding, *Ardea* **68**:225.

Eisenberg, J. F. (ed.), 1979, *Vertebrate Ecology in the Northern Neotropics,* Smithsonian Institution Press, Washington, D.C.

Emlen, S. T., and Oring, L. W., 1977, Ecology, sexual selection, and the evolution of mating systems, *Science* **197**:215.

Ficken, M. S., and Witkin, S. R., 1977, Responses of Black-capped Chickadee flocks to predators, *Auk* **94**:156.

Fjeldså, J., 1977, *Guide to the Young of European Precocial Birds*, Scarv Nature Publications, Tisvildelje, Denmark.

Foster, M. S., 1974a, A model to explain molt–breeding overlap and clutch size in some tropical birds, *Evolution* **28**:182.

Foster, M. S., 1974b, Rain, feeding behavior, and clutch size in tropical birds, *Auk* **91**:722.

Frederickson, L. H., 1969, An experimental study of clutch size of the American Coot, *Auk* **86**:541.

Furrer, R. K., 1976, Magpie *Pica pica* defending its nest becomes victim of Carrion Crows *Corvus c. corone*, *Ornithol. Beob.* **73**:240.

Gibson, F., 1971, The breeding biology of the American Avocet (*Recurvirostra americana*) in central Oregon, *Condor* **73**:444.

Gibson, F., 1978, Ecological aspects of the time budget of the American Avocet, *Am. Midl. Nat.* **99**:65.

Goodman, D., 1974, Natural selection and a cost ceiling on reproductive effort, *Am. Nat.* **108**:247.

Goransson, G., Karlsson, J., Nilsson, S. G., and Ulfstrand, S., 1975, Predation on birds' nests in relation to antipredator aggression and nest density: An experimental study, *Oikos* **26**:117.

Graul, W. D., 1973, Adaptive aspects of the Mountain Plover social system, *Living Bird* **12**:69.

Graul, W. D., 1975, Breeding biology of the Mountain Plover, *Wilson Bull.* **87**:6.

Grubb, T. C., Jr., 1977, Discrimination of aerial predators by American Coots in nature, *Anim. Behav.* **25**:1065.

Hagar, J. A., 1966, Nesting of the Hudsonian Godwit at Churchill, Manitoba, *Living Bird* **5**:5.

Hall, K. R. L., 1959, A study of the Blacksmith Plover *Hoplopterus armatus* in the Cape Town area. I. Distribution and breeding data, *Ostrich* **30**:117.

Hall, K. R. L., 1964, A study of the Blacksmith Plover *Hoplopterus armatus* in the Cape Town area II. Behavior, *Ostrich* **35**:3.

Hamilton, R. B., 1975, Comparative behavior of the American Avocet and the Black-necked Stilt (Recurvirostridae), *AOU Ornithol. Monogr.* **17**.

Hildén, O., and Vuolanto, S., 1972, Breeding biology of the Red-necked Phalarope *Phalaropus lobatus*, *Ornis Fenn.* **49**:57.

Hills, S., 1980, Incubation capacity as a limiting factor of shorebird clutch size, *Am. Zool.* **20**:774.

Högstedt, G., 1974, Length of the pre-laying period in the Lapwing *Vanellus vanellus* L., in relation to its food resources, *Ornis Scand.* **5**:1.

Höhn, E. O., 1967, Observations on the breeding biology of Wilson's Phalarope (*Steganopus tricolor*) in central Alberta, *Auk* **84**:220.

Hopcraft, J. B. D., 1968, Some notes on the chick-carrying behavior in the African Jacana, *Living Bird* **7**:85.

Howe, H. F., 1978, Initial investment, clutch size, and brood reduction in the Common Grackle (*Quiscalus quiscula* L.), *Ecology* **59**:1109.

Howe, M. A., 1975, Behavioral aspects of the pair bond in Wilson's Phalarope, *Wilson Bull.* **87**:248.

Howe, M. A., 1982, Social organization in a nesting population of eastern Willets (*Catoptrophorus semipalmatus*), *Auk* **99**:88.

Hudson, W. H., 1920, *Birds of La Plata*, Vol. II, Dent, London.

Hussell, D. J. T., and Page, G. W., 1976, Observations on the breeding biology of Black-bellied Plovers on Devon Island, N. W. T., Canada, *Wilson Bull.* **88**:632.

Inglis, I. R., 1977, The breeding behaviour of the Pink-footed Goose: Behavioural correlates of nesting success, *Anim. Behav.* **25**:747.

Ingram, C., 1978, Carriage of the young and related adaptations in the anatomy of the woodcock *Scolopax rusticola, Ibis* **120**:67.

Jackson, F. J., 1938, *The Birds of Kenya Colony and the Uganda Protectorate,* Vol. I, Gurney & Jackson, London.

Jayakar, S. D., and Spurway, H., 1965a, The Yellow-wattled Lapwing, a tropical dry-season nester (*Vanellus malabaricus* Boddaert, Charadriidae). I. The locality, and the incubatory adaptations, *Zool. Jahrb, Abt. Syst. Oekol. Georg. Tiere* **92**:53.

Jayakar, S. D., and Spurway, H., 1965b, The Yellow-wattled Lapwing, *Vanellus malabaricus* (Boddaert), a tropical dry-season nester. II. Additional data on breeding biology, *J. Bombay Nat. Hist. Soc.* **62**:1.

Jayakar, S. D., and Spurway, H., 1968, The Yellow-wattled Lapwing *Vanellus malabaricus* (Boddaert), a tropical dry-season nester, *J. Bombay Nat. Hist. Soc.* **65**:369.

Jehl, J. R., Jr., 1973, Breeding biology and systematic relationships of the Stilt Sandpiper, *Wilson Bull.* **85**:115.

Jehl, J. R., Jr., 1975, *Pluvianellus socialis*: Biology, ecology and relationships of an enigmatic Patagonian shorebird, *Trans. San Diego Soc. Nat. Hist.* **18**:25.

Jenni, D. A., and Betts, B. J., 1978, Sex differences in nest construction, incubation, and parental behaviour in the polyandrous American Jacana (*Jacana spinosa*), *Anim. Behav.* **26**:207.

Johnsgard, P. A., 1981, *The Plovers, Sandpipers, and Snipes of the World,* University of Nebraska Press, Lincoln.

Kendeigh, S. C., 1952, Parental care and its evolution in birds, *Ill. Biol. Monogr.* **22**:1.

Kistchinski, A. A., 1975, Breeding biology and behaviour of the Grey Phalarope *Phalaropus fulicarius* in east Siberia, *Ibis* **117**:285.

Klomp, H., 1951, Over de achteruitgang van de Kievit, *Vanellus vanellus* (L.), in Nederland en gegevens over het legmechanisme en het eiproductie-vermogen, *Ardea* **39**:143.

Klomp, H., 1954, De terreinkeus van de Kievit, *Vanellus vanellus* (L.), *Ardea* **42**:1.

Klomp, H., 1970, The determination of clutch-size in birds, a review, *Ardea* **58**:1.

Lack, D., 1947, The significance of clutch-sizes, I, *Ibis* **89**:302.

Lack, D., 1948, The significance of clutch-sizes, III, *Ibis* **90**:25.

Lack, D., 1954, *The Natural Regulation of Animal Numbers,* Oxford University Press (Clarendon), London.

Lack, D., 1968, *Ecological Adaptations for Breeding in Birds,* Chapman & Hall, London.

Laven, B., 1941, Beobachtungen über Balz and Brut beim Kiebitz (*Vanellus vanellus* L.), *J. Ornithol. Erganzungsband III* **89**:1.

Lenington, S., 1973, Parental behavior in the Killdeer, M.Sc. thesis, University of Minnesota.

Lenington, S., 1980, Bi-parental care in Killdeer: An adaptive hypothesis, *Wilson Bull.* **92**:8.

Little, J. de V., 1967, Some aspects of the behavior of the Wattled Plover *Afribyx senegallus* (Linnaeus), *Ostrich* **38**:259.

Maclean, G. L., 1967, The breeding biology and behaviour of the Double-banded Courser *Rhinoptilis africanus* (Temminck), *Ibis* **109**:556.

Maclean, G. L., 1969, A study of seedsnipe in southern South America, *Living Bird* **8**:33.

Maclean, G. L., 1972a, Problems of display postures in the Charadrii (Aves: Charadriiformes), *Zool. Afr.* **7**:57.

Maclean, G. L., 1972b, Clutch size and evolution in the Charadrii, *Auk* **89**:299.

Maclean, G. L., 1973, A review of the biology of the Australian desert waders, *Stiltia* and *Peltohyas, Emu* **73**:61.

Maclean, G. L., 1977, Comparative notes on Black-fronted and Red-kneed Dotterels, *Emu* **77**:199.

Maxson, S. J., and Oring, L. W., 1980, Breeding season time and energy budgets of the polyandrous Spotted Sandpiper, *Behaviour* **74**:200.

Mayfield, H. F., 1973, Black-bellied Plover incubation and hatching, *Wilson Bull.* **85**:82.

Mayfield, H. F., 1979, Red Phalaropes breeding on Bathurst Island, *Living Bird* **17**:7.

Meinertzhagan, R., 1954, *Birds of Arabia,* Oliver & Boyd, London.

Miller, E. H., 1979a, Functions of display flights by males of the Least Sandpiper, *Calidris minutilla* (Vieill.) on Sable Island, Nova Scotia, *Can. J. Zool.* **57**:876.

Miller, E. H., 1979b, Egg size in the Least Sandpiper *Calidris minutilla* on Sable Island, Nova Scotia, Canada, *Ornis Scand.* **10**:10.

Miller, E. H., and Baker, A. J., 1980, Displays of the Magellanic Oystercatcher (*Haematopus leucopodus* Garnot), *Wilson Bull.* **92**:149.

Morony, J. J., Jr., Bock, W. J., and Farrand, J., Jr., 1975, *Reference List of the Birds of the World,* American Museum of Natural History, New York.

Mueller, H. C., 1976, Reactions of quail to flying vultures, *Condor* **78**:120.

Murphy, G. I., 1968, Patterns in life history, *Am. Nat.* **102**:391.

Myers, J. P., 1978, One deleterious effect of mobbing in the Southern Lapwing (*Vanellus chilensis*), *Auk* **95**:419.

Myers, J. P., 1980, The pampas shorebird community: Interactions between breeding and non-breeding members, in: *Migrant Birds in the Neotropics: Ecology, Behavior, Distribution, and Conservation* (J. A. Keast and E. S. Morton, eds.), Smithsonian Institution Press, Washington, D.C.

Nie, N. H., Hull, C. H., Jenkins, K. S., and Bent, D. H., 1975, *Statistical Package for the Social Sciences,* 2nd ed., McGraw–Hill, New York.

Nisbet, I. C. T., 1978, Dependence of fledgling success on egg-size, parental performance and egg-composition among Common and Roseate Terns, *Sterna hirundo* and *S. dougallii, Ibis* **120**:207.

Norton, D. W., 1973, Ecological energetics of Calidridine sandpipers breeding in northern Alaska, Ph.D. thesis, University of Alaska.

Orians, G. H., and Willson, M. F., 1964, Interspecific territories of birds, *Ecology* **45**:736.

Oring, L. W., and Knudson, M. L., 1972, Monogamy and polyandry in the Spotted Sandpiper, *Living Bird* **11**:59.

Osborne, D. R., and Bourne, G. R., 1977, Breeding behavior and food habits of the Wattled Jacana, *Condor* **79**:98.

Owen, D. F., 1977, Latitudinal gradients in clutch size: An extension of David Lack's theory, in: *Evolutionary Ecology* (B. Stonehouse and C. Perrins, eds.), pp. 171–179, University Park Press, Baltimore.

Paluden, K., 1952, Contributions to the breeding biology of *Larus argentatus* and *Larus fuscus, Vidensk. Medd. Dan. Naturhist. Foren. Khobenharn* **114**:1.

Parmelee, D. F., 1970, Breeding behavior of the Sanderling in the Canadian high Arctic, *Living Bird* **9**:97.

Parmelee, D. F., and Payne, R. B., 1973, On multiple broods and the breeding strategy of Arctic Sanderlings, *Ibis* **115**:218.

Parmelee, D. F., Greiner, D., and Graul, W., 1968, Summer schedule and breeding biology of the White-rumped Sandpiper in the central Canadian Arctic, *Wilson Bull.* **80**:5.

Parsons, J., 1976, Factors determining the number and size of eggs laid by the Herring Gull, *Condor* **78**:481.

Payne, R. B., 1974, The evolution of clutch size and reproductive rates in parasitic cuckoos, *Evolution* **28**:169.

Perrins, C. M., 1977, The role of predation in the evolution of clutch size, in: *Evolutionary Ecology* (B. Stonehouse and C. Perrins, eds.), University Park Press, Baltimore.
Perrins, C. M., and Moss, D., 1975, Reproductive rates in the Great Tit, *J. Anim. Ecol.* **44**:695.
Phillips, R. E., 1980, Behavior and systematics of New Zealand plovers, *Emu* **80**:177.
Pienkowski, M. W., and Green, G. H., 1976, Breeding biology of Sanderlings in north-east Greenland, *Br. Birds* **69**:165.
Pitelka, F. A., Holmes, R. T., and MacLean, S. F., Jr., 1974, Ecology and evolution of social organization in Arctic sandpipers, *Am. Zool.* **14**:185.
Rahn, H., Paganelli, C. V., and Ar, A., 1975, Relation of avian egg weight to body weight, *Auk* **92**:750.
Ramia, M., 1967, Tipos de sabanas los llanos de Venezuela, *Bol. Soc. Venez. Cie. Nat.* **27**:264.
Raveling, D. G., 1979, The annual cycle of body composition of Canada Geese with special reference to control of reproduction, *Auk* **96**:234.
Ricklefs, R. E., 1970, Clutch size in birds: Outcome of opposing predator and prey adaptations, *Science* **168**:599.
Ricklefs, R. E., 1974, Energetics of reproduction in birds, in: *Avian Energetics* (R. A. Paynter, Jr., ed.), pp. 152–292, Nuttall Ornithology Club, Cambridge, Mass.
Ricklefs, R. E., 1980, Geographic variation in clutch size among passerine birds: Ashmole's hypothesis, *Auk* **97**:38.
Ridpath, M. G., 1972, The Tasmanian Native Hen, *Tribonyx mortierii*. I. Patterns of behaviour, *CSIRO Wildl. Res.* **17**:1.
Rinkel, G. L., 1940, Waarnemigen over het gedrag van de Kievit (*Vanellus vanellus* (L.)) gedurende de broedtijd, *Ardea* **29**:108.
Ross, H. A., 1979, Multiple clutches and shorebird egg and body weight, *Am. Nat.* **113**:618.
Royama, T., 1966, Factors governing feeding rate, food requirements and brood size of nestling Great Tits *Parus major, Ibis* **108**:313.
Royama, T., 1969, A model for the global variation of clutch size in birds, *Oikos* **20**:562.
Ruthke, P., 1977, Futtertragen und Dunenlkeid des Odins-wassertreters (*Phalaropus lobatus*), *Vogelwelt* **98**:229.
Ryder, J. P., 1970, A possible factor in the evolution of clutch size in Ross' Goose, *Wilson Bull.* **82**:5.
Safriel, U. N., 1975, On the significance of clutch size in nidifugous birds, *Ecology* **56**:703.
Schifferli, L., 1973, The effect of egg weight on the subsequent growth of nestling Great Tits *Parus major, Ibis* **115**:549.
Sheldon, W. G., 1967, *The Book of the American Woodcock,* University of Massachusetts Press, Amherst.
Shipley, F., 1980, Four on the shore, *Bird Watch* **8**:1.
Simmons, K. E. L., 1956, Territory in the Little Ringed Plover *Charadrius dubius, Ibis* **98**:390.
Skead, C. J., 1955, A study of the Crowned Plover, *Ostrich* **26**:88.
Skeel, M. A., 1978, Vocalizations of the Whimbrel on its breeding grounds, *Condor* **80**:194.
Slagsvold, T., 1975, Hypotheses on breeding time and clutch-size in birds, *Nor. J. Zool.* **15**:219.
Sordahl, T. A., 1981, Predator-mobbing behavior in the shorebirds of North America, *Wader Study Group Bull.* **31**:41.
Stearns, S. C., 1976, Life-history tactics: A review of the ideas, *Q. Rev. Biol.* **51**:3.
Stearns, S. C., 1977, The evolution of life history traits: A critique of the theory and a review of the data, *Annu. Rev. Ecol. Syst.* **8**:145.

Thomas, D. G., 1969, Breeding biology of the Australian Spur-winged Plover, *Emu* **69:**81.
Troth, R. G., 1979, Vegetational types on a ranch in the central llanos of Venezuela, in: *Vertebrate Ecology in the Northern Neotropics* (J. F. Eisenberg, ed.), pp. 17–30, Smithsonian Institution Press, Washington, D.C.
Väisänen, R. A., 1977, Geographic variation in timing of breeding and egg size in eight European species of waders, *Ann. Zool. Fenn.* **14:**1.
von Haartman, L., 1971, Population dynamics, in: *Avian Biology* (D. S. Farner and J. R. King, eds.), Vol. 1, Academic Press, New York.
Walters, J., 1979, Interspecific aggressive behaviour by Long-toed Lapwings (*Vanellus crassirostris*), *Anim. Behav.* **27:**969.
Walters, J. R., 1980, The evolution of parental behavior in lapwings, Ph.D. thesis, University of Chicago.
Walters, J. R., 1982, Parental behavior in lapwings (Charadriidae) and its relationships with clutch sizes and mating systems, *Evolution* **36:**1030.
Walters, J., and Walters, B. F., 1980, Co-operative breeding by Southern Lapwings *Vanellus chilensis*, *Ibis* **122:**505.
Webster, J. D., 1941, The breeding of the Black Oystercatcher, *Wilson Bull.* **53:**141.
Weidmann, U., 1956, Observations and experiments on egg-laying in the Black-headed Gull (*Larus ridibundus* L.), *Anim. Behav.* **4:**150.
Welty, J. C., 1982, *The Life of Birds*, 3rd ed., Saunders, Philadelphia.
Western, D., 1973, The structure, dynamics, and changes of the Amboseli ecosystem, Ph.D. thesis, University of Nairobi.
Wetmore, A., 1926, Observations on the birds of Argentina, Paraguay, Uruguay, and Chile, *U.S. Natl. Mus. Bull.* **133:**1.
Wilcox, L., 1959, A twenty year banding study of the Piping Plover, *Auk* **75:**129.
Williams, G. C., 1966, Natural selection, the costs of reproduction, and a refinement of Lack's principle, *Am. Nat.* **100:**687.
Wilson, E. O., 1975, *Sociobiology*, Belknap Press, Cambridge.

Chapter 8

ANTIPREDATOR BEHAVIOR: AGGRESSIVE AND DISTRACTION DISPLAYS OF SHOREBIRDS

Michael Gochfeld

Department of Environmental and Community Medicine
UMDNJ–Rutgers Medical School
Piscataway, New Jersey 08854

I. INTRODUCTION

All organisms face two major tasks: survival and successful reproduction. For the most part, survival depends on obtaining adequate food and on avoiding predation. Similarly, successful reproduction involves obtaining suitable mates and breeding areas, providing adequate food for the young (or leading them to it), and protecting them from predators. Such care requires appropriate habitat and nest site selection, efficient foraging, vigilance and appropriate response to the threat of predation. A preliminary list of categories of antipredator behavior is presented in Table I.

This chapter examines the responses of adult shorebirds or waders to potential egg or brood predators or intruders near the nest. Murphy (1926) labeled this "nest protection" behavior. Responses to predators vary among shorebird species and among individuals, and variation is influenced by the intruder and its behavior, and by many other factors. Nest protection behavior includes cryptic behavior throughout the cycle (Gochfeld, 1979), aggressive behavior, and certain conspicuous behavior patterns apparently suited to lure predators away from vulnerable eggs or young. The last, variously referred to as "distraction displays" (Nice, 1943) or diversionary behavior (Armstrong, 1949), has long been of interest. In examining the literature of the past half-century on avian distraction displays, one is struck by how this literature reflects the history of ethological thought. Perhaps unique among ethological problems, the

Table I. Categories of Nest Protection Behavior

- Habitat and nest site selection
  - Protected site
    - Avoidance of ungulate herds or human activity
  - Cryptic site
  - Spacing out
- Cryptic nest-building and parental care
  - Incubation and feeding of young
  - Cryptic young hide or remain immobile
- Predator-induced nest protection behavior
  - Nest-departure behavior
    - Early surreptitious departure
      - Fly-away trick
    - Sitting tight
    - Special behaviors
      - Egg and chick covering
      - Chick carrying
  - Distraction displays
    - Directly related to departure
      - Explosive departure
      - Impeded flight and incapacitation displays
      - Running behaviors
        - Crouched run
        - Upright or rapid run
        - Wing-out run
        - Rodent run
    - Other distraction displays
      - Injury-feigning or broken wing
      - Tail flagging
      - Agitated circling flight
      - Erratic fluttering
    - Displacement activities
      - False brooding
      - False feeding
      - False swimming
      - Pseudo-sleeping
      - False maintenance behavior
      - False reproductive behavior
  - Aggressive behavior
    - Approach
    - Aggressive circling and scolding
    - Mobbing and attacking
    - Ungulate display
  - Parent(s) leads young away

subject of distraction displays has rarely been investigated systematically. It is difficult to think of another problem where the ratio of anecdotal accounts to scientific studies is so great. I attribute this in part to the difficulty in describing such behavior in detail.

In reviewing accounts of shorebird antipredator behavior, I will indicate those aspects that are readily amenable to more rigorous field investigation and will point out certain descriptive measurements that can be recorded including specific types of behavior, postures, transitions, and the effect of the intruder's behavior.

A. Categories and Nomenclature of Nest Protection Behavior

Some nest protection behavior is exceedingly cryptic; some is dramatically conspicious. The manner in which the parent bird leaves the nest or young, the behavior given on departure, and the reliance on cryptic, deceptive, and aggressive behaviors must all be examined. This chapter focuses on the deceptive behavior involved in so-called distraction displays and on aggressive behavior aimed at the intruder. Aggression and distraction may alternate, and have evolved together to enhance the overall antipredator effectiveness of parent birds.

It is, of course, possible to overclassify behavior patterns such that the essential features are obscured rather than clarified. Conversely, a categorization such as Table I does not convey adequate information, because, for example, injury-feigning behavior varies substantially among species and situations, involving differing motor patterns for use of body, legs, and wings, either symmetrically or asymmetrically, with either slight vibratory movements or violent flailing. Also, in addition to its threatening role, mobbing behavior is conspicuous and attracts the attention of intruders, and therefore may subserve the dual function of intimidation and deception.

B. Response to an Approaching Predator

The approach of a predator to a nest is often a gradual process, with the predator or intruder starting out at a distance and moving either directly or circuitously toward the nest. The approach may be nest-directed, in the case of a hunting predator, or the intrusion may be incidental. An intruder may be a potential predator that is not hunting, or may not be a potential predator at all. For birds nesting in dense cover, or in the case of an aerial intruder, the threat may appear suddenly close at hand, allowing less flexibility in nest-departure behavior. The large majority of shorebirds, however, nest in open habitats (Johnsgard, 1981) where the incubating bird, or at least its mate on the territory, can visually scan for danger over a long distance.

Although one may question whether incubating birds make "decisions," it is nonetheless useful to consider a model that incorporates decision making (Krebs and Davies, 1978). At every moment during the approach of the intruder, the incubating bird has the option to decide (Fig. 1) whether to depart from the nest or to remain on it. If it chooses departure, it decides how to leave the nest among several options. The first decision is influenced by its appraisal of the intruder and its approach. Is it an organism that appears threatening? Will its path bring it sufficiently close to the nest to jeopardize it (see Burger and Gochfeld, 1981)? Is the intruder a threat to the adult itself, or only to the nest contents? Can the intruder detect the departing adult readily? Not all of this information is actually available to the decision-maker. The decision of whether or not to leave the nest is also influenced by the species, by the individual's previous experience (Burger and Gochfeld, 1983), and by the stage in the breeding cycle.

The second decision, upon leaving the nest, is to choose among alternative behaviors: departure and disappearance, aggression, and various types of conspicuous behavior (e.g., distraction displays) that focus the attention of the intruder on the adult. The variety of these displays is substantial, but among them the displays that reveal the adult as apparently injured ("injury feigning"), perhaps suffering a broken wing ("broken wing display"), are the best known examples.

In the decision-making model, one can also invoke the bird's appraisal of its own investment, either thus far in the cycle (Dawkins and Carlisle, 1976) or in the remainder of the cycle or lifetime. However, most investigators find it difficult to visualize how a bird makes such an appraisal proximately. However, assuming that relative investment is important either proximately or ultimately, one can at least hypothesize that birds with long breeding seasons and the opportunity to renest after a failure (low temperate and tropical species) may respond differently to threats than Arctic species, which, lacking the opportunity for renesting in a given year, have more at stake in the first brood. It may thus be significant that distraction displays figure most prominently among birds of the high northern latitudes (Armstrong, 1954).

C. Avian Distraction Behavior

Many species of birds exhibit characteristic and dramatic behavior when incubation or brooding is interrupted by the approach or proximity

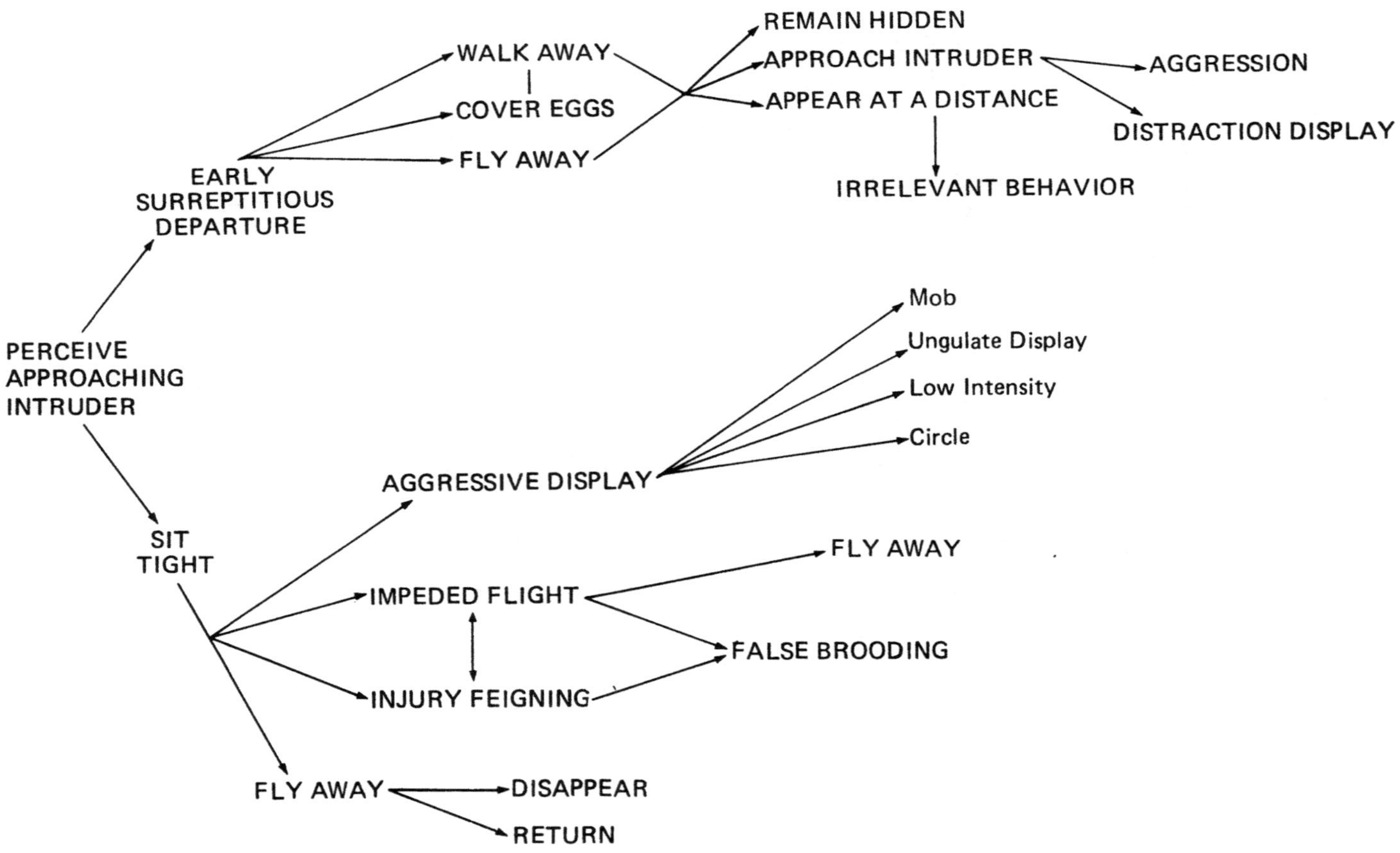

Fig. 1. Flow chart showing options and transitions among antipredator behavior patterns.

of a potential predator. Such behavior that appears to focus the predator's attention on the adult, distracting it from a more profitable quest, the eggs and young, would be adaptive if it actually reduced predation on eggs or young, without resulting in adult mortality. The kinds and intensity of behavior manifested vary widely among species and depend on many factors, but they are often sufficiently prominent (though not necessarily stereotyped) within a species to warrant the term "display."

1. Nomenclature of Distraction Behavior

Williamson (1949) credits Nice (1943) with coining the term "distraction display," reflecting their apparent function. Although many authors use this term, it has been repudiated by some as anthropomorphic. Murphy's term "nest protection displays" (1926) covers aggressive and distractive behaviors, without providing implicit judgements concerning their origin or how predators perceive them. Jourdain (1936) used the term "injury feigning" in a major review of such behavior. At that time many writers believed that such displays were highly stereotyped, perhaps even reflexive, and the use of the term "feigning" implied a certain deliberateness that was not generally acceptable. In Williamson's (1949, 1952) view, distraction display included active defensive behavior and passive behaviors, the latter of which Williamson preferred to call "lure displays," rather than "injury-feigning" or "broken wing" displays. He used the term "mobile" and "static lure" depending on whether the bird was moving or remaining in one position.

Armstrong (1949) preferred the term "diversionary" behavior instead of broken wing, lure display, decoy display, parental ruse, distraction display, etc. He notes that although the term "injury feigning" was in popular use, authors frequently commented that a particular display bore little resemblance to an injured bird. He objected to this term (Armstrong, 1949, p. 89), arguing that "simulation" had less of a connotation of purposiveness than "feigning," a distinction I question. However, as Armstrong (1949) emphasized, injury feigning is only one form of distraction display. He divided diversionary behavior into two subcategories: distraction displays (including injury simulation, chick simulation, eccentric comportment), and displacement behavior, and even coined Greek terms for each.

Thompson (1964) used "distraction display" for elaborate, stereotyped behavior (such as injury feigning and rodent running), and used "diversionary" behavior as a broader term including less stereotyped behaviors as well. This chapter argues that such a distinction is not useful.

In recent years the term "distraction display" seems to predominate, while its many forms are named separately. The term "injury feigning" likewise seems to be in favor, and I use it here, begging the question of whether the performing bird is actually feigning anything at all. The following sections will examine in greater detail the contexts in which different antipredator behaviors are encountered, the motor patterns used in such behavior, and how such behaviors arise.

2. Taxonomic and Ecological Distribution of Distraction Displays

Distraction displays are characteristic of the various shorebird families (Charadriiformes, in part) and many other groups, predominating in ground-nesting birds regardless of their phyletic heritage (Armstrong, 1954). This reflects the vulnerability of ground nests to intruders. A second generalization (Armstrong, 1954) is that colonial species do not use distraction displays, although there are exceptions (e.g., stilts *Himantopus* spp.; pratincoles *Glareola* spp.; Black Skimmers *Rynchops niger*).

Among other Charadriiformes, very few gulls or terns show distraction displays, the exceptions including Sabine's Gull [*Xema sabinii* (Brown *et al.*, 1967)], Franklin's Gull [*Larus pipixcan* (Burger, 1974a)], Brown-hooded Gull [*L. maculipennis* (Burger, 1974b)], Common Gull [*L. canus* (Ruttledge, 1944)], and Least Tern [*Sterna antillarum* (Gochfeld, 1974)]. In contrast, the Black Skimmer and jaegers (*Stercorarius* spp.) have well-developed distraction displays. Among the shorebirds, distraction displays occur in most species, particularly the smaller ones, whereas large species (e.g., avocets, lapwings) characteristically resort to mobbing or threat behavior. Where species nesting in open habitats can detect intruders at a great distance, a sitting bird may leave the nest early before the intruder can detect it, but if the sitting bird is surprised at the nest or has chosen to sit tight, it may give distraction displays when the intruder flushes it.

II. THE CONTEXT IN WHICH NEST PROTECTION BEHAVIOR OCCURS

To better understand distraction displays and antipredator aggression, it is necessary to consider the spatial and temporal contexts and habitats in which they occur as well as the kinds of stimuli that elicit them.

Table II. Some Species for Which There Is a Reported Change in Distraction or Aggression between the Egg Phase and the Chick Phase[a]

Species	Egg phase	Chick phase	Source
Variable Oystercatcher		Increased aggression	Jones (1979)
Black-winged Stilt	Circle with plaintive cries	Mob vigorously	Arnold and Ferguson (1962)
		Great agitation at hatching	Witherby *et al.* (1940)
Eurasian Avocet	No DD	DD given	Brown and Lynn-Allen (1979)
Egyptian Plover	FB is common	More varied DD	Howell (1979)
	No IF	IF occurs	
Australian Pratincole		Displays more intense	Maclean (1973) MacGillivray (1924)
Collared Pratincole	Motionless, wing-up display	Frantic IF	Yeates (1948)
Northern Lapwing	No IF	Occasional IF	Ennion (1949)
Long-toed Lapwing	Low aggression	Attack more, f > m	Walters (1979)
Black-headed Plover	Mob occasionally	Greet and mob	North (1937)
Crowned Plover	Undemonstrative	Very aggressive	Skead (1955)
	Low aggression	Increased aggression	Ade (1979)
Wattled Plover	Circle humans	Mob humans	Little (1967)
Brown-chested Plover	Early departure	Occasional IF	Serle (1956)
Masked Plover	Leave early, hide	Mobbing is common	Thomas (1969)
Black-bellied Plover		Increased intensity	Bent (1929)
		Increased intensity	Seebohm (1901)
Greater Golden Plover	Inconspicuous	Scolding increased	Ratcliffe (1977)
Killdeer		Increase DD	Bent (1929)
		m increase DD at hatching	Mundahl (1982)
Piping Plover	Increase IF during incubation	Lower display distance	Gochfeld (unpublished)
		DD intense at hatching	Cairns (1982)
Kittlitz's Plover	Leave early, covers eggs	Often IF	Martin (1972)
Three-banded Plover	Leave early, flies away	Sometimes IF	Martin (1972)
White-fronted Sandplover	Leave early, covers eggs	May IF	Martin (1972)

[a] Abbreviations: m, male; f, female; EEP, early egg phase; LEP, late egg phase; ACF, agitated circling flight; BW, broken wing display; DD, distraction display; EF, erratic fluttering; FB, false brooding; IF, injury feigning; RR, rodent run; H, hatching.

Table II. *(Continued)*

Species	Egg phase	Chick phase	Source
Red-capped Dotterel	IF occasional	IF routine	Hobbs (1972)
Eurasian Dotterel	Left early	Vigorous IF	Davidson (1954)
Diademed Sandpiper-plover		Adults run more anxiously	Johnson and Goodall (1965)
Northern Jacana	Display infrequent EEP: Higher LEP: Declines	Display 30% of time Peaks at hatching	Jenni and Betts (1978)
Wattled Jacana	Male guards Female flies early	Vigorous IF May carry chicks	Hudson (1920)
Green Sandpiper	Both very quiet	Agitated circling	Davidson (1954)
Wood Sandpiper		Increased anxiety	Bent (1929)
Willet	EEP: May leave early, flies away	Male swoops and mobs	This chapter
Spotted Sandpiper	Infrequent IF	Frequent IF; rapid run	Miller and Miller (1948)
	Occasional IF	Frequent RR and EF	Mousley (1937)
Red-necked Phalarope	Leave early	Sit tight, then IF	Hohn (1968)
		DD peaks after hatch	Nice (1943)
Red Phalarope	Leave early	Circle and call	Meltofte (1979)
	No aggressive approach	Aggressive approach	Kapitonov (1962)
Eurasian Curlew	May leave early	Generally sit tight	Dement'ev *et al.* (1969)
Black-tailed Godwit	Scold	Attack	Lind (1961)
Hudsonian Godwit	Sit tight Guarding increased	Attack Aggressive circling	Hagar (1966)
Bar-tailed Godwit		Mobbing increased	Bent (1927)
Marbled Godwit	Does not dive-bomb	Usually dive-bomb	Nowicki (1973)
Red Knot	Sit tight, then fly far away	Circle continuously	Dement'ev *et al.* (1969)
	Fly very far away LEP: Become more tolerant	Sit tighter	Portenko (1981)
Semipalmated Sandpiper	EEP: Leave early LEP: Sit tight	Noisy circling	Kuzyakin (1959)
Temminck's Stint	EEP: Leave early LEP: Sit tighter		Davidson (1954)
Purple Sandpiper	EEP: Leave early and short RR LEP: Sit tighter and longer RR		Bengtson (1970)
Broad-billed Sandpiper	Flew then dropped	IF	Davidson (1954)
	Chick-feign	No chick-feign	Flint (1973b)

(Continued)

Table II. *(Continued)*

Species	Egg phase	Chick phase	Source
Ruff	EEP: Leave early H: circle intruder	May distract	Dement'ev *et al.* (1969)
Common Snipe	No display	IF	Lye (1948)
	Incomplete DD	More complete DD	Faris (1942)
	Briefer DD	Longer DD	Williamson (1950b)
	Usually leave	Impeded flight	Mousley (1939)
	Peak IF early and late, lower in middle		

A. Temporal and Spatial Factors and Habitat

1. Temporal and Spatial Factors

Rarely mobbing or distraction behavior may occur outside the breeding season. For many species, either the likelihood or the intensity of such displays increases during incubation, peaks around hatching or early brooding, and (in precocial species) declines as chicks become independent (see Table II). In altricial species, the protective displays continue beyond fledging (Barash, 1975) reflecting the protracted period of vulnerability. Aggressive or distraction displays are given in the vicinity of the nest or young. Some displays are given at maximal intensity when the intruder is actually at the nest. Other displays are given at a distance from the nest either to divert an approaching intruder or to "suggest" that the nest is somewhere else (e.g., "false brooding" behavior).

2. Habitat

Shorebird responses to the threat of predation are influenced by nesting habitat. Birds nesting in flat, open habitats can see approaching intruders at a long distance. Where there is vegetative cover, a bird departing from its nest may use such cover to avoid being identified by the intruder, or may disappear into the cover, and reappear at a point away from the nest, luring the intruder to that point. Some shorebirds also use semiconcealment in their distraction behavior, displaying when partially concealed by irregularities in the terrain or by vegetation. To some extent such partial concealment would seem to reduce the conspicuousness of distraction behavior and this will be discussed below.

B. Appropriate Stimuli

The usual stimulus eliciting antipredator aggression or distraction display is the approach of an intruder recognizable as a potential predator. For mobbing behavior, Curio (1976) found that birds could distinguish a threatening intruder from an innocuous one, and Burger and Gochfeld (1981) found that gulls could distinguish whether an intruder was approaching a nest directly or tangentially. Williamson (1946) reported that Whimbrel (*Numenius phaeopus*) recognized his purposive movement toward nests, as opposed to casual intrusion. With rare exceptions, shorebirds do not give distraction displays to conspecifics or other shorebirds (but see Armstrong, 1954; Clay, 1967), though they may respond with aggression. Distraction displays are given to intruders that appear to be hunting and that might be induced to follow an apparently injured bird, or that might be deluded by other behavior and diverted from vulnerable eggs or young. Occasionally such displays occur when a mate, rather than offspring, is in danger (Male, 1950).

Shorebirds respond to potential predators in a variety of ways. Even a single individual may respond differently to human, carnivore, ungulate, or avian intruder. For example, many of the large plovers respond to ungulates (which pose the threat of trampling a nest) by sitting tight, then lunging or charging suddenly at the intruder's face, with legs and wings extended (this will be referred to below as an "ungulate display"), a posture characteristic of threat behavior in many species. Mountain Plover [*Charadrius montanus* (Walker, 1955)], Common Redshank (*Tringa totanus*), St. Helena Plover [*C. sanctaehelenae* (Pitman, 1965)], and many other species give categorically different responses to human, cattle, carnivore, and avian intruders.

Armstrong (1956) reviews several reports where species were more effective than humans in eliciting distraction displays. The stimulus eliciting a response may have little to do with the factors responsible for its evolution. Thus, Armstrong (1954) argues that although Purple Sandpipers (*Calidris maritima*) exhibit rodent run displays when approached by humans, the display (in which they are reputed to show a resemblance to a fleeing mouse or lemming) probably evolved to lure rodent-hunting foxes.

III. DESCRIPTIONS OF NEST PROTECTION BEHAVIOR AND DISPLAYS

Shorebirds illustrate a broad spectrum of motor patterns among their nest protection behaviors. This section will describe certain common patterns based on field experience and on published accounts. In reality, many displays change rapidly and even an injury-feigning display starts, stops, and constantly changes in both quantitative (intensity) and qualitative aspects of the display behavior. Thus, published accounts usually capture the predominant or most dramatic (peak intensity) display, and many low-intensity variants are lost. Therefore, it is often difficult to ascribe characteristic displays to a species based on published accounts.

A. Nest-Departure Behavior and Response Distance

There are many times and ways to leave a nest. At the two extremes, a bird may leave a nest surreptitiously when an intruder is at such a great distance that it is unaware of the incubating bird's presence at all. At the other extreme, an incubating bird may sit tight relying on a cryptic nest site and coloration, and may not depart the nest until it is actually touched. I use the term "response distance" (= "flushing distance") to denote the distance between the intruder and the nest at the point when the parent bird flushes.

Although many species can be characterized either as sitting tight or as leaving early, there is substantial individual variation related to breeding cycle and habitat. If the incubating bird can see an intruder, or if it has a mate standing "guard" to warn it in adequate time, it may leave early and surreptitiously. However, the same bird surprised by the sudden appearance of a nearby threat sits tight. Although it is the conspicuous departures that most often lead into distraction displays, the departure behavior should be viewed as a continuum. From discussion in Skutch (1976), one can infer that response distance tends to be bipolar, with far more birds either tight sitters or early leavers, than there are with intermediate response distances. A bird with an intermediate response distance loses the benefits of both behavioral extremes.

Once the bird begins its departure, there is a spectrum of options open to it, and it may draw on several, often in rapid succession. The behavior off the nest often follows logically from the kind of nest-leaving behavior. Figure 1 provides a flow sheet for different categories of nest protection behavior.

1. Early Surreptitious Departure

Many shorebirds, particularly those nesting in open habitats, leave their nests at the first indication of an intruder approaching, even at a great distance. Walking or running stealthily, usually in a crouched posture, they escape detection as they disappear into vegetation, over a hill, or simply walk a long distance before standing erect, taking flight or calling. The bird may appear "casual," may engage in other activities such as feeding or sleeping, may approach the intruder and either mob or begin a distraction display. The apparent feeding, preening, or sleeping behaviors given in a context of escape from the nest, apparently represent displacement activities and have been labeled "false feeding," "false preening," and "false" or "pseudo-sleeping" (see p. 317). The stimulus eliciting an early surreptitious departure may be the sight of the approaching intruder, or the call of a mate standing "guard" near the nest as reported for lapwings (Ennion, 1949; Bates and Lowther, 1952), Killdeer [*Charadrius vociferus* (Gabrielson, 1922)], Greater Yellowlegs (*Tringa melanoleuca*), phalaropes, Dunlin (*Calidris alpina*), and Long-billed Curlew [*Numenius americanus* (Bent, 1927, 1929)].

Egg-Covering. The covering of eggs (and chicks) by sand or vegetation as a bird leaves its nest is a relatively unusual phenomenon manifested by few taxa of birds (e.g., coots, grebes, and some plovers). Kittlitz's Plover [*Charadrius pecuarius* (Maclean, 1974)] and White-fronted Plover [*C. marginatus* (Summers and Hockey, 1981)] regularly manifest such behavior; at the approach of a predator the bird stands, shuffles sand over its eggs, then departs surreptitiously. Double-banded Coursers [*Rhinoptilus africanus* (Maclean, 1974)] and Egyptian Plovers [*Pluvianus aegyptius* (Howell, 1979)] use their bills to brush sand over their eggs. Kittlitz's and St. Helena Plovers cover their eggs when humans approach on foot, but not when cows approach (Pitman, 1965). Hall (1959) suggests that egg-covering evolved primarily in response to humans, and Howell (1979) attests to its effectiveness.

Chick Carrying. Some species [e.g., jacanas (Hopcraft, 1968) and woodcocks (Bent, 1927; Witherby *et al.* (1940)] transport young more or less regularly. Some other species do so rarely. Woodcocks that have been frightened from the nest will return and carry the young, one by one, between the legs, to another place (Bent, 1927).

Options Subsequent to Early Departure. There are several sequences of behavior that may ensue following early surreptitious departure (see Fig. 1):

1. Bird remains secretive and hidden (some species cover eggs) "Fly away trick"
2. Bird reappears at a distance from the nest
   a. Irrelevant behaviors: standing, feeding, sleeping
   b. Distraction behavior
      1. False brooding, feeding, or sleeping behavior
      2. Injury feigning

3. Bird approaches intruder
    a. Aggressive behavior
        1. Circles noisily, often with several other birds
        2. Dives at, mobs, or strikes intruder
    b. Distraction behavior

"Fly Away Trick." One special case of early departure was described by Meltofte (1977) who called attention to certain species that leave the nest in a long flight, ending far from the intruder. They eventually return to the nest by flying close to the ground from another direction. This is characteristic of the Knot (*Calidris canutus*). He recognized that there are multiple ways of leaving the nest and suggested that the "fly-away trick" was a form of surreptitious departure. Williamson (1952) reports that Eurasian Oystercatchers (*Haematopus ostralegus*) whose chicks had been captured flew off into the distance with a conspicuous "butterfly" type flight with deep, slow, exaggerated wing beats. Sometimes the birds would give the same flight pattern but circle the intruder, dipping low, and touching the ground. Burger (1963) describes a Spotted Sandpiper (*Actitis macularia*) that, after a sequence of distraction display, would fly noisily away, high over a river, and then, minutes later, would fly back quietly to the vicinity of the nest.

Surreptitious departure may be a satisfactory option available to incubating shorebirds, particularly early in incubation when prolonged exposure of eggs to chilling represents relatively little hazard. In the tropics, however, where exposure to extreme heat is a hazard throughout incubation, or at high latitudes, where chilling of eggs near hatching may be fatal, early departure incurs a significant cost.

2. Sitting Tight

Many shorebirds are cryptically colored and can remain nearly invisible while incubating. Consequently, many individuals sit tight, not flushing until the intruder is but a meter away. In some species there is substantial intraspecific as well as individual variation, with some individuals flushing early while others sit tight, and some individuals sitting tight at one point and flushing early at another. Some species specialize in sitting tight. For example, the Australian Dotterel (*Peltohyas australis*) is one of the most perfectly camouflaged ground-nesting birds, and even when not incubating may remain immobile when approached (Maclean, 1973). Similarly, snipes and woodcocks as a rule sit tight (Bent, 1927; Witherby *et al.*, 1940), while beach-nesting plovers, on the other hand, tend to leave early (Johnsgard, 1981).

If the bird does not employ early surreptitious departure, it is likely to be observed by the intruder when it does leave the nest, and it is no longer sufficient to simply rise and walk slowly away. Thus, birds that do sit tight often employ a very conspicuous or explosive departure followed immediately by some other conspicuous behavior pattern. The explosive departure may be accompanied by vocalizations or even noisy wing-beats and may reveal a conspicuous plumage pattern. Common outcomes of sitting tight and late conspicuous departure are outlined below:

1. Aggressive behavior
   a. Approach
   b. Aggressive circling or scolding
   c. Mobbing or attack
   d. Upright threat (usually directed at ungulates)
2. Distraction behavior
   a. Impeded flight and related incapacity displays
   b. Injury feigning
   c. False brooding or feeding, etc.
3. Disappearance

B. Distraction Displays and Related Behaviors

A discussion of distraction behavior must involve arbitrary subdivisions. The various motor patterns displayed generally share the common feature of conspicuousness, but the distinctions among them are often blurred. Williamson (1952) emphasized the distinction between moving and stationary displays, but there are other and equally important distinctions, for example the attitude of the body during locomotion, and whether the wings and/or tail are displayed in a particular fashion or are folded. For example, a running bird may crouch with wings folded ("crouched run"), hold wings slightly abducted with feathers ruffled ("rodent run"), or extend and flutter the wings ("injury feigning"). A stationary bird may crouch with wings folded ("false brooding"), or lie with wings partly or fully extended (incapacitation or "exhausted bird"). Figures 2 and 3 illustrate the spectrum of body, wing, and tail positions in the injury-feigning display of one Piping Plover (*Charadrius melodus*).

I have organized the following discussion largely based on the mode of locomotion, though other presentations would be equally satisfactory. An incubating shorebird leaving its nest can either fly, run, or walk. Flying is quickest, assures safety from a terrestrial predator, but renders the adult (and perhaps the nest) more conspicuous, and may be hazardous if

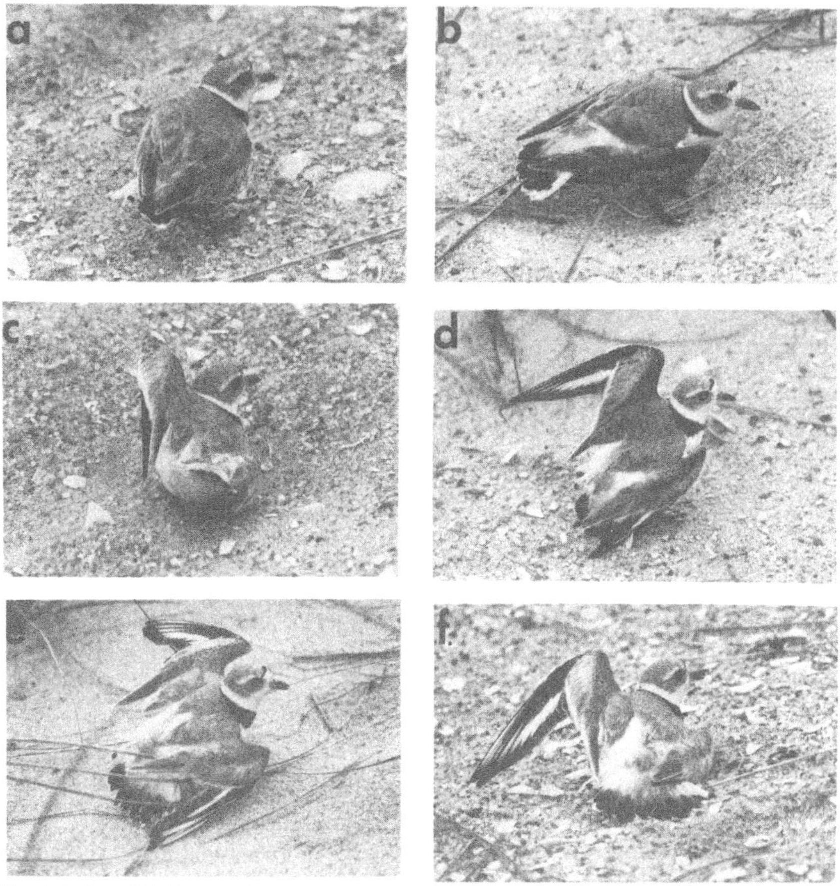

Fig. 2. Injury-feigning displays of the Piping Plover *Charadrius medodus* (lower intensity patterns). (a) Body inclined 20° to the right, wings adducted, tail folded; (b) right wing 10% abducted, tail 33% fanned; (c) legs flexed (crouch), left humerus 80° elevated but wing not extended, tail 10° elevated and folded; (d) legs neutral (normal stance), left humerus 70° elevated, wing 33% extended, tail 50% fanned; (e) similar with right humerus depressed 45°; (f) legs flexed, tail 80% fanned and ground into substrate, right wing abducted, left wing 40% extended with elevated humerus. [Photographs by J. Burger.]

the intruder is an aerial predator. In fact, many shorebirds simply flatten themselves on their nest when an aerial predator is nearby (e.g., Little, 1967; Hussell and Page, 1976; Maclean, 1976). Both flying and running displays serve to distract, confuse, or intimidate predators. Such displays may be given immediately upon leaving the nest or subsequently as a bird returns to engage the intruder.

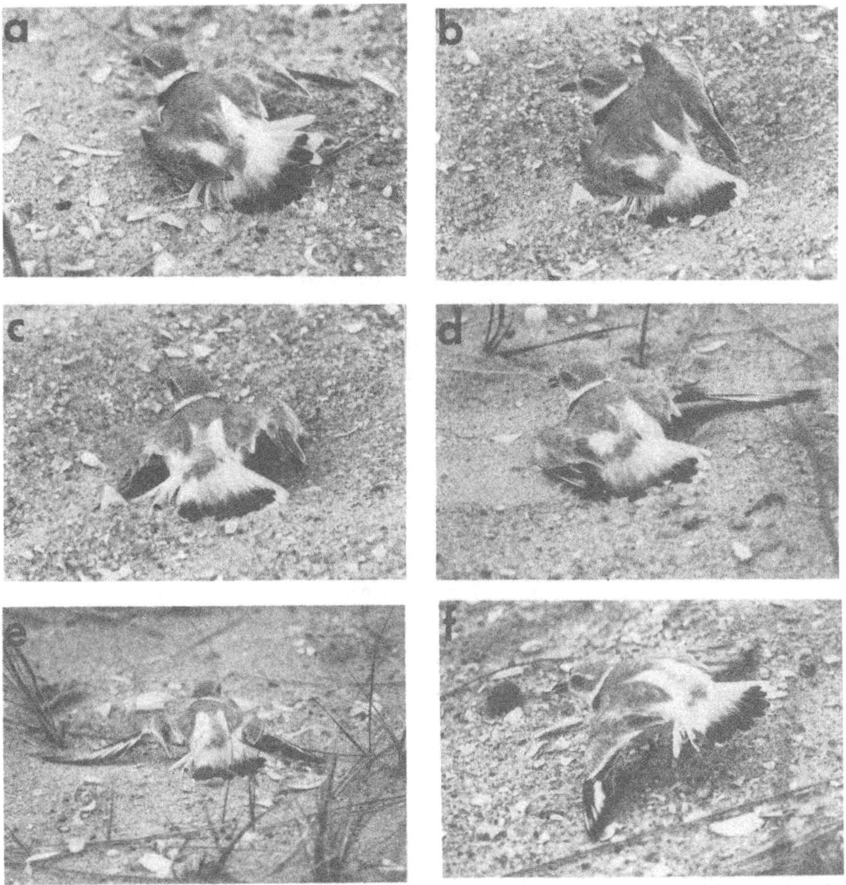

Fig. 3. Injury-feigning displays of the Piping Plover *Charadrius melodus* (higher intensity patterns). (a) Similar to Fig. 2f but legs more completely flexed, and right wing 50% extended; (b) crouch with elevated right humerus, depressed left humerus, and fanned tail; (c) full crouch with wings about 33% extended and tail 80% fanned; (d) asymmetrical extension of wings 80% on right and 20% on left; (e) nearly symmetrical wing extension with breast pressed to substrate; (f) wing-out run. [Photographs by J. Burger.]

1. Standing

A nesting shorebird may stand still at some distance from its nest. Standing may be a brief transition in a train of distraction displays. Brown (1962) emphasized the upright "alert" posture of Western Sandpiper (*Calidris mauri*). Vigorous head bobbing characterizes some species and is described in detail for the Kentish or Snowy Plover, *Charadrius alexandrinus* (Simmons, 1951), and for the Pacific Golden Plover (Sauer,

1962). Magellanic Oystercatchers (*H. leucopodus*) stand and tail-flag (Miller and Baker, 1980).

2. Flying Activities

A bird flying from its nest may fly away from or toward the intruder, fly a short distance or far, and may or may not engage in specialized displays. The following displays usually are given after "sitting tight" (i.e., when the predator is virtually at the nest).

Explosive Departure. By flushing suddenly and noisily underfoot, a bird attracts the attention of the startled predator which may not, for the moment, recognize that it has flushed an incubating bird. The sudden departure may induce the predator to follow immediately, and by the time the predator decides that it is unable to capture the adult, it may be too far from the nest to discover it. This is, of course, an anthropomorphic interpretation that requires experimental examination. This behavior is often seen in conjunction with conspicuous plumage patterns, for instance in the Willet (*Catoptrophorus semipalmatus*). Explosive departures are characteristic of many species that sit tight.

Impeded Flight and Other Incapacitation Displays. These displays are given by some birds when they flush from the nest (Table III). The bird flies for a short distance either with wings fluttering vigorously and legs dangling, seeming to barely navigate, or it flies directly away for a short distance and then drops vertically to the ground. Later the bird may flutter near the intruder or rise and fly away. Common Snipe (*Gallinago gallinago*), flushed from the nest, usually fly off, but occasionally "drop like a stone to the ground as if losing the power of flight" (Mousley, 1939; Spencer, 1948). This behavior differs in appearance from most "impeded flight" for the bird appears to have died suddenly. This variant is better called an "aborted flight."

Certain kinds of flight seem to me analogous to impeded flight, although the effect is different. Lowe (1972) describes an impeded flight display by a European Woodcock (*Scolopax rusticola*) in which the legs dangled below the bird as it flew slowly, giving Lowe the impression that it was carrying a chick, a behavior also reported for the species. Spotted Sandpipers with chicks will flutter erratically all about the intruder, "just out of arm's reach" (Miller and Miller, 1948). This circling behavior could represent a transition to an impeded flight display or to injury feigning. The aborted and impeded flights seem analogous to the static lure, exhausted bird, or death feigning discussed below.

Of 27 willets I flushed from the nest during incubation, 14 gave a form of impeded flight, dropping low over the marsh a few meters ahead

Table III. Some Species for Which Impeded Flights Have Been Described

Species	Behavior	Source
Eurasian Stone Curlew	Rarely: flutter, legs dangling	Witherby *et al.* (1940)
Eurasian Avocet	Low flight, retarded wing-beats, leg trailing through water	Witherby *et al.* (1940)
Wattled Plover	Skim low over ground before dog	Little (1967)
Greater Golden Plover	Skim low over ground, trailing wings and tail over grass	Ratcliffe (1977)
Eurasian Dotterel	Rarely flush with shivering wings and drooped tail	Witherby *et al.* (1940)
Willet	Drop briefly to ground 1–3 m from nest, then fly low, with shallow wing-beats and dangling legs	This chapter
Upland Sandpiper	Low flight with quivering wing-beats	Rockwell (1912)
Red Knot	At one nest, adult flew low over ground as though wounded	Portenko (1981)
Western Sandpiper	"Injury flight"	Brown (1962)
Rock Sandpiper	Leave nest with characteristic jerky flight close to ground for a few meters, then injury feign	Hanna (1921)
Dunlin	Fly limply low over ground with fluttering wings and typical trill	Portenko (1981)
Buff-breasted Sandpiper	Approach with legs dangling	Sutton (1967)
Common Snipe	Flush, then drop vertically	Mousley (1939) Spencer (1948)
Eurasian Woodcock	Flew with dangling legs as if carrying a chick	Lowe (1972)
	Labored flight; dangling legs	Witherby *et al.* (1940)

of me with wings and tail fanned and flailing, continuing their flight low over the salt marsh then out over the bay on weak, shallow wing beats with the legs dangling down. In six of these birds, there was a brief hesitation with the bird landing for 1–2 sec about 2 m from the nest, holding wings briefly aloft then taking flight. This behavior was first noted in 1983, and is sufficiently subtle to have escaped my attention in previous years. This flight, accompanied by a plaintive call, gave the impression that the birds were barely capable of flight. Often, however, they continued for as much as 1 km, over the water where the intruder (author) could hardly be expected to follow them. Thus, one must be careful in interpreting impeded flight as a lure, although it may focus the attention of the intruder on the adult and lead it a short distance.

Diversionary Circling. Birds frequently circle intruders either at close range or at distances of many meters, either silently or with vocalizations. Is the bird circling in order to return to the nest, only to be thwarted by the continued presence of the intruder? Is the circling a way of monitoring the intruder's activity? Or is the circling itself diversionary? There is perhaps a continuum between circling behavior and mobbing, and it is discussed in Section C.1 below as "agitated circling behavior." In many cases, circling is apparently part of distraction behavior. For example, Sutton (1967) describes a Buff-breasted Sandpiper (*Tryngites subruficollis*) that repeatedly flew up in front of him, circled around, and landed in a different direction. There was no accompanying injury-feigning display, the bird alternating bouts of false feeding (see below) between the short circling flights. A Solitary Sandpiper (*Tringa solitaria*) flushed from its nest flew from tree to tree with erratic, conspicuous swallow-like flight (Street, 1923). Williamson (1952) describes Eurasian Oystercatchers circling intruders with slow, conspicuous wing-beats, occasionally flying low and actually touching the ground.

In a terrestrial variant seen in various species, birds circle the intruder, fluttering and calling often in very close proximity to the intruder. Such behavior is usually given only at the nest or when the intruder has captured a chick. Black-headed Plovers (*V. tectus*) with young run to greet intruders with loud cries, and walk around with head held low, and crest depressed, engaging in false feeding (North, 1937).

3. Running Activities

A running bird may vary its speed, its posture, and the position and motion of its wings and tail. Walking or running from an intruder may reflect simple escape behavior, but the variety of walks and runs discussed below involve a mixture of cryptic and conspicuous postures, and I interpret them as distraction behavior. Various terminology has been used. Thus, the "rodent run" of Erard (1963) equals the "crouch run" of Simmons (1955).

Crouched Run. The significance of the crouched run is that it reduces rather than enhances the conspicuousness of the departing bird. Crouched run is characteristic of many plovers, e.g., Killdeer, Piping Plover [*Charadrius melodus* (personal observation)], Ringed Plover [*C. hiaticula* (Williamson, 1947)], Little Ringed Plover [*C. dubius* (Armstrong, 1952)], and Snowy or Kentish Plover (Simmons, 1951). Haverschmidt (1949) considers it characteristic for oystercatchers to leave a nest furtively and run in a stooped posture to avoid detection. He considered this behavior to be highly stereotyped.

Sanderlings (*Calidris alba*) flushed from the nest do a crouched run and then squat at a distance from the nest (Parmelee, 1970). The first reaction by Little Ringed Plover to intruders consists of running away with body held low. Armstrong (1954) considers this equivalent to the rodent run discussed by Williamson (1950a). Armstrong notes that the bird is "being inconspicuous conspicuously," and he considers it a distraction display, noting, however, that it is impossible to make a rigid distinction between a purely escape role and a distraction display.

My observations of Killdeer and Piping Plover suggest that the crouched run serves initially as a surreptitious departure, allowing the bird to move away from the nest before "advertising" its presence with another behavior pattern. The bird may continue its crouched departure for many meters. However, if the intruder is within about 10 m of the nest, the crouched run is usually interrupted, and the bird pauses to stand upright or engage in injury feigning, accompanied by plaintive calls. No evidence can yet be adduced to distinguish an escape function. Conversely, when the crouched run is given by a bird that has returned to the intruder and then ruffs its feather and runs directly ahead, one can see a clear distraction component. Perhaps an important distinction lies in the bird's position with respect to the intruder. If it remains a certain distance ahead, stopping when the intruder stops, the display component seems to be paramount. If it continues to run, remaining well ahead of the intruder, escape would be paramount. Detailed field study of this behavior would be valuable, for distraction displays have traditionally (see p. 359) been interpreted as arising from a behavioral conflict between an escape tendency and a parental tendency.

Upright or Rapid Run. The distinction between different forms of running is not always clear. Several authors, e.g., Miller and Miller (1948), refer to an upright rapid running behavior, which does not include the crouched posture characteristic of either crouched run or rodent run yet is more rapid than the normal departure speed. Williamson (1952) refers to the rapid, furtive running by Eurasian Oystercatchers which move about 75 m from the nest, waiting for the intruder to pass. A rapid run, characteristic of Piping Plovers and Killdeer, is frequently interrupted by injury feigning (see Table XI).

Wing-Out Run. Some shorebirds run with their wings widely spread, and with legs either flexed or fully extended. This distinctive behavior is described for Long-billed Curlew (*Numenius americanus*) by Graul (1971) and illustrated for Piping Plover in Fig. 3f.

Rodent Run. The widely discussed rodent run is a topic of particular interest. Several authors, notably Williamson (1950a) and Duffey *et al.* (1950), were impressed by certain running behaviors in which the departing shorebird resembled a small rodent. This distraction display, known as the rodent run, is typically a rapid running in a crouched position, with the head tilted downward, the body plumage fluffed, and the wings slightly extended and drooped. It is often accompanied by squeaking. In some reports the term includes running with wings outstretched and fluttered as in injury feigning, whereas Armstrong (1952) treats this as crouched run.

Duffey *et al.* (1950) describe this behavior in detail for the Purple Sandpiper. The incubating sandpiper sits tight, then jumps from the nest and runs off with a piping call. Body plumage is fluffed, the wings are drooped, and the head is carried low and extended. Wings are often quivered, and the bird usually runs in a zigzag path, stopping occasionally as if to ascertain whether it is being followed. Bent (1929) mentions female Ruffs (Reeves) (*Philomachus pugnax*) running with every feather erect, looking like a small mammal.

The rodent run is believed to derive its adaptive value from the departing adult's resemblance to a rodent, an appearance that some think is heightened by the dark median stripe on the upper tail coverts that resembles the back pattern of Collared Lemmings (*Dicrostonyx* spp.). As such, the behavior should be effective with predators such as the Arctic fox [*Alopex lagopus* (Larson, 1960)].

Many species give a rodent run, including Dunlin (*Calidris alpina*), Pectoral Sandpiper (*C. melanotus*), Sanderling (*C. alba*), Ringed Plover (*Charadrius hiaticula*), Eurasian Dotterel (*C. morinellus*), and Greater Golden Plover [*Pluvialis apricaria* (Williamson, 1950a)]. Parmelee (1970) concludes that early reports that Sanderlings did not rodent run resulted from the difficulty of discovering their nest. He found that on the approach of an intruder, a sanderling flushed from its eggs creeps away like a lemming (Parmelee, 1970). The bird's hunched appearance makes it look larger than normal.

Williamson (1950a) points out that the Purple Sandpiper apparently confines its distraction display actions to the rodent run; however, Bengtson (1970) describes this species running in a crouched posture with wings elevated or flapping, the tail fanned and depressed, uttering squeaks. Although Bengtson considers this a rodent run, I interpret it as more similar to injury feigning. Burger (1963) describes a Spotted Sandpiper that, instead of injury feigning, ran for about 7 m, with wings spread, back hunched, and tail fanned. Are such displays rodent run or injury feigning, or does the existence of apparently intermediate motor patterns render the distinction artificial?

Piping Plovers (personal observation) alternate crouched running and injury feigning, and occasionally the running, although silent, is accompanied by drooping and fluttering wings (see Fig. 2) suggestive of a rodent run. In such circumstances one can see rodent running arising as a display combining components of two other behavior patterns, such a display becoming the predominant mode of distraction display under certain selective regimes, e.g., for Arctic species such as the Purple Sandpiper, where foxes may be the primary predator. Clearly it would be desirable to examine populations of Purple Sandpipers breeding in areas where foxes are absent, to determine if relaxation of a specific form of predation pressure results in a different antipredator behavior regime. To some extent Pitman (1965) discusses this for the St. Helena Plover, an island isolate of Kittlitz's Plover.

In the case of large shorebirds, the term "rodent run" seems misapplied. The behavior described by Webster (1941) for Black Oystercatchers running in a semicrouching position with nape feathers raised and tail lowered resembles descriptions of the rodent run. However, the running displays of curlews, involving widely outspread wings (Witherby *et al.*, 1940; Graul, 1971), are clearly different. It is difficult to imagine any predator mistaking an oystercatcher or large curlew for a mouse, lemming, or even a hedgehog, regardless of how the wings were held. Although Williamson (1950a) proposed that the expression "rodent run" be retained for the type of behavior that "clearly owes its biological success to its resemblance of a small mammal running away," the common feature of the rodent run in Calidridini sandpipers and the wing-out running of the curlews, lies in the conspicuousness and attractiveness to the intruder.

That a fox may be more interested in pursuing an earth-bound lemming than it would be in a potentially volant adult shorebird may enhance the selective value of the display. In the case of the larger shorebirds, the running behavior may delude the intruder into believing that the bird is incapable of flight. Thus, there is a communality between the rodent run and the injury-feigning displays to be described below. Therefore, I prefer to use more descriptive terms such as "upright run" to describe rapid running in an upright posture, "crouched run" for the rapid running with wings folded, and "wing-out run," the latter displays blending into the injury-feigning displays described below. As yet there has been no experimental investigation of whether predator responses to a shorebird rodent run in any way resemble the response to a running rodent as opposed to a running shorebird. Experiments, probably with captive predators, could distinguish among responses to the various stimuli.

Chick Feign. For a predator searching for flightless chicks, a chick-like display should certainly attract their attention. Ficken and Ficken (1962) described parulid warbler distraction displays resembling a begging fledgling. Flint (1973b) interpreted certain shorebird distraction displays as "chick feigning," in which the adults ran in a crouched posture, weakly moving the wings, and giving a call like a helpless chick. Not having personally seen the displays he described in Broad-billed Sandpiper (*Limicola falcinellus*), Pectoral Sandpiper, and Temminck's Stint (*C. temminckii*), I find it difficult to distinguish such behavior from the rodent run. Clearly the resemblance of an adult to a helpless chick is analogous, and perhaps even more seductive, than resemblance to a running rodent. However, the display only requires success in attracting the attention of the predator, and does not need to appear "true" to the human eye.

Distinctions among Running Behaviors. It is not always apparent how one can classify a particular kind of run. Brown (1962) concluded that in the Western Sandpiper (*Calidris mauri*), there was a continuum of behavior from an injury flight to a rodent run to a crouched run. The difference is partly one of intensity, yet is also qualitative, in that many species that do not show rodent run do have crouched run. As it must be more efficient for birds to use flight for escape, rather than running from terrestrial intruders, I suspect any running behavior of having possible distraction display function.

4. Injury-Feigning and Related Displays

These are the displays that first come to mind when one thinks of distraction displays. The common denominator is that the adult bird acts as if it were partly or wholly incapacitated, and as if the intruder need only follow it a few steps further to be sure of a meal. To a certain extent the terms "injury feigning" and "broken wing" act have been used synonymously with "distraction display." Skutch (1976, p. 406) emphasizes the striking difference between the rapidly moving wings and tail and the slow forward locomotion, noting that it is this "disparity" that connotes injury. He also emphasizes that many so-called injury-feigning displays do not particularly suggest injury to the human observer. This section will indicate the breadth of motor patterns covered by these terms. Detailed illustrations are provided by Brown (1962) and Erard (1963).

There is a broad spectrum of behavior from displays in which a bird runs off with one wing slightly outstretched and perhaps dragging on the substrate, to a bird that collapses and appears dead. The impeded and aborted flight displays described above under nest-departure are directly analogous. Notably, although recent writers emphasize the broken wing aspects of displays, earlier writers spoke of limping and feigning lameness [e.g., for the Black-bellied Plover; Upland Sandpiper, *Bartramia longicauda*; Rock Sandpiper, *Calidris ptilocnemis*; White-rumped Sandpiper,

C. fuscicollis; Dunlin; Common Snipe (Audubon, 1839; Seebohm, 1901; Eaton, 1910; Bent, 1927)].

Broken Wing Acts. This is the archetype of injury feigning, often used synonymously with it. A few published descriptions are paraphrased below. The Surfbird (*Aphriza virgata*) gives an injury feign (Dixon, 1927) that is characteristic of many calidridine sandpipers (e.g., the Knots). When flushed from the nest, it runs with wings half spread, and with the fanned tail dragging on the ground (Jehl, 1968). Killdeer extend one wing over the back, beating the other in the dust, with the tail spread and the breast and tail pressed to the ground (Gabrielson, 1922; Deane, 1944). Thick-billed Plovers (*Charadrius wilsonia*) display with spread tail, creeping forward, beating one or both wings on the ground (Tomkins, 1944), characteristic of the *Charadrius* plovers. The display of the Three-banded Plover (*C. tricollaris*) is described for a bird that approached the intruder to within a meter, calling excitedly, then shuffled away in a crouch, jerking its head, with the tail raised (reminiscent of woodcocks), and its wings extended to a varying degree (Tyler, 1978). The Spotted Sandpiper shows an apparently helpless fluttering over the ground with its wings partially extended and its tail fanned and dragging the ground. Occasionally, like a Killdeer, it spreads its wings fully, waves them, and rolls from side to side. Sanderlings creep away from the nest with the tail down, beating fully spread wings on the ground (Parmelee, 1970).

Without illustration or rigorous description, it is difficult to determine whether these are different ways of describing the same display, or whether species actually differ in displays. Written descriptions clearly capture the peak of display activity, and seldom do writers emphasize the rapid qualitative and quantitative transitions in behavior (see among others Nethersole-Thompson, 1973, for one exception).

In the injury feigning of many species, the deployment of the wings is often asymmetrical. Symmetrical wing extension is seen, for example in the wing-out run, at certain moments in *Charadrius* plovers (see Figs. 3e and f and Erard, 1963); in Pacific Golden Plover (Sauer, 1962); and in pratincoles (*Glareola pratincola*; Fig. 4b).

The exposure of contrasting wing and tail patterns enhances conspicuousness, and its importance has been emphasized by several writers including Williamson (1952) for Eurasian Oystercatcher and Erard (1963) for Kentish Plover. Figures 2 and 3 illustrate contrasting patterns in the Piping Plover. The Killdeer's injury feigning displays its striking buff tail pattern. The Australian Pratincole (*Stiltia isabella*) flutters and stumbles away from the intruder in a crouched posture with its half-spread wings beating the ground at intervals. This reveals the white leading edge of the

Fig. 4. (a) Injury-feigning display of Greater Golden Plover in England; (b) symmetrical, collapsed or exhausted bird posture of Collared Pratincole at Natal, South Africa; (c) Red Phalarope injury feigning on water at Churchill, Manitoba. [Photographs by J. Burger.]

wing, the white upper tail coverts, and the white tail spots (Maclean, 1973).

Erratic Fluttering. For want of a better term, one can characterize certain displays as "erratic fluttering," for birds run, fly, engage in jump flights, with rapid changes in direction, and rapidly beating wings. Unlike the apparent deliberateness of many injury-feigning displays, the birds seem to be in panic and to be making convulsive, uncontrolled movements. The fact that birds may engage in such fluttering, then pause and follow the intruder a short distance, then repeat the display, belies the lack of control. The behavior has the appearance of great excitation, and it may be analogous to "agitated circling flight."

This behavior apparently impressed earlier writers with its convulsiveness, and led others to postulate it as an outcome of conflicting motivations to brood and to escape. In view of the close approach of the adult to the intruder, it would appear to represent thwarted threat behavior. In some cases the adult shorebirds appear very agitated, sometimes with apparently uncontrolled movements that have been characterized as "seizures" and may show a transition to more typical injury feigning. Such behavior is probably common among sandpipers, particularly after hatching when the intruder is near or has captured the chicks. It has been well described for the Spotted Sandpiper (Mousley, 1937; Miller and Miller, 1948). Wandering Tattlers (*Heteroscelus incanus*) respond similarly when the chicks are captured (Dixon, 1933) or even when their alarm call is imitated (Murie, 1924). Even larger species such as the Whimbrel may behave this way (Williamson, 1946).

Exhausted Bird. This term was used by Williamson (1952) to describe situations where birds paused in their injury feigning or other displays, collapsing in one spot, either laying still or weakly moving extremities. Other terms used for similar displays include "static lure display" (Simmons, 1951), "incapacity display" (Armstrong, 1952), and "death feigning" (Sargeant and Eberhardt, 1975). Such displays may represent simply a pause in the more typical, running distraction display, but Williamson (1952) treated it as a distinct display characteristic of Ringed Plovers, Eurasian Oystercatchers, and many other waders. The Collared Pratincole (Fig. 4b) illustrates this display.

Other Conspicuous Displays. Many distraction displays do not fit readily into the above categories, yet are characterized by heightened conspicuousness. The theoretical importance of examining such displays is that they appear in the context of distraction displays, possess conspicuousness to a high degree, yet to the human observer do not resemble "injury" (e.g., Skutch, 1976, p. 403). Hamilton (1975) provides a detailed description of the "tightrope display" in Black-necked Stilts (*Himantopus mexicanus*) and American Avocets (*Recurvirostra americana*), in which birds face an intruder, standing erect, with one wing elevated, extended, and waving, the other wing extended, but drooping downward. The "one-wing lift" (Fig. 5n) is another display wherein the bird stands erect, with one wing elevated over its back. It is given in a variety of contexts, and Bengtson (1970) reported a male Purple Sandpiper pausing in its rodent

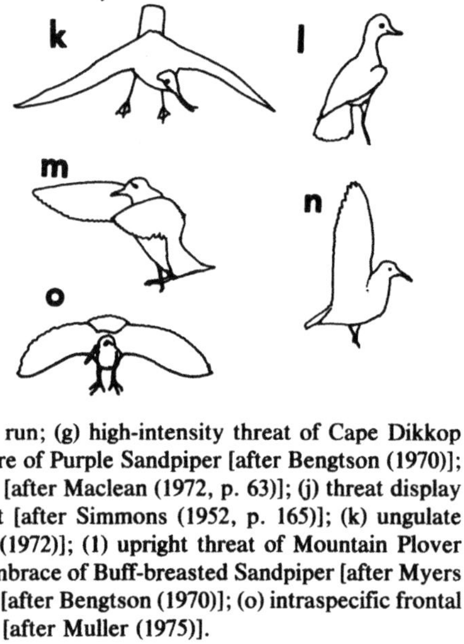

Fig. 5. Diagrams of distraction and aggressive displays. (a) Tail down rush of Mountain Plover [after Graul (1973, p. 75)]; (b) head down threat of Mountain Plover [after Graul (1973, p. 75)]; (c) injury feigning of Semipalmated Plover to human [after Simmons (1955, p. 166)]; (d) mutual display resembling false brooding of Magellanic Oystercatcher [after Maclean (1972, p. 67)]; (e) broken wing display of Greater Golden Plover [after Johnsgard (1981, p. 133)]; (f) horizontal threat of Piping Plover [after Cairns (1982, p. 535)] showing similarity to rodent run; (g) high-intensity threat of Cape Dikkop [after Maclean (1972, p. 65)]; (h) erect posture of Purple Sandpiper [after Bengtson (1970)]; (i) intraspecific threat of Collared Pratincole [after Maclean (1972, p. 63)]; (j) threat display of Ringed Plover to human intruder at nest [after Simmons (1952, p. 165)]; (k) ungulate display of Eurasian Avocet [after Maclean (1972)]; (l) upright threat of Mountain Plover [after Graul (1973, p. 74); (m) double-wing embrace of Buff-breasted Sandpiper [after Myers (1979)]; (n) one-wing lift of Purple Sandpiper [after Bengtson (1970)]; (o) intraspecific frontal aggressive display of Greater Painted Snipe [after Muller (1975)].

run to give a one-wing lift display. These displays presumably represent threat behavior, and additional examples are given below (Section III.C.4).

The Eurasian Oystercatcher (Williamson, 1952) has both static and mobile distraction displays, usually creeping rapidly in a crouch with tail down, wings half-spread, arched, and beating rapidly, displaying the white feathers of the secondaries, lower back, and tail. Miller and Baker (1980) illustrate conspicuous tail-flagging displays of the Magellanic Oystercatcher, a display not known for either the American or Eurasian Oystercatchers. Williamson (1950b) describes a Common Snipe flying from its nest, then running with tail widely fanned, and tilted upwards. It opened and closed its tail displaying the contrasting patterns, and did a crouched run with head retracted and bill pointing downward.

5. Displacement-Type Distraction Displays

Many shorebirds, after leaving the nest, engage in behavior that seems unrelated to breeding or defense of the nest. Several behavior patterns given in contexts that characteristically elicit distraction behavior, have been termed "false brooding," "false feeding," "false" or "pseudo-sleeping," "false preening," and "false swimming." "False preening" and "false swimming" are probably misnomers, for the birds in question are actually preening and swimming, but the behaviors are analogous to false brooding, false feeding, and pseudo-sleeping in that the bird gives the impression of involvement in activities irrelevant to the threat of predation. The birds give no indication that they have a nest nearby, and many writers have published their frustration at trying to find nests of such birds (e.g., Hanna, 1921).

These behaviors are often interpreted as arising as displacement activities (e.g., Armstrong, 1949) that have become ritualized and serve to delude the intruder into "believing" that the individual does not have eggs or young nearby. Displacement in this ethological sense implies an irrelevant behavior arising in a situation of motivational conflict (Tinbergen, 1952), and it is premature to account for all of these distraction behaviors in this manner. Whether certain individuals are more susceptible to displacement behavior or show preferences for one or another category of displacement remains to be studied. Of the behavior patterns listed below, false brooding is the most common and widespread.

False Brooding. Table IV lists some species for which false brooding has been described. Many plovers and sandpipers will squat on the ground away from the nest, and appear to be incubating. They sit watching the intruder, and may make settling movements that enhance the illusion. Such behavior is quite elaborate in the Eurasian Oystercatcher, for example, particularly on the Faeroes, where they may settle in a hollow, and shuffle their bodies as if adjusting the brood patches on the clutch. Unlike many species that interrupt a crouched run to squat briefly as if on a nest, the oystercatcher remains there when approached, and displays typical nest-leaving behavior when departing from the false nest. The peak of this behavior early in incubation (Williamson, 1952) suggests a derivation from nest-scraping behavior, and the tendency for false brooding peaks early in incubation before injury feigning.

Although most distraction displays of Wilson's Phalarope (*Phalaropus tricolor*) are given by males, the female does perform false brooding

Table IV. Some Species for Which False Brooding (FB) Has Been Reported

Species	Comment	Source
Eurasian Oystercatcher	Peaks in EEP; common in Holland, rare in Britain	Williamson (1952)
Magellanic Oystercatcher		Miller and Baker (1980)
American Oystercatcher	Weak and infrequent	Miller and Baker (1980)
American Avocet		Hamilton (1975)
Black-necked Stilt		Hamilton (1975)
		ffrench (1973)
Egyptian Plover	FB is common	Howell (1979)
Temminck's Courser		Maclean (1967)
Milky Pratincole	Squat with open wings as if on nest	Ali and Ripley (1969)
Wattled Lapwing		Little (1967)
Crowned Plover		Ade (1979)
Southern Lapwing	FB if it cannot lure humans	Hudson (1920)
Black-bellied Plover	FB is frequent	Drury (1961)
Ringed Plover	FB on tern nest	Smith and Smith (1974)
Little Ringed Plover	Squat with shifting movements	Armstrong (1952)
Killdeer	Infrequent FB	This chapter
Piping Plover	Sometimes complete collapse	Bent (1929)
	> 80% of birds show FB	This chapter
Kittlitz's Plover	FB and false nest-making	Pitman (1965)
St. Helena Plover		Pitman (1965)
Snowy Plover		Erard (1963)
Eurasian Dotterel	FB	Nethersole-Thompson (1973)
Northern Jacana	Males may FB if not followed	Jenni and Betts (1978)
Wilson's Phalarope	Females may FB	Hohn (1967)
Surfbird		Gabrielson and Lincoln (1959)
Sanderling	Squats in depression	Parmelee (1970)

(Hohn, 1967). American Avocets and Black-necked Stilts both engage in false brooding although other conspicuous distraction and threatening behaviors predominate in the performances of most individuals (Hamilton, 1975). If injury-feigning male Wattled Jacanas (*Jacana jacana*) are not followed by the intruder, the bird may engage in false brooding.

Certain pauses in behavior sequences given in response to intruders may resemble false brooding. For example, African Wattled Plovers seldom give distraction displays, but Little (1967) saw a male run from a nest in a hunched posture with the neck extended, then squat flat on the ground (seemingly a form of false brooding), then fly off. Sanderlings may run from the nest, then squat at a distance (Parmelee, 1970). There is a resemblance between the false brooding described above and the "exhausted bird," the latter display performed with the wings extended by Collared Pratincoles (Fig. 4b) and Milky Pratincoles [*Glareola lactea* (Ali and Ripley, 1969)].

Although one is tempted to interpret false brooding as a reflection of a strong incubation or brooding tendency, Brown (1962) provides evidence that false brooding is probably a distraction display unrelated to the immediate incubation tendency. I interpret my data on Piping Plover similarly, for false brooding displays are not less common in the first 5 min after a plover leaves the nest, than 30 min later when, traditional Lorenzian ethology (see Gould, 1982, p. 186) would lead us to believe, that the brooding tendency would have increased.

Nest Betrayal. An interesting variant of false brooding is described for a Ringed Plover that ran from an intruder into a nearby nesting area of Little Tern (*Sterna albifrons*), and there stood, bobbing next to one of the tern nests. When approached there it flew away. The behavior was repeated at weekly intervals, the plover seemingly advertising the presence of a nest not its own (Smith and Smith, 1974). I have seen a similar display by a Least Tern, and there are occasional anecdotes of birds incubating or brooding at other nests when disturbed. One must distinguish such behavior arising from the momentary "confusion" engendered by the intruder, from "betrayal" as a specific adaptation that accomplishes not only predator distraction but interference competition.

False Feeding. Table V lists some species for which episodes of false feeding have been reported in the context of distraction displays. Three variants are seen commonly. In one case a bird that has left the nest early will appear either on a characteristic feeding area (e.g., a nearby mudflat) or in a part of its territory away from the nest, and will begin feeding; in some cases food is actually obtained. Alternatively, the bird simply pecks at the ground in a manner reminiscent of grass-pulling in gulls, the classic example of a displacement activity. In other cases a bird engaged in a distraction display will make one or more prefunctory or convincing pecks at the ground.

Pseudo-sleeping. In some species, a bird that has left its nest early will often appear to be sleeping, showing little evidence of interest in a nest. Williamson (1950c) describes the pseudo-sleeping behavior of oystercatchers, noting that although the bill is tucked under the scapulars, the eyes are open and apparently fixed on the intruder. Just how long an individual bird can perform this behavior, and how such birds respond when the intruder gets close to the nest, remain to be examined. Williamson (1952) considers this a passive distraction display.

Table V. Some Species for Which False Feeding (FF) and Other Displacement Activities Occur in the Context of Distraction Displays

Eurasian Oystercatcher	FF	Williamson (1952)
Red-necked Avocet	False sleeping 3 m from nest	Bryant (1947)
American Avocet	FF	Hamilton (1975)
Egyptian Plover	Sand-tossing is common as if egg-covering; also does vigorous FF including stone-turning	Howell (1979)
Northern Lapwing	Runs and "artlessly pretends to feed"	North (1937)
Black-headed Plover	FF, also false courtship: one bird crouched, mate mounted	North (1937)
Crowned Plover	FF	North (1937)
	FF interspersed with mobbing	Skead (1955)
Wattled Lapwing	FF	North (1937)
Brown-chested Plover	Stands motionless and FF	Serle (1956)
Masked Plover	FF and occasional injury feigning interspersed with mobbing	Thomas (1969)
Black-bellied Plover	FF frequent	Drury (1961)
American Golden Plover	FF infrequent	Drury (1961)
Semipalmated Plover	FF if lure is unsuccessful	Drury (1961)
Kittlitz's Plover	FF	Pitman (1965)
Snowy Plover	FF	Erard (1963)
Upland Sandpiper	High aerial display	Rockwell (1912)
Little Curlew	Displacement tossing of buds	Labutin *et al.* (1982)
Red-necked Phalarope	Rapid FF on water surface	Davidson (1954)
Buff-breasted Sandpiper	FF interspersed with circling	Sutton (1967)
Stilt Sandpiper	Pebble tossing	Jehl (1973)
Short-billed Dowitcher	FF	Kistchinski and Flint (1973)

False Swimming. American Avocets frequently land in the water and swim or flutter in circles around a nest threatened by an intruder (Hamilton, 1975). I make a distinction between swimming that appears to be an irrelevant activity, from behavior causing a conspicuous disturbance, for many species of typically aquatic birds [e.g., ducks, Red Phalaropes (J. Burger, personal communication) (see Fig. 4c), and Northern Jacana (Miller, 1931)] perform dramatic injury-feigning displays on water.

False Preening or Other Maintenance Behavior. This is one activity that shorebirds engage in while remaining at a distance from their nest. The significance lies in the fact that such maintenance behavior occurs in a context of threat to nest or young, when the adult would not usually engage in such behavior. The resulting air of nonchalance might suggest to the intruder that there are no offspring about.

Other Apparent Displacement Activities. A variety of other behavioral avenues are open to birds faced by an intruder near the nest. Little Curlews (*Numenius minutus*) pick up and toss buds (Labutin *et al.*, 1982). Red Knots attempt courtship (Portenko, 1981). Semipalmated Sandpipers perform grass-pulling, one of the classic displacement activities (Ashkenazie and Safriel, 1979). White-rumped Sandpipers begin carrying eggshells (Parmelee *et al.*, 1968).

C. Aggressive Antipredator Behavior

In addition or as an alternative to distraction displays, shorebirds may act aggressively toward intruders. Although various behavior patterns are recognized by field workers as "aggressive," one cannot assume that they arise out of motivations similar to those that prompt aggressive interactions among conspecifics. Aggressive behavior may intimidate or discourage an intruder, but certain aggressive displays also render the adult bird highly conspicuous and rivet the attention of the intruder on the bird rather than on its nest. Thus, aggression may have a dual benefit, and it is important to investigate the relative contributions of intimidation and distraction. I will discuss aggressive behavior under four headings.

1. Approach or Greeting Flight

Many shorebirds, whether they leave the nest early or sit tight, may subsequently approach an intruder. In some cases birds that leave early may fly out as if to greet the intruder at distances of over 500 m from the nest. Such long-distance greeting is mentioned for Black-necked Stilts (Bent, 1927), Northern Lapwing (Ennion, 1949), Willet (personal observation; Vogt, 1938), Greater Yellowlegs (Bent, 1927), phalaropes (Bent, 1927; Portenko, 1981), godwits (Bent, 1927; Lind, 1961; Hagar, 1966), Whimbrel (Williamson, 1946), Dunlin, Baird Sandpiper (*Calidris bairdii*), Rock Sandpiper, and dowitchers (Bent, 1927; Kistchinski and Flint, 1973).

Approach may result in attack or mobbing, circling behavior (either aerial or on foot), or some other display. The approaching bird may veer off and land at a distance, engaging in a different behavior. For example, when an intruder approaches a Piping Plover nest, the bird may leave the nest surreptitiously, but if the intruder actually gets near the nest, the bird will return to it with conspicuous behaviors such as direct attack (occasionally approaching to within 1 m), agitated circling, crouched run-

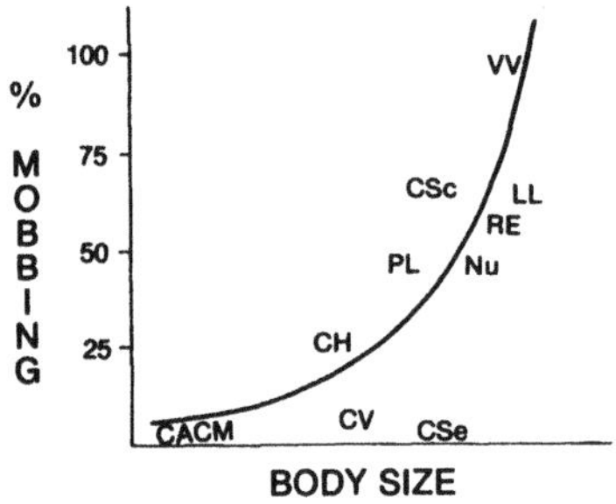

Fig. 6. Relationship between tendency to mob and body size among shorebird taxa. CA, Snowy Plover; CH, Stilt Sandpiper; CM, Piping Plover; CSe, Willet (egg phase); CSc, Willet (chick phase); CV, Killdeer; LL, Black-tailed Godwit; Nu, curlews (*Numenius*); PL, American Golden Plover; RE, American Avocet; VV, Northern Lapwing

ning, or injury feigning. Brown (1962) notes that in Western Sandpipers, approach and retreat alternated in 35 of 52 cases.

2. Mobbing and Attack

Typical mobbing behavior involves a group of birds approaching, circling, swooping at, or even striking a potential predator. "Attack" has various connotations. Nice (1943) defined "attack" as involving actual strikes; Jenni and Betts (1978) use "attack" for approach and aerial diving, while Walters (1979) uses "attack" for spectacular threat displays without contact. Whether physical contact is actually made depends on the relative strengths of the tendency to attack and of fearfulness. Only occasionally do shorebirds actually strike the intruder (Williamson, 1952), although strikes are common in other Charadriiformes groups (e.g., gulls, terns, skuas). The mobbing attacks are repeated steadily and are often accompanied by loud vocalizations. Mobbing may arise in the same context as distraction display, may occur as an alternative to distraction display, and may even alternate with distraction display. In general, mobbing characterizes the larger species of shorebirds (see Fig. 6), particularly lapwings (*Vanellus*), godwits (*Limosa*), and curlews (*Numenius*). Sordahl (1981) found that mobbing occurred in most North American shorebird families, but was usually restricted to the largest species. Notably, none of the small plovers mob, but both American Golden and Black-bellied Plovers may mob.

Mobbing is characteristic of lapwings. Walters (1979) notes that Long-toed Lapwings (*V. crassirostris*) are usually aggressive toward intruders. They make swooping threat displays without actually touching the intruder. Crowned Plovers (*V. coronatus*) mob only in the breeding season and actually strike intruders, including humans (Skead, 1955). They are undemonstrative during incubation, but mob vigorously when they have chicks. Australian Spur-winged Plover (*V. m. novaehollandiae*) also mob vigorously, in the chick phase, though they rarely strike (Thomas, 1969).

Although there are many reports of mobbing of human intruder, mobbing is not restricted to them. The Wattled Lapwing (*Vanellus senegallus*) attacks mongooses and dogs by running toward them in a crouched threat posture, then flying up at them with a loud call (Little, 1967). Skead (1955) describes Crowned Plovers "teasing" a dog, fluttering over it, leading it on, then mobbing it repeatedly. However, there are qualitative differences in response to different intruders, which will be discussed below. Southern Lapwings (*V. chilensis*) occasionally mob humans (Hudson, 1920), and frequently mob avian predators, occasionally succumbing in the attack (Myers, 1978).

Black-tailed Godwits (*Limosa limosa*) typically attack avian intruders, flying above and dive-bombing them, followed by either a gradual "attack soar" or sudden vertical "escape soar," prior to the next attack (Lind, 1961). Godwits did not attack mammals or gulls on the ground, but hovered over them calling, which may represent a low-intensity manifestation of the mobbing tendency. Lind (1961) interprets hovering as a balance between the attack and escape "drive." Similarly, American Avocets perform a dive-bombing display near their nest.

Eurasian Oystercatchers typically make aggressive flights toward humans, usually swerving aside at the last moment, but occasionally striking the intruder (Williamson, 1952). Variable Oystercatchers (*Haematopus unicolor*) mob dogs and humans, diving and striking the head and back with wings or with bill and feet (Jones, 1979). Willets rise to meet human intruders as soon as the intruder lands on the island (Vogt, 1938). The Willets circle, calling loudly until the whole "colony" joins in (Bent, 1929). Sordahl (1979) includes this as mobbing, although from my experience with Atlantic coast populations, I would distinguish this circling behavior which occurs commonly during incubation, from the higher intensity mobbing which some willets perform when they have young.

3. Scolding and Agitated Circling

Scolding. Loud vocalizations allow an intruder to locate the calling bird. The vocalizations given by shorebirds when an intruder is present could function either as threats to discourage a closer approach, or as conspicuous signals that focus the intruder's attention on the adult, thereby enhancing the effectiveness of distraction displays. Many shorebirds give loud characteristic "scolding" calls in the presence of an intruder. These calls may accompany mobbing, or may be given as the bird circles, stands on the ground, or perches at a "safe" distance. Scolding is very common among the Tringini sandpipers, many of which fly from tree to tree calling loudly, as an intruder moves through their territory. Lind (1961) describes a "quivering scolding flight" of the Black-tailed Godwit, which flies in circles around intruders with the wings quivered shallowly. Wilson's Phalaropes will circle around an intruder, often with quivering wing tips (Hohn, 1967).

Agitated Circling. Agitated circling is characterized by birds circling an intruder, usually in flight with loud, apparently anxious calls. This may represent low-intensity mobbing, but in many species it appears to be a characteristic display that does not develop into mobbing; hence, I have separated it from mobbing and scolding. American Avocets (Hamilton, 1975) have an aerial flight display, circling noisily and fluttering in circles near the nest. Willets characteristically make loud alarm-type calls as they circle an intruder, usually birds from several pairs participating in the act (Bent, 1929). Both adult willets, particularly the male, defend young by circling intruders and calling loudly (Vogt, 1938; Howe, 1982). However, I have found that in some willet pairs (7 of 32), the nonincubating mate tends to call loudly and fly away from the nest vicinity, rather than engage in agitated circling flight. Southern Lapwings that are not habituated to humans circle and fly around noisily without actually mobbing humans (personal observation). In New York and New Jersey, American Oystercatchers will circle an intruder giving loud piping calls (personal observation). They remain at a distance of 20 to 50 m, seldom approaching more closely.

4. Ungulate Displays and Related Threat Displays

Ungulates represent a significant hazard for many ground-nesting birds. Mason and MacDonald (1976) report 25% of Common Snipe nests lost by trampling. Many species give different responses to ungulates than to other intruders. They typically sit tight, then suddenly stand erect, with neck and legs fully extended, raise their wings fully (Fig. 5g), and call loudly, often lunging or charging at and sometimes pecking the intruder. Table VI summarizes some species for which "ungulate displays" have been reported. The ungulate displays closely resemble those used in intraspecific aggression (see Fig. 5 and illustrations in Maclean, 1972).

Many species of Lapwings give ungulate displays. Northern Lapwings display toward ungulates near the nest, by standing suddenly, and waving their raised wings. They will charge toward sheep, calling persistently, and displaying their white underwings with dark edges. They actually strike lambs, and then may feign injury. Southern Lapwings stood on the nest and vigorously threatened approaching sheep, calling loudly

Table VI. Some Species for Which a Characteristic Display toward Ungulates Has Been Described

Species	Description	Source
Eurasian Stone Curlew	Wings and tail spread	Witherby *et al.* (1940)
Double-striped Thick-knee	Upright UD[a]	Freese (1975)
Double-banded Courser	Full UD, wings and tail spread, head down	Maclean (1967)
Australian Pratincole	Fluffs, flutters, lunges	Maclean (1967)
Northern Lapwing	Displays and attacks	Ennion (1949)
	Charges sheep waving raised wings	Nethersole-Thompson (1940)
Black-headed Plover	Stands, shrieks, and pecks at legs	North (1937)
Crowned Plover	Stands and pecks	North (1937)
Wattled Lapwing	Stands and pecks	North (1937)
Southern Lapwing	Always aggressive, typical UD	Hudson (1920)
Masked Plover	Aggressive upright, wings out, tips touch ground, tail fanned	Thomas (1969)
Killdeer	Spreads wing, ruffs feathers, runs toward	Bent (1929)
	Sits tight, lunges with spread wings	Stone (1937) Taverner (1936)
Three-banded Plover	Upright aggression to dogs	North (1937)
Red-breasted Dotterel	Flutters upward into animal's face	McKenzie (1953)
Mountain Plover	Flys up into face of cattle	Walker (1955)
	Wing-out rush posture	Graul (1975)
Common Redshank	Typical upright UD with wings up	Witherby *et al.* (1940)
Surfbird	Explodes in face	Taverner (1936)

[a] UD, ungulate display.

all the while (eventually attracting shepherds who took the eggs; Hudson, 1920). African Wattled Lapwings may stand and call if a sheep is nearby, or hold wings vertically and peck at the sheep's head. The same birds moved off silently if a human approached (Little, 1967). Ungulates accounted for much nest loss in the Masked Plover (Barlow *et al.*, 1972), which gives an upright display with wings out and tail fanned (Thomas, 1969). Common Redshanks have typical ungulate displays. Double-banded Coursers will rush at ungulates with wings and tail fully spread and head down in a full threat display (Maclean, 1967). A similar ungulate display is given by Double-striped Thick-knees (*Burhinus bistriatus*).

Although such threat displays may be widespread among the larger plovers, they are not characteristic of smaller species. Killdeer, intermediate in size, perform typical ungulate displays (Stone, 1937; Taverner, 1936). Mountain Plovers threaten cattle by flying in their faces, yet give injury feigning to dogs and humans (Walker, 1955). Surfbirds explode into the face of Mountain Sheep (Hall, 1959).

In some cases ungulate displays effectively turn the intruder aside, whereas in others they actually attract the attention of curious animals (e.g., sheep), and may result in the loss of the nest contents to trampling, or to a hungry shepherd (Hudson, 1920). There are rather few accounts of injury-feigning displays to ungulates [see Hall (1959) for one case in Kittlitz's Plover]. The Australian Pratincole does a distraction display to sheep, fluttering its wings and fluffing its feathers while calling loudly, or it makes short sharp flights, which can divert sheep (Maclean, 1973).

Other Terrestrial Threat Displays. Characteristic ungulate displays are occasionally given to humans [see Plate V in Simmons (1952)]. Thomas (1969) reports Australian Spur-winged Plovers displaying to human intruders with an upright posture with wings extended, tips touching the ground. At maximum intensity the wings were quivered, the body bobbed up and down, and the tail was fanned. Hall (1959) describes a Kittlitz's Plover that, after covering its eggs with sand, turned and ran toward him crouching low with wings and tail widely spread, stood about 0.5 m away, and lunged at him, making "angry" noises. J. Burger (personal communication) describes a Southern Lapwing that ran toward human intruders in a crouched, hunched posture, calling loudly.

Hamilton (1975) describes a "tightrope" display of American Avocets and Black-necked Stilts, in which the bird adopts an upright posture, with wings fully extended symmetrically, the tail fully spread and depressed. Unlike ungulate displays, the neck is usually retracted, and this display is directed toward humans as well as other intruders. The bird faces the intruder, calling continuously, walks slowly toward the intruder, tipping its wings from side to side, but not fluttering them. This display involves motor patterns somewhat different from those of lapwing ungulate displays.

D. Response, Approach, Display, and Luring Distances

An important feature of antipredator displays is the distance between the displaying bird, the intruder, and the offspring. Remarkably few anecdotes mention any of the distances discussed below.

Response Distance. This is the distance between the intruder and the nest when the adult initiates its response. This may be many hundreds of meters in the case of early surreptitious departure or less than 1 m for birds that sit tight. The factors that influence variability in displays, influence the response distance as well. Although there is much inter- and

Table VII. Approach Distances for Shorebirds with the Intruder Standing or Sitting at the Nest (Mobbing and Attack Excluded)[a]

Species	Distance[b]	Behavior	Source
Australian Pratincole	2–3 m	IF	MacGillivray (1924)
Pacific Golden Plover	Very close	(If human is sitting)	Portenko (1981)
Ringed Plover	8 m	Leave early, return	Portenko (1981)
Snowy Plover	Very close	Leave early, return	Take-Tsukasa (1967)
Piping Plover	2–30 m	Variable	Gochfeld (unpublished)
Three-banded Plover	1 m	Called excitedly, CR	Tyler (1978)
Willet	~20 m	During egg phase	Gochfeld (unpublished)
Buff-breasted Sandpiper	10 m	ACF late incubation	Sutton (1967)
Dunlin	3 m	Return to brood	Taka-Tsukasa (1967)

[a] Abbreviations: IF, injury feigning; CR, crouched run; ACF, agitated circling flight.
[b] This is the closest distance to which a shorebird approaches when the intruder is at the nest.

intraspecific variation in response distance, in practice there is a dichotomy between early departure and sitting tight.

Analysis of intraspecific variation in response distance is needed. For example, Brown (1962) found that in Western Sandpipers, birds that sit tight are more likely to give crouched run and to alternate approach with departure, while birds with a response distance greater than 10 m are less likely to alternate and are more likely to perform injury feigning.

Approach Distance. This is the closest point that the displaying bird approaches the intruder, whether mobbing or giving distraction displays. Table VII lists some approach distances reported when a human stood at the nest or captured the young. Species that mob vigorously have low approach distances, whereas more fearful species may not approach closer than 50 m. There is a striking decrease in approach distance after chicks are captured. Willets that approached no closer than 15 m, came to within 6 m when a chick I had captured began giving distress notes. Piping Plovers and Red Phalaropes approached to within 2 m (personal observation; Bengtson, 1968), and Least Sandpipers (*Calidris minutilla*) to within 0.7 m (Austin, 1932). Purple Sandpipers sat on the hand that was collecting eggs from its nest (Portenko, 1981).

Display Distance. This is the usual distance between the bird and the intruder when distraction displays are given. Some species distract within 2 m of the intruder, others may remain at approximately 15–20 m. In my experience, Killdeer tend to exemplify the latter category, but Deane

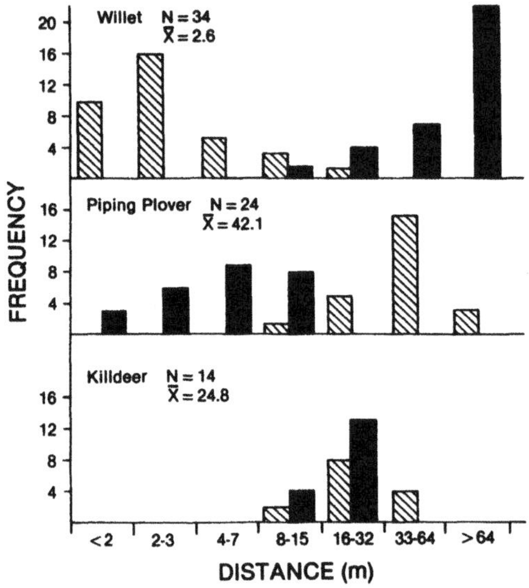

Fig. 7. Histograms comparing response distance and display distance for three shorebird species. Response distance indicated by hatched bars; display distance indicated by solid bars.

(1944) reported that the Killdeer display distance to humans was about 10 m during incubation and decreased to 3 m at hatching, while to dogs the display distance was 2 m. Figure 7 compares the response and display distances for three species.

Luring Distance. Some species will lure an intruder for long distances away from the nest before flying away or circling back to the nest. Some examples of luring humans include 50–100 m for the Piping Plover and Killdeer (personal observation; Gabrielson, 1922), 200 m for the Red Knot (Portenko, 1981), several hundred meters for the Rock Sandpiper (Bent, 1927), and up to 1200 m for the Purple Sandpiper (Bengtson, 1970). Lynes (1910) reported a lapwing luring a stoat (*Mustela erminea*) 400 m.

E. Motor Patterns or Acts

There is a great variety in the motor patterns used by shorebirds in antipredator behavior. The nominal categories used above provide guidelines for describing behavior but are not, themselves, adequate for all species or for all behavior. Some motor patterns are difficult to classify. Moreover, changes in intensity of display may result in different classifications. Particularly in a comparative study it is necessary to provide useful descriptions of the motor patterns. In reviewing the evolution of

Table VIII. Motor Patterns Employed in Shorebird Antipredator Behavior[a]

Body part	Motion	Intensity
	Primary motor patterns	
Body orientation	Inclination	Degree or angle of inclination from horizontal
Neck orientation	Inclination	Degree or angle of inclination from horizontal
Neck	Extended vs. retracted	Percent of full extension
Head and bill orientation	Inclination	Degree or angle of inclination from horizontal
Tail orientation	Inclination Cocked vertically vs. depressed to substrate	
Tail	Fanned vs. closed	Percent of full expansion
Legs (left and right)	Extension vs. flexion	Percent of full flexion or extension
Wings (left and right)	Abducted vs. adducted	Percent of full abduction or extension
Wings: Humerus	Inclination	Degree or angle of inclination from horizontal
	Rotation Undersurface forward (+) vs. dorsum forward (−)	Degree or angle of inclination from horizontal
	Flexion (forward) vs. extension	
	Locomotion	
Walk		
Run	Includes "charging"	
Fly		
Lunge	Move forward less than one body length	
	Secondary motor patterns[b]	
Head bobbing	Rotation of tail	
Body bobbing	Rocking of body downward	
Bill opening and closing	Rocking of body from side to side	
Eyes: Open or closed	Droop wing tips	
Active extension or retraction of neck	Walk toward or away or circle	
Active extension or retraction of legs (standing vs. crouching)	Alternate movements rapidly with fluttering wings	
	Run toward or away or circle	
Active abduction or adduction of wings (flapping)	Fly toward or away of circle	
Fanning of tail	Fly with downward swoop and/or upward soar	

[a] This is not an exhaustive list of possible movements. Vocalizations are excluded from this preliminary list.
[b] Secondary motor patterns are similar to "act systems" of Carpenter and Ferguson (1977).

stereotyped behavior in reptiles, Carpenter and Ferguson (1977) provided long lists of "acts" (specific movements) or "act systems" (specific behavior patterns) for the major taxa. Table VIII lists certain behavioral acts or motor patterns that appear important in shorebird antipredator behavior. Table IX provides an analysis of orientation and deployment for selected examples from Figs. 2, 3, and 5.

Table IX. Analysis of Motor Patterns for Selected Shorebird Nest Protection Displays[a]

Orientation	Eurasian Avocet threat (Fig. 5k)	Greater Painted Snipe threat (Fig. 5o)	Buff-breasted Sandpiper threat (Fig. 5m)	Piping Plover distraction display Fig. 2d	Fig. 3d	Fig. 3f
Body inclination	−20	−30	75	60	20	0
Neck inclination	0	40	45	45	30	10
Neck extension	+	0	−	0	0	+
Head inclination	−30	−30	0	0	−10	−20
Tail inclination	90	30	10	−30	−30	−20
Tail fanning	20	90	?	25	90	90
Leg extension	90	90	50	50	10	80
Wing abduction (L/R)	90	100	90	50/0	10/90	80/90
Humerus elevation	−30	20	30	30/−45	−20/−10	−10
Humerus rotation	0	−45	80	−10	−10/0	0

[a] Positive value = upward inclination, extension, or abduction, rotated up. Negative value = downward inclination, flexion, or adduction, rotated down. Inclination, elevation, and rotation measured in degrees. Extension, fanning, abduction measured as percent of full range of motion.

F. Stereotypy

Many authors (e.g., Friedmann, 1934; Armstrong, 1954; Skutch, 1955) have discussed the extent to which distraction displays are stereotyped. No quantitative studies are available on shorebird distraction display stereotypy, but it appears that the highest intensity of an injury-feigning display (see for example Fig. 3e) may be stereotyped. On the other hand, shorebirds show seemingly endless flexibility in the sequence and intensity of behaviors performed, resulting in the lack of overall stereotypy which impressed Skutch (1976). Detailed study is needed to examine the consistency of motor patterns from one performance or from one bird to another, as well as the transitions among behaviors, determining under what circumstances one transition becomes more likely than another.

Remarkably few previous observers have emphasized a constant stimulus (see Brown, 1962, for an important exception). As many injury-feigning displays involve asymmetrical deployment of the wings and tail, it is valuable to examine whether there is any consistent preference for greater extension of one wing or the other among particular individuals, whether alternation of side occurs, or whether this is determined by the position of the intruder. Carpenter and Ferguson (1977) illustrate how stereotyped behavioral acts can be incorporated into highly variable behavior.

G. Reentrapment

One of the important features of distraction displays is reentrapment, the tendency of the performing bird to keep observing the responses of the intruder, and periodically returning toward a recalcitrant intruder as if to recapture its attention. Bengtson (1970) notes that a Purple Sandpiper that is not followed stops its display and returns closer to the intruder to try to "hook" him again. Hamilton (1975) notes that American Avocets move away from an intruder, decreasing the paces of their display when followed, but if not followed the bird approaches and even follows the intruder, increasing the intensity of its display. Reentrapment has also been reported for Australian Pratincole (Maclean, 1973), American Golden Plover (Bent, 1929), Dunlin (Hortling and Baker, 1932), Red Knot (Bent, 1927), and Wilson's Phalarope (Bent, 1927). Reentrapment is prominent in the Piping Plover (personal observation). When I remained standing at the nest, the birds repeatedly performed injury feigning or crouched run, moving away from the nest, then returning to repeat the display, often initiating a new display within 2 m of me.

Matthiesen (1967) was impressed that the reentrapment practices by shorebirds after performing a rodent run represented a flaw, for the predator would recognize that it was not dealing with a fleeing rodent, and could use the reentrapment behavior as a clue to locating the nest.

H. Transitions among Distraction Display Components

Shorebirds responses to intrusion include alternations and transitions among several distinct motor patterns, sometimes in response to a stimulus change. For example, Duffey *et al.* (1950) describe a Purple Sandpiper that performed a rodent run but switched to close agitated circling when a chick was captured. Brown (1962) reported Western Sandpipers (*Calidris mauri*) switching rapidly between injury-feigning and rodent run-type displays, with some birds incorporating motor patterns that were intermediate. Townsend and Allen (1907) describe how a Spotted Sandpiper alternately lured and attacked dogs.

Nethersole-Thompson (1973) describes rapid shifts between rodent run and sudden stops analogous to false brooding in the Eurasian Dotterel. He describes the bird disturbed from the nest moving forward, then running back toward the nest, stopping, starting, circling, disappearing momentarily, then rushing forward toward the intruder, briefly pausing to snap at a fly (false feeding), rapidly racing away, then sitting (false brooding), then shuffling away, flailing its wings and trailing its white-tipped

Table X. Comparison of Responses to the Author Given by Three Species of Shorebirds (Up to 10 Behavior Patterns Recorded for Each Individual)

Response	Piping Plover	Killdeer	Willet
Number of nests approached	17	6	27
Number of approaches	24	14	34
Early departure[a]	96%	71%	0%[e]
Intermediate departure[b]	4%[d]	29%	12%
Late departure and stayed away[c]	0%	0%	88%
Dropped to ground near nest	0%	0%	46%[f]
Impeded flight away from nest	0%	0%	52%[g]
Upright run	38%	71%	0%
Crouched run	92%	50%	0%
Repeated crouched run	25%	29%	0%
False brooding	38%	36%	0%
Deliberate upright walk	0%	0%	3%
Circle around	53%	7%	9%
Injury feign	82%	86%	0%
Injury feign and attack	4%	0%	0%
Mobbing	0%	0%	—[h]

[a] Response distance (RD) > 25 m.
[b] RD between 5 and 25 m.
[c] RD < 5 m.
[d] One bird nesting at base of dune was surprised on the nest at 5 m.
[e] Other nesting willets were believed to have departed early but their nests were not found.
[f] Based on 13 events in 1983.
[g] Based on 27 events.
[h] Mobbing does occur in chick phase.

tail, and squeaking like a small mammal in distress. The author's frenetic prose captures the rapid changes between crouched run, aggressive approach, false feeding, injury feigning, and rodent run. My observations of Killdeer and Piping Plover also show rapid transitions between running, injury feigning, and false brooding.

Eurasian Avocets (*Recurvirostra avosetta*) rapidly alternate aggressive circling and mobbing with a terrestrial aggressive display and occasionally with broken wing displays (Olney, 1970). False-feeding bouts are often interspersed with other distraction displays as for example with mobbing in the Crowned Plover (Skead, 1955) and with crouched run and mobbing in the Masked Plover (Thomas, 1969). Figure 1 diagrams some common sequences of antipredator behavior, and the loops indicate common behavioral transitions and alternations.

Piping Plover. Using a transition matrix I examined behavioral transitions in the Piping Plover. I observed 26 Piping Plovers at 19 nests. Where two birds at a nest were studied, at least one was banded. The

Table XI. Transition Matrix for Distraction Behavior Given by 20 Adult Piping Plovers to an Intruder 5 m from the Nest[a]

Activity	Walk or run[b]	Fly[c]	False brood	Injury feign
Walk or run	8	8	7	11
Fly	7	5	3	6
False brood	8	0	3	4
Injury feign	14	3	6	6

[a] $\chi^2 = 9.1$, $df = 9$, $p > 0.30$.
[b] Walking or running including crouched run, upright run, rodent run.
[c] Flying including flying toward or away and agitated circling flight.

sequence of motor patterns was recorded. For each bird I used only the first five behavioral transitions (see Table XI) given while I stood 5 m from the nest. Behaviors were grouped as walking or running (including crouched, rapid, and rodent runs), flying, false brooding, and injury feigning. The most prominent activity was walking or running (38%), with injury feigning next (28%). However, the transitions did not differ significantly from random ($\chi^2 = 9.1$, $df = 9$). This should not be construed as evidence that birds behave randomly, but rather that there is no fixed sequence of activities. Further analysis will require examination of transitions under controlled conditions to determine whether particular behavioral changes can be predicted and whether dyadic or triadic sequences can be identified.

IV. FACTORS INFLUENCING NEST PROTECTION BEHAVIOR

It is apparent that there is great inter- and intraspecific variability in avian distraction behavior (Skutch, 1976). Several classes of factors must be examined: (1) the state of the respondent, (2) the nature of the stimulus, and (3) the conditions under which the stimulus and response occur. The individual factors are shown in Table XII. Although the type and behavior of the intruder are of paramount importance, there is virtually no information on shorebird responses to natural predators. This unfortunate gap requires extrapolation and speculation based on experience (often anecdotal) with the human as intruder. Experimental studies based on controlled intrusion are important.

Table XII. Factors Influencing Qualitative and Quantitative Variation in Antipredator Behavior of Shorebirds

Characteristics of individual
Interspecific variation
Spatial variations: geography and populations
Temporal variations: season and cycles
Sex
Habituation and learning
Age and experience
Individual variation
Characteristics of intruder
Type of intruder
Nature of intruder's approach and posture
Influence of intruder's response to display
Characteristics of the context
Habitat
Weather
Observer variability

A. Characteristics of Individual

1. Interspecific Variation

Table X and Fig. 7 compare the responses of three species that I studied. The categories (Table X) are arranged in an approximate order of increasing intensity of distraction display. Piping Plovers almost invariably left early (response distance exceeding 25 m), while Willets almost invariably sat tight (response distance usually less than 3 m). Killdeer showed a greater tendency to perform upright runs, whereas the Piping Plover tended to perform crouched runs. Frequency of injury feigning was similar in both plovers, although the motor patterns differed. Willets, on the other hand, almost invariably flew from the nest, often with a brief pause or impeded flight. Only one Willet performed a slow, deliberate "upright walk." Willets nesting in sandy areas in Common Tern colonies usually leave early and circle noisily (personal observation). The response distances and display distances (Fig. 7) reveal three types of behavior, early departure and close displays in the Piping Plover; sitting tight and remote displays in the Willet; and intermediate response and display distances in the Killdeer. Studies of these species at other localities will reveal whether these categorizations are representative of the species, or are modified in other habitats.

Table XIII compiled from Drury (1961) compares distraction display activities among three Arctic Plovers. One can infer that in certain re-

Table XIII. Comparison of Nest Protection Behavior by Three Species of Plovers[a]

Black-bellied Plover	American Golden Plover	Semipalmated Plover
	Low intensity	
Head up alarm	Head up alarm at 75 m	Flies to greet intruder
False feeding	False feeding	False feeding
Runs with head down	Runs with head up	Runs with head down
Tail barely fanned	Tail barely fanned	Tail fanned
Tail tilted toward intruder		
Wings drooped		Near wing slightly extended and dragged
Ruffled back feathers	Ruffled back feathers	
Rushes forward	Stands upright	Faces away from intruder
	High intensity	
Prostrate on ground	Breast on ground	Wriggles away on ground tilted on side
Bill thrust forward		
Tail cocked, fanned	Tail slightly cocked and fanned	
Wings 90% spread	Wings partly open	Beats one wing up in air
Feet stamping	Feet stamping	
Faces to side of intruder	Faces intruder	
	Response to pursuit	
Runs off rolling to side	Crouched run	Crouched run
Wings folded, lowered	Wings open, down at side	Wings folded
Tail spread	Tail canted downward, partly fanned	Tail raised and folded

[a] After Drury (1961).

spects Black-bellied Plover behavior closely resembles that of American Golden Plover; in other cases it more closely resembles Semipalmated Plover. The interspecific differences may be exaggerated, as the text captures the more prominent features of each species' displays.

There is, however, substantial taxonomic variation among the shorebirds in qualitative and quantitative aspects of distraction behavior. Significant interspecific variability, therefore, must be examined among closely related species. For most taxa comparative studies remain to be done. Many anecdotal accounts lead to the impression of remarkable similarity of displays among closely related species, for example, members of *Charadrius* or *Calidris*. Portenko (1981) remarks that Temminck's

Stint, Red Knot, Ruddy Turnstone (*Arenaria interpres*), Ringed Plover, and large species of *Calidris* tend to be similar in distraction behavior, but Rufous-necked Stint (*C. ruficollis*), Little Stint, Curlew Sandpiper, and Dunlin formed a group of species with different displays. Investigations of interspecific differences in distraction displays among closely related species would have to examine habitat-related differences as well as differences in the stimuli eliciting displays.

2. Spatial Variations: Geography and Populations

Some shorebird species show substantial geographic variation in antipredator behavior. The northern race of Greater Golden Plover has well-developed injury feigning in the chick phase, contrary to the southern race (Williamson, 1948). Even within Britain, plover nest-departure behavior varies from place to place, some populations being characterized by sitting tight, others by leaving early. This may be partly influenced by density [birds in dense groups tend to leave earlier (Ratcliffe, 1977)]. Similarly, Red Phalaropes in Greenland (Manniche, 1910) and on Bathurst Island (Mayfield, 1979) perform distraction displays, but Parmelee *et al.* (1967) did not see distraction displays on Victoria Island. Distraction displays are common in the American form of the Common Snipe (Bent, 1927; Mousley, 1939), and on the Faeroe Islands (Williamson, 1950b), but not elsewhere in Britain or Europe. Similarly, Greater Golden Plovers and Eurasian Oystercatchers in Britain (Ellison and Johnston, 1946; Cawkell, 1948) and The Netherlands (Haverschmidt, 1946) seldom give distraction displays, while such behavior is more frequent on offshore islands (e.g., Benson, 1948; Williamson, 1943, 1952). In Costa Rica, injury feigning is common in the Northern Jacana population at El Viejo but not at Turrialba (Jenni and Betts, 1978).

McCaffery (1982) found that American Golden Plovers mob jaegers at Franklin Bluffs but not at Barrow, Alaska, perhaps due to the presence of the Pomarine Jaeger (*Stercorarius pomarinus*) at the latter site. In Alaska, Whimbrel tend to be inconspicuous during the nesting season unlike birds at Churchill (Gabrielson and Lincoln, 1959) or on the Faeroes (Williamson, 1946).

3. Temporal Variations: Seasons and Cycles

In many species there is a marked change in the types of nest protection behavior late in incubation or early in the chick phase. In general the tendency to distract is low early in the breeding season. During incubation the tendency may increase gradually, or it may increase dra-

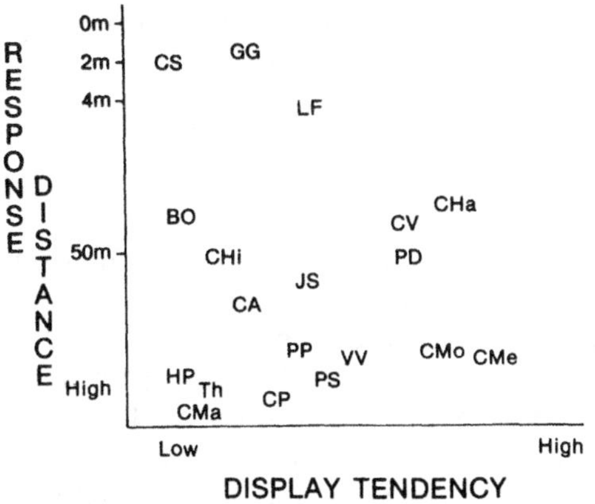

Fig. 8. Relationships between response distance and display intensity for selected taxa. BO, Stone Curlew; CA, Dunlin; CHa, Ringed Plover; CHi, Stilt Sandpiper; CMa, White-fronted Sandplover; CMe, Piping Plover; CMo, Mongolian Plover; CP, Kittlitz's Plover; CS, Willet; CV, Killdeer; GG, Common Snipe; HP, American Oystercatcher; JS, Northern Jacana; LF, Broad-billed Sandpiper; PD, American Golden Plover; PP, Ruff; PS, Black-bellied Plover; Th, seedsnipes (*Thinocorus*); VV, Northern Lapwing.

matically just before or just after hatching. Figure 8 illustrates how the intensity and display distance might vary with stage of breeding cycle, Fig. 9 indicates alternative models for display change with cycle, and Table II lists some examples of quantitative and/or qualitative changes in either frequency or intensity of nest protection behavior. While chicks are dependent, distraction behavior may be very vigorous, but it generally decreases either gradually as chicks grow or dramatically when they fledge. Some species give distraction displays even when young are flying

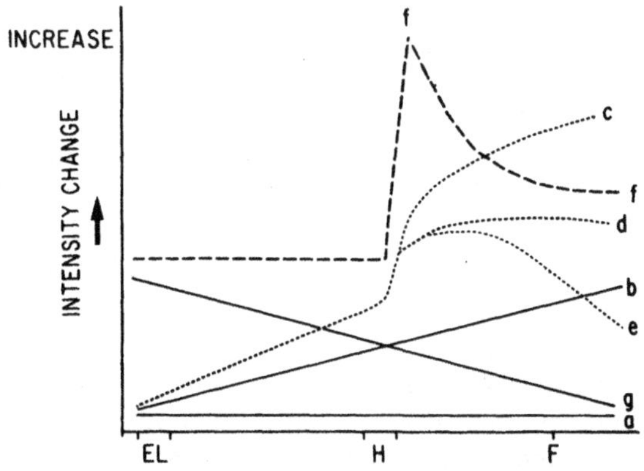

Fig. 9. Models of variation in intensity of antipredator displays as a function of the breeding cycle. a, no change with time; b, gradual and steady increase; c–e, increase during incubation with sudden increase near hatching; c, increase continues after hatching; d, intensity plateaus after hatching; e, intensity declines after hatching; f, no change during incubation with sudden peak at hatching, followed by decline (or plateau, not shown); g, decline during the reproductive cycle.

[e.g., Eurasian Dotterel (Armstrong, 1952)]. Rarely distraction displays continue even after young have died (Williamson, 1951). Mobbing may sometimes occur outside of the breeding season.

4. Sex

One would predict that sex would be an important factor influencing departure tendencies, attack behavior, and distraction displays. Despite the numerous anecdotal accounts, it is difficult to interpret information on sex roles in aggression and distraction displays. In most species it is difficult to distinguish sexes in the field. Some can be sexed at close range only by an experienced field observer. Many older papers ascribe certain roles to females apparently on the assumption that only females did most or all incubation, with sex being verified in some instances by collected specimens. Recent information on the complex array of shorebird mating systems (particularly scolopacids) reveals that in some species only one bird of a pair may be in attendance at a nest [e.g., Sanderlings (Parmelee, 1970)], while in others the male plays the predominant role (e.g., phalaropes). Table XIVa is a partial list of accounts that appear to reliably ascribe different roles to males and females; accounts of relatively equal roles are listed in Table XIVb.

5. Habituation and Learning

Habituation is the reduction in the frequency or intensity of a response upon repeated presentation of the stimulus. It differs from learning in that cognitive processes are presumed not to be involved. These processes play important roles in modifying all behavior, particularly in so-called higher organisms, and a discussion of these would be out of place in this chapter. Nonetheless, it is worth mentioning examples where distraction display modification upon repeated exposure has been described. There are few accounts in the literature, yet such modification occurs quite rapidly upon repeatedly flushing a sitting bird from a nest. The apparent habituation may be very rapid, with birds responding very differently on a second visit than on the previous visit to the territory. One difficulty in studying habituation is that behavior changes anyway with time during the course of the breeding cycle. Moreover, changes in fearfulness may occur at the same time as habituation, resulting in unpredictable responses. Mobbing behavior is also subject to habituation (e.g., McNicholl, 1973), and this remains to be examined in shorebirds.

Sanderlings that usually leave the nest early may sit tight after repeated exposure to a human intruder (Parmelee, 1970). Observing indi-

Table XIVa. Accounts of Sex Differences in Nest Protection Behavior[a]

Species	Female	Male	Source
Gray-breasted Seed-snipe	—[b]	RR and jump flight	Maclean (1969)
Eurasian Stone Curlew	Steals away	Stands guard, distracts	Dement'ev *et al.* (1969)
Variable Oystercatcher	More aggressive	$N = 1$	Jones (1979)
Northern Lapwing	99% incubation	ACF	Ennion (1949)
	Leaves silently	—[b]	
	Mobs		
	Most care	Guards; some care	Witherby *et al.* (1940)
Black-bellied Plover		Anxious m > f	Johnson and Goodall (1965)
	Deserts earlier	Remains with brood	Hussell and Page (1976)
Little Ringed Plover	CR f > m		Armstrong (1952)
Killdeer		DD m > f	Mundahl (1982)
Red-breasted Dotterel		Main nest protector	Johnsgard (1981)
Eurasian Dotterel	Little tending	Incubates	Witherby *et al.* (1940)
Diademed Sandpiper-plover	Hides	Approaches and DD	Johnson and Goodall (1965)
Pheasant-tailed Jacana	Abandons	Tends chicks, IF	Johnsgard (1981)
Bronze-winged Jacana	—[b]	Tends chicks	Mathew (1964)
Northern Jacana	Abandons	All care	Miller (1931)
	Some defense	± all incubation	Jenni and Betts (1978)
	No DD	Broods and DD	
Wattled Jacana		All incubation	Osborne and Bourne (1977)
Greater Painted Snipe	—[b]	Mainly male care	Komeda (1983)
Spotted Redshank	—[b]	Incubates, tends	Witherby *et al.* (1940)
Common Greenshank	—[b]	Wing-quivering	Witherby *et al.* (1940)
Green Sandpiper	Most incubation	Most care of young	Witherby *et al.* (1940)
Wood Sandpiper	—[b]	Guarding m > f	Witherby *et al.* (1940)
Willet	Guards early, then abandons	Guards throughout	Howe (1982)
Terek Sandpiper	Distracts	Flys out to scold	Dement'ev *et al.* (1969)
Wilson's Phalarope	Occasional DD	Usual DD even in LEP	Hohn (1967)
Red-necked Phalarope	—[b]	Most or all care	Bent (1927), Witherby *et al.* (1940)

(Continued)

Table XIVa. (*Continued*)

Species	Female	Male	Source
Red Phalarope		Most or all care	Bent (1927), Witherby *et al.* (1940)
		All incubation	Ridley (1980)
Slender-billed Curlew	Incubates	Stands guard	Witherby *et al.* (1940)
Hudsonian Godwit	Sits tight	Stands guard	Hagar (1966)
Bar-tailed Godwit		More demonstrative	Witherby *et al.* (1940)
Ruddy Turnstone	Incubation f > m	Care of young m > f	Witherby *et al.* (1940)
Red Knot	Does not incubate	Incubates, DD	Dement'ev *et al.* (1969)
		Great anxiety	Witherby *et al.* (1940)
Western Sandpiper	Both incubate	m > f posthatch	Holmes (1971)
Least Sandpiper		Care m > f	Bent (1927)
White-rumped Sandpiper	Only attendant	Abandons	Parmelee *et al.* (1968)
Baird's Sandpiper		Incubates and IF	Portenko (1981)
Pectoral Sandpiper	Incubates	Abandons	Flint and Kistchinski (1972)
Purple Sandpiper		All care?	Witherby *et al.* (1940)
Rock Sandpiper	Incubates	Incubates m > f	Bailey (1943)
Broad-billed Sandpiper	—[b]	Incubates and tends	Flint (1973a)
Buff-breasted Sandpiper	Incubates and tends	Abandons	Sutton (1967)
Ruff	All parental care	Abandons	Bent (1929)
Short-billed Dowitcher	Mainly EEP, then deserts	More in LEP	Kistchinski and Flint (1973)
	Main incubation	Most care	Witherby *et al.* (1940)
Common Snipe	Most incubation	Some incubation	Witherby *et al.* (1940)
Eurasian Woodcock	Incubates, tends young	—[b]	Witherby *et al.* (1940)

[a] Abbreviations: m, male; f, female; EEP, early egg phase; LEP, late egg phase; ACF, agitated circling flight; CR, crouched run; DD, distraction display; IF, injury feigning; RR, rodent run.
[b] The sex contributes little or none to the particular activity.

Table XIVb. Species for Which Relatively Equal Participation by Males and Females Has Been Reported[a]

Species	Behavior	Source
Double-striped Thick-knee	Both exhibit early departure	Freese (1975)
Eurasian Oystercatcher	Both incubate, tend young	Witherby *et al.* (1940)
American Oystercatcher	Both distract	Oberholser (1938)
Black-winged Stilt	Both incubate	Bates and Lowther (1952)
Eurasian Avocet	Both scold and mob	Olney (1970)
Egyptian Plover	Both cover eggs and chicks	Howell (1979)
Collared Pratincole	Both incubate, tend young	Witherby *et al.* (1940)
Yellow-wattled Lapwing	Both incubate	Ali and Ripley (1969)
Wattled Plover	Both incubate, tend young	Little (1967)
Sociable Plover	Both incubate, tend young	Witherby *et al.* (1940)
Southern Lapwing	Both tend young	Walters and Walters (1980)
Red-wattled Lapwing	Both incubate	Ali and Ripley (1969)
Black-bellied Plover	Both participate equally	Johnsgard (1981)
	Both incubate, tend young	Witherby *et al.* (1940)
	Both incubate, DD	Hussell and Page (1976)
Greater Golden Plover	Both approach, scold	Ratcliffe (1977)
	Both approach, DD	Witherby *et al.* (1940)
Pacific Golden Plover	Both upright run, whistle	Portenko (1981)
Ringed Plover	Both incubate, IF	Johnsgard (1981)
Semipalmated Plover	Both incubate	Peters and Burleigh (1951)
Little Ringed Plover	Both incubate, IF	Ali and Ripley (1969)
Killdeer	Both incubate, IF, tend young	Bent (1929)
	Both incubate, tend young, IF	This chapter
	Both incubate and tend equally	Lenington (1980)
Piping Plover	Both incubate, IF	This chapter
	Both IF	Cairns (1982)
Snowy Plover	Both incubate, IF	Ali and Ripley (1969)
Kentish Plover	Both incubate, tend young	Witherby *et al.* (1940)
Double-banded Dotterel	Both DD	Cunningham (1973)
Two-banded Plover	Both incubate, good IF	Johnsgard (1981)
Mongolian Plover	Both incubate, IF	Ali and Ripley (1969)
Greater Sand Plover	Both tend nestlings	Dement'ev *et al.* (1969)
Rufous-chested Dotterel	Both call	Johnsgard (1981)
Eurasian Dotterel	Both guard chick	Nethersole-Thompson (1973)
Northern Jacana	Both guard young	Jenni and Betts (1978)
Common Redshank	Both incubate, IF	Ali and Ripley (1969)
	Both tend young	Witherby *et al.* (1940)
Common Sandpiper	Both tend young	Witherby *et al.* (1940)
Red-necked Phalarope	Both guard young	Austin (1932)
Upland Sandpiper	Both protect boldly	Oberholser (1938)
Little Stint	Both distract	Dement'ev *et al.* (1969), Bent (1927)
Dunlin	Both incubate, tend young	Witherby *et al.* (1940)

(Continued)

Table XIVb. (*Continued*)

Species	Behavior	Source
Whimbrel	Both incubate, tend young	Witherby *et al.* (1940)
Long-billed Curlew	Both distract	Bent (1927)
Black-tailed Godwit	Both demonstrate	Witherby *et al.* (1940)
Semipalmated Sandpiper	Both sit tight	Kuzyakin (1959)
	Both incubate and IF	Bent (1927)
Western Sandpiper	Both IF	Holmes (1973)

[a] Many authors (particularly before 1960) assumed that the bird standing guard is the male and the incubating bird is the female; such records are excluded from the above table.
[b] Abbreviations: DD, distraction display; IF, injury feigning.

vidually marked Piping Plovers at three nests, I found that when a bird was flushed three times in a 2-day period, the injury-feigning display, which was intense after the first visit, was not manifest by the third. By 3 days, at least one of the birds had "recovered." Alternatively, by standing for 30 min next to Piping Plover nests one can see the decline in intensity of behavior in terms of the proportion of time the bird is displaying and the proximity to the intruder (Gochfeld, unpublished data). Portenko (1981) describes multiple visits to nests of Red-necked Phalarope and Pectoral Sandpiper, with a marked decrease in the response distance of the former and in the approach distance of the latter. In studies of habituation one must be prepared to ask whether a change in behavior represents habituation or a new strategy.

6. Age and Experience

Although age and experience influence several aspects of reproductive biology, I have found no comparison of variation in shorebird antipredator behavior in relation to either factor. Skutch (1976) suggests that birds whose behavior successfully lures a predator and saves the nest, may learn from this experience and reuse the behavior, perhaps with embellishments, the next time a potential predator approaches.

7. Individual Variation

In the final analysis of most behavior there remains a substantial proportion of unexplained variability that must be ascribed to individual variation. Skutch (1976, p. 411) was particularly impressed with individual variation in distraction display. Thus, after taking into account membership in populations with certain characteristics, stage of breeding cycle,

Table XV. Descriptions of Individual Variation in Antipredator Behavior

Species	Description	Source
	Variations in nest departure	
Double-banded Courser	Variable response distance	Maclean (1967)
Masked Plover	If good visibility, leaves early If surprised, flies with alarm call	Thomas (1969)
Greater Golden Plover	Related to density?	Ratcliffe (1977)
Pacific Golden Plover	Some run or fly early; others creep away	Portenko (1981)
Eurasian Dotterel	Some leave early, some late	Nethersole-Thompson (1973)
Spotted Sandpiper	Sometimes sits tight	Mousley (1939)
Red Phalaropes	Variable response distance	Mayfield (1979)
Semipalmated Sandpiper	Variable response distance	Ashkenazie and Safriel (1979)
Pectoral Sandpiper	Variable response distance and IF	Flint and Kistchinski (1972)
Sanderling	IF or RR, very variable	Parmelee (1970)
	Variations in distraction behavior	
Crowned Plover	Most aggressive, gave best IF	Ade (1979)
Little Ringed Plover	Only one IF seen at several nests	Parrinder (1948)
Three-banded Plover	In crouched run, wings sometimes extended	Tyler (1978)
Temminck's Stint	4 of 30 cases chick-feign (N = 20)	Flint (1973b)
Dunlin	Variable behavior, no details	Portenko (1981)
Sanderling	RR, IF	Parmelee (1970)
	Low variability	
Semipalmated Plover	Each of 30 pairs gave DD	Austin (1932)
Killdeer	Consistent IF posture	Friedmann (1934)
Purple Sandpiper	Always sits tight	Larson (1960)
Broad-billed Sandpiper	Chick-feign in 21 of 21 cases at 6 nests	Flint (1973b)

[a] Abbreviations: DD, distraction display; IF, injury feigning; RR, rodent run; N, number of nests.

age and experience, sex, previous experience with intruder, and habituation or learning, individual birds will still show behavioral differences. Some examples of individual variation are given in Table XV.

Individual variation in response distance is described for Double-banded Coursers (Maclean, 1967), Greater Golden Plovers (Ratcliffe, 1977), and many other species. Individual Red-capped Dotterels (*Char-*

adrius ruficapillus) seemed to behave differently vis-à-vis distraction displays, with no regard to season or other identifiable factors (Hobbs, 1972).

I examined individual variation in three species during incubation (Table X). At 23 Willet nests, individuals sat tight (response distance < 5 m) in 88% of cases. Piping Plovers conversely left early in 95% of cases. Killdeer also left early in all cases, but their response distance averaged much lower than that of the Piping Plover. About 60% of Willets departed with a flight low over the marsh and/or water, with their legs dangling and with shallow rapid wing-beats, which I interpret as a low-intensity impeded flight. The two plovers showed strong tendencies to perform injury feigning, but not all individuals did so, even late in incubation. For those individuals that were marked, there was a consistently lower likelihood of seeing injury feigning on my subsequent visits to the nest.

B. Characteristics of Intruder

Various characteristics of the intruder must be examined to see how such stimulus variation induces response variation among breeding shorebirds. Obviously, birds should differ in response to an intruder, depending upon whether it is recognized as a potential threat or as benign (Curio, 1976). This subject is one of the most challenging aspects of antipredator behavior, yet little detailed information exists at this time. C. Ristau and D. Griffin (personal communication) propose to investigate this using remote-controlled model predators.

1. Type of Intruder

Shorebirds respond differently to different kinds of intruders (Armstrong, 1954; Skutch, 1955, 1976). For example, Little (1967) reports that African Wattled Plovers crouch when a raptor is seen, aggressively approach dogs and mongooses, seemingly ignore ungulates, yet often run off silently or alternatively mob at the approach of a human. Hobbs (1972) reports that a person approaching on foot elicits early departure and crouch run from Red-capped Dotterels, while approach in a car results in sitting tight; foxes elicit a rodent run. The dotterels vary in response to avian predators, running rapidly about 50 m when corvids approach, but flattening themselves on the nest when small falcons fly over. Black-bellied Plovers show little responses to gulls, flatten on the nest to Long-tailed Jaegers (*Stercorarius longicaudus*), chase Parasitic Jaegers (*S. par-*

asiticus), and greet, mob, and distract human intruders (Hussell and Page, 1976).

Red Phalaropes crouch or run from a human, but give aerial attacks to avian predators such as jaegers (Mayfield, 1979). Clay (1967) reports a Ringed Plover repeatedly giving injury feigning to a conspecific male in a territorial context, and such behavior can be elicited experimentally in the species by presenting a stuffed Ringed Plover (Edwards *et al.*, 1948). Bent (1929) reports Killdeer using broken wing displays to humans and carnivores, while the species gives a threat display to ungulates. Walker (1955) reports Mountain Plover using ungulate displays to sheep and injury feigning to the sheep dogs accompanying them. Stone Curlews give injury feigning to dogs but rarely to humans (Witherby *et al.*, 1940).

Avian Intruders. Many shorebirds simply flatten themselves on the nest when an avian predator flies nearby [e.g., Wattled Plover (Little, 1967), Black-bellied Plover (Hussell and Page, 1976), Australian Dotterel (Maclean, 1976)]. Mobbing is a common response [e.g., in Long-toed Lapwing (Walters, 1979) and Masked Plover (Barlow *et al.*, 1972)]. Red Phalaropes fly to greet approaching jaegers then engage in a luring flight (Mayfield, 1979). Occasionally, distraction displays are given in response to avian intrusion; for example, Page *et al.* (1983) report Snowy Plovers attempting to distract California Gulls (*L. californicus*).

The Human Intruder. Most anecdotal information on antipredator behavior concerns avian responses to the human intruder. Simmons (1955) believes that historically humans have been important predators on shorebirds or their nests, thereby selecting for certain behavior patterns. Thus, the shorebird responses to humans do not represent responses to novel stimuli but to traditional predators. Armstrong (1956) argued that until recently human populations were so small as to represent relatively insignificant predation pressure on noncolonial shorebirds. Moreover, he believes that humans would have quickly learned that distraction displays could be used as clues to the whereabouts of nests, and that such behavior would become dysgenic and would be selected against. Thus, in Armstrong's (1956) view, birds continue to give distraction displays to humans simply because the displays are capable of being elicited by generalized stimuli.

Automobiles. A special case frequently mentioned in the literature is that birds that are very wary of human intruders are generally tolerant of approaching automobiles. This is widely established in the folklore of natural history, and has been specifically mentioned with respect to incubating shorebirds in several cases, e.g., Australian Pratincoles (Bailey,

1935), Greater Golden Plovers (Ratcliffe, 1977), and Red-capped Dotterels (Hobbs, 1972).

2. Nature of Intruder's Approach and Posture

The tendency to perform distraction displays, and the type and intensity of the distraction display vary depending on where the intruder is in the territory, both with respect to the nest and with respect to the adult. Responses to rapidly moving intruders may be different than to intruders that move slowly or aimlessly. Spotted Sandpipers respond differently depending on the intruders position vis-à-vis young. If not near the young, the sandpiper runs rapidly away, repeating this until the intruder is lured far from the young. If the young are approached closely, the bird begins to flutter erratically all about the intruder (Miller and Miller, 1948). Burger (1963) reports that Killdeer initially responded to her approach by flying noisily around her, then switched to distraction displays when she came within 2 m of the nest. Burger and Gochfeld (1981) showed that gulls respond differently depending on the nature of one's approach toward the nest. Establishing eye contact may be a clue to a predatory intent. The intruder's posture also influences behavior, and Armstrong (1954) describes different responses to sitting and standing humans. Birds may recognize individual humans by characteristic garments, which may influence their responses.

3. Influence of Intruder's Response to Display

Many species vary their distraction displays in response to the intruder's behavior. For example, if a displaying adult is followed, it may continue to move away from the nest, often for long distances [e.g., up to 1200 m in the Purple Sandpiper (Bengtson, 1970)]. If the intruder does not follow, the displaying bird may circle back, land directly in front of the intruder, and behave in a manner that may elicit following behavior (reentrapment). Eurasian Oystercatchers change behavior if they are not followed (Williamson, 1952). Spotted Sandpipers doing a crouched run away from the nest stopped and returned closer to intruders that did not follow (Miller and Miller, 1948). Burger (1963) made a similar observation on Killdeer. Reentrapment is widespread among shorebirds that give distraction displays, and indicates that birds constantly monitor the responses of the intruder and modify their distraction displays accordingly. Mousley (1939) recognized this and chose to stop the moment he flushed a bird, and stand still at that point, while he recorded the bird's behavior. In my study I chose instead to stand at a fixed distance from the nest.

Both represent attempts to standardize the impact of intruder behavior on shorebird behavior.

C. Characteristics of the Context

1. Habitat

Armstrong (1954) discussed the influence of habitat on avian distraction noting that distraction behavior (specifically injury-simulation) is accentuated (1) in open, unprotected situations, (2) for ground-nesting species with open accessible nests, (3) for solitary species, (4) for species subject to diurnal predation, and (5) in high northern latitudes. Among shorebirds there are additional considerations. Species that tend to leave the nest early when an intruder is sighted can do so more readily in open habitats where there is good visibility from the nest than in "closed" habitats wherein the intruder may be close to the nest before its presence is discovered. It is in the latter situation that distraction displays arise. Some shorebirds characteristically perform distraction displays in cover, periodically appearing and disappearing as they run behind a bush or crouch in a depression. Such behavior is substantially reduced in habitats with uniform physiognomy.

One must also look at microhabitat or intraterritorial variability in the likelihood that an intruder will elicit a certain response. Meltofte (1977) identified certain areas in a territory at which a distraction display was more likely to be stimulated. These were not simply dependent on distance from the nest.

2. Weather

Larson (1960) noted that for many Arctic species there is an increase in the intensity of distraction displays during periods of bad weather. I suggest that where nest contents are potentially heat-stressed, incubating birds show a strong tendency to return to shelter the nest, and this may reduce the tendency for certain behavior. To the extent that extreme weather conditions favor sitting tight, injury-feigning behavior close to the nest may be favored.

D. Observer Variability

Different observers may be impressed by different aspects of a shorebird's aggressive or distraction displays. This may lead to observer bias in what behaviors are recorded, some workers, for example, emphasizing the behavior patterns given at peak intensity. Attempts to objectively describe behaviors are important, and cinematography or videotaping will be extremely useful.

E. Discussion of Variation

Although some of the inter- and intraspecific behavioral variation can be explained by the above factors, there will always remain unexplained individual variation. The above synopsis indicates a variety of factors influencing the kind, sequence, and intensity of antipredator behavior and distraction displays. These factors, superimposed on individual variability, make it difficult to predict what any individual shorebird will do when danger threatens; however, by examining and quantifying factors, one can construct behavioral models to account for a greater part of the variability than is currently possible by intuitive means.

A great gap in our knowledge involves the role of the predator. The author's field studies indicate a strong ability to influence diversionary behavior depending on how the nest is approached, where one stands in relation to the nest, whether one follows the displaying bird, and whether one returns to the nest. The behavior of a hunting predator involves a complex mixture of all these stimuli. Moreover, species and individuals respond differently to humans, carnivores, avian predators, and ungulates. Even a detailed study of a few shorebird species exposed to a variety of predators would greatly further understanding of these antipredator behaviors.

V. SOCIAL ASPECTS OF NEST PROTECTION BEHAVIOR

It has been recognized that colonial species are less likely to engage in distraction displays than are solitary ground nesters (e.g., Armstrong, 1954), yet many shorebirds, including some with well-developed distraction displays, may nest in clusters, sometimes with such small internest distances that they must be considered colonial (e.g., certain stilts, pratincoles). In such groups, antipredator behavior usually is social. However, even in noncolonial breeders, it is not unusual to have pairs on adjacent territories join in displaying to an intruder. The following discussion examines social interactions in both mobbing and distraction behavior, and interspecific as well as intraspecific participation (Table XVI).

A. Mobbing Behavior: Social Aspects

Although the term "mobbing" connotes a large group of birds swooping, diving, or attacking an intruder, the same behavior can be given by a single bird. Sordahl (1981) noted that mobbing is characteristic of larger species of shorebirds. Marbled Godwit (*Limosa fedoa*) nests are frequently clustered, and Nowicki (1973) found up to 18 birds joining in mobbing him when he intruded near the nest. Similarly, Lind (1961) reported up to 16 Black-tailed Godwits mobbing mammals near the nest. After hatching, adult Hudsonian Godwits mob vigorously with up to 10 birds participating (Hagar, 1966).

Vogt (1938) noted that Willets will mob intruding dogs or crows and that pairs on several territories would join in, making the mobbing more effective. The contagious nature of the Willet song flight used in circling intruders has impressed many ornithologists (Wilson and Bonaparte, 1832; Vogt, 1938; Sordahl, 1981). Even nonbreeding Willets join the mobbing. Makkink (1936) found that Eurasian Avocets engaged in group mobbing that united the birds into a semblance of a colony. Mobbing increased in frequency and intensity as the season progressed. African Spur-winged Plovers also may show clumped nesting and mob as a group (Helversen, 1963). In Britain, Greater Golden Plovers generally nest widely spaced. However, when a pair with chicks flies out to scold and circle a human intruder, it is often joined in this effort by two or three nearby pairs (Ratcliffe, 1977). Ade (1979) notes that Crowned Plovers nesting in close proximity formed a communal early warning system, such that the alarm call of any sentinel caused incubating birds to leave their scrapes before the intruder was near at hand. Nechaev (1980) reported that the scolding behavior given when an intruder enters the territory of Spotted Greenshanks (*Tringa guttifera*) attracts neighbors, and up to 22 birds have been counted participating in the scolding.

Interspecific Social Mobbing. Often adults of two or more shorebird (and occasionally other) species join in circling and mobbing intruders. In Manitoba, Hagar (1966) found Whimbrels, Lesser Yellowlegs (*Tringa flavipes*), Short-billed Dowitchers (*Limnodromus griseus*), and Stilt Sandpipers (*Calidris himantopus*) joining Hudsonian Godwits, while in The Netherlands, Lind (1961) mentions Redshanks joining Black-tailed Godwits. See Table XVI for other examples.

Table XVI. Social Aspects of Mobbing and Distraction Behavior of Shorebirds

Species	Mobbing	Participants	Source
Eurasian Avocet	Circle and call	Whole colony	Hutson and Bannerman (1931)
Northern Lapwing	Circle and call	Several pairs	Ennion (1949)
Crowned Plover	Early warning	Loose colony	Ade (1979)
Greater Golden Plover	Circle and scold	2–3 pairs	Ratcliffe (1977)
American Golden Plover	Aggressive circling flight	8 adults	Sutton and Parmelee (1956)
Wilson's Phalarope	Circling and scolding	Group	Bent (1927)
Black-tailed Godwit	Scold mammals	Up to 16	Lind (1961)
Hudsonian Godwit	Mob vigorously	Up to 10	Hagar (1966)
Marbled Godwit	Mobbing	Up to 16	Nowicki (1973)
Common Redshank	Mobbing	Several	Lind (1961)
Spotted Greenshank	Scolding	Up to 22	Nechaev (1980)
Willet	Mob dogs and crows	Several pairs	Vogt (1938)
Western Sandpiper	Circle intruder	Group	Bent (1927)
Rock Sandpiper	Fly to greet human	Up to 12	Bent (1927)

Species	Distraction behavior	Participants	Source
Black-winged Stilt	Upright stance, wing waving	Many, colony	Armstrong (1954)
Eurasian Avocet	Broken wing display; wing raising	Colony of 30	Ali and Ripley (1969)
Milky Pratincole	Mobbing and distraction display	Many	Ali and Ripley (1969)
Long-billed Curlew	Outcry, try to lure	5–6 males	Bent (1929)
Marbled Godwit	Distraction display	Neighbors	Nowicki (1973)
Pectoral Sandpiper	Chick-feign	3–4 adults	Flint (1973b)
Rock Sandpiper	Circle and lure	Up to 12	Bent (1927)
Dunlin	Distraction display	Adjacent pairs	Meltofte (1979)
Broad-billed Sandpiper	Chick-feign	Other species	Flint (1973b)

Species participating in interspecific mobbing	Source
Black-winged Stilt, Black-tailed Godwit, Northern Lapwing	Drycz (1981)
Short-billed Dowitcher, Lesser Yellowlegs, Hudsonian Godwit, Whimbrel	Hagar (1966)
Black-tailed Godwit, Common Redshank	Lind (1961)
Cream-colored Courser, White-crowned Plover, Egyptian Plover	Hudson and Bannerman (1931)
Spoon-billed Sandpiper (joined by other stints)	Bent (1927)
Black-bellied Plover, Red Phalarope	Mayfield (1979)

B. Distraction Behavior: Social Aspects

The Milky Pratincole nests in colonies of over 100, and birds typically display by landing and squatting on the ground in front of the intruder with open wings. Ali and Ripley (1969) mention that they seem to be brooding, but the description appears to represent a form of injury-feigning display, rather than false brooding. Social distraction display is also seen in Dunlin when adjacent pairs converge on an intruder (Meltofte, 1979). Adult Marbled Godwits not only mob socially, but Nowicki (1973) reported that when an adult was engaged in distraction behavior, neighboring godwits joined in with distraction displays, although they performed at much lower intensity. In contrast, in American Avocets, which nest socially, several pairs may circle intruders, but Hamilton (1975) reported that only birds whose nests are actually approached give distraction displays. Both Black-necked and Black-winged Stilts wave their wings and give short jump flights, and several birds, widely spread over the breeding ground, may display simultaneously (Wilson and Bonaparte, 1832; Armstrong, 1954).

C. Discussion of Social Aspects

It is apparent that social interactions are prominent in those species that engage in mobbing, and sporadic in species giving distraction displays. Mobbing aggregations are more conspicuous than are individuals. It remains to be tested whether shorebirds mobbing or distracting in a group are more effective in discouraging or distracting predators than are solitary birds. Among the Charadriiformes, skimmers, pratincoles, and certain other species provide the opportunity to examine such behavior in truly colonial species, while many shorebirds provide opportunities to investigate social distraction at varying densities.

VI. EFFICACY AND COSTS OF VARIOUS TYPES OF ANTIPREDATOR BEHAVIOR

Insofar as predation is a major hazard faced by virtually all species, effective antipredator behavior must represent a crucial aspect of life history strategies. The ornithologist as an intruder, virtually omniscient with respect to shorebird ecology and behavior, represents a poor evaluator of the efficacy of antipredator behavior. Naive observers, more readily thwarted or distracted, are probably better subjects (see also Armstrong, 1956). Williamson (1950b) describes a case wherein a Faeroe islander was completely deceived by a distraction display and gave chase.

A. Nest-Departure Behavior: Efficacy

Many species of shorebirds enjoy a reputation for the excellent concealment of their nests. Many authors remark on the extreme difficulty in discovering nests, particularly for species exhibiting early surreptitious departure [e.g., Greater Golden Plovers (Ratcliffe, 1977)]. Early departure is such a widespread phenomenon among shorebirds that its relative effectiveness will not be examined here. I find no reports specifically on the efficacy of explosive departures or abortive flights.

B. Aggressive Behavior: Efficacy

Vogt (1938) describes Willets effectively deterring dogs and crows by mobbing, often in groups. When a predator was near, the mob followed it. Variable Oystercatchers subject to severe harassment and predation by dogs attacked an intruding dog by swooping at it and striking it on the head, causing it to retreat (Jones, 1979). American Avocets deter deer and coyotes (Gibson, 1971). Tarboton and Nel (1980) in South Africa found a sandbar area completely covered by buffalo tracks except for a 50-cm circle around a White-crowned Plover (*V. albiceps*) nest, indicating that the incubating bird had successfully repelled the large ungulates. Deane (1944) reports a Killdeer successfully deflecting an approaching cow from its nest using a lunge and aggressive upright display. The Australian Pratincole makes short flights with fluffed feathers and loud calls, which can divert sheep from the nest site (Bailey, 1935; Maclean, 1973).

Even humans can be discouraged. Bryant (1948) describes a pair of avocets (*Recurvirostra novaehollandiae*) that dove incessantly at a camera placed near its nest. After several hours with about 300 dives/h, and no evidence of habituation, the camera was finally removed. The efficacy of mobbing behavior in deterring predation has been demonstrated for gulls and terns by Kruuk (1964) and Lemmetyinen (1971).

C. Distraction Behavior: Efficacy

That so many species employ distraction displays as prominent components of their parental care repertoires attests to the efficacy of these displays in protecting progeny.

Canine Predators. The uniformly effective decoying of dogs by many bird species led Skutch (1955) to conclude that dogs can hardly resist a fluttering bird. Many Arctic shorebirds are exposed to Arctic fox predation. The most effective behavior is sitting tight, but when a nest is discovered, distraction display plays a role in diminishing fox predation (Jourdain, 1936; Larson, 1960). Hohn (1958) noted that American Golden Plover and ptarmigan (*Lagopus* spp.) were completely effective in eliciting pursuit from a sled dog, and believed that the dog's responses to avian distraction display were similar to those of wild wolves or foxes. Duffey *et al.* (1950) and Bengtson (1970) reported that a husky dog would immediately chase a Purple Sandpiper that gave a rodent run, and was completely fooled. Taverner (1936) reported the effectiveness of Killdeer distraction displays in fooling domestic dogs. A Red-capped Dotterel slightly distracted a fox with a rodent run display (Hobbs, 1972). Witherby *et al.* (1940) reported a Stone Curlew (*Burhinus oedicnemus*), a species that rarely feigns injury to humans, luring a dog across a field.

Other Carnivores. Lynes (1910) reported a Northern Lapwing successfully decoying a weasel from its chicks by screaming loudly and flapping outstretched wings.

D. Costs of Antipredator Behavior

Although a broken wing display appears to put the displaying bird at substantial risk of predation, the actual risks are apparently small. Many observers report that a bird engaged in distraction display is constantly alert, probably even in a state of hyperalertness. Certainly if attempts to lure predators away from young resulted in the demise of the adult, selection would act strongly to terminate luring behavior or to enhance vigilance. Relatively few anecdotes describe the failure of antipredator behavior. Whereas a bird that departs early and remains away from the nest may provide no clue as to the nest's location, the performance of vigorous aggressive or distraction behavior may provide an initiated predator, particularly (but not only) humans, with a clue as to the nest's location.

In some cases the behavior seems ill-contrived and maladaptive. Thus, Rock Sandpipers (Hanna, 1921) and Northern Lapwings (Ennion, 1949) sometimes lure humans toward, rather than away from the nest. Hudson (1920) describes how Argentine shepherds located Southern Lapwing nests when the birds gave persistent, noisy, conspicuous displays to ungulates. Jourdain (1936) reports a displaying Red Knot captured by a dog. There are also occasions when mobbing birds come too close to their target and are captured. Myers (1978) reports the capture of a Southern Lapwing by the Crested Caracara (*Polyborus plancus*) it was mobbing. J. Burger (personal communication) watched a Peregrine (*Falco peregrinus*) snatch one of the Common Terns (*Sterna hirundo*) that was mobbing it over a colony. I have captured by hand mobbing Common Terns, and presume that an agile canid could do likewise (which may explain why terns infrequently mob canids).

That self-immolation is avoided attests to the constant monitoring of the intruder or hyperalertness. As self-sacrifice during the period when the young are still dependent would most likely result in the loss of current as well as future reproductive potential, one cannot invoke kin selection to account for the evolution of such behaviors.

VII. ORIGINS OF PREDATOR-INDUCED AGGRESSION AND DISTRACTION DISPLAYS

Predation is one of the powerful selective pressures confronting virtually all organisms. Antipredator behavior is shaped to minimize the threat of predation and to maximize survival of self and offspring when predation is imminent. Aggression and distraction are often successful in thwarting predation, and the opportunity for genetic fixation of such behavior is substantial.

A generation ago, ethologists (e.g., Tinbergen, 1952) focused attention on the form and origin of displays. More recent studies tend to emphasize factors influencing variability in displays. The literature on mobbing and distraction displays is replete with speculations regarding the origin and causes of such displays. It is not always apparent whether an author is discussing the proximate factors underlying a particular display in a particular circumstance, or the ultimate factors influencing the evolution and ritualization of the display in a particular species. Proximate factors are those that influence the likelihood of giving a display and the kind and intensity of display given in a particular situation. These include the nature of the stimulus, the context (habitat, season, etc.), and endogenous factors that result in certain tendencies or motivations. Ultimate factors are those that account for the evolution of particular patterns of behavior and the propensity to behave in certain ways. The proximate factors triggering certain antipredator behavior may bear no apparent relation to the factors accounting for its evolution.

It is also essential to distinguish behavior that is or appears stereotyped from behavior modified by experience. Table XVII summarizes

Table XVII. Interpretations of the Origins of Distraction Displays

Author	Interpretation
Selous (1927)	Suggested that conflict between fear and other drives resulted in distorted activity.
Dewar (1928)	Argued that birds cannot know how a wounded bird would act. It is unlikely that many families would converge on same behavior; birds often give distraction display (DD) inappropriately. They would try to keep in view of the predator.
Friedmann (1934)	Emphasized conflict between personal fear and parental care.
Swarth (1935)	Questioned Friedmann's conflict theories, noting that Lesser Yellowlegs attack rather than injury feign; does this mean that plovers are more fearful than Yellowlegs?
Chisholm (1936)	Reasoned Friedmann's ideas must be wrong as birds that have moved away from the nest will return and try again to lure if not followed. Invoked "semiparalysis."
Cruickshank (1936)	Doubted that excessive fear plays a role.
Jourdain (1936)	Rejected Friedmann's assertions.
Nice (1943)	Disagreed with Friedmann. Coined term "distraction display."
Deane (1944)	Suggested training and intelligence rather than reflex accounts for distraction display.
Williamson (1950b)	Conflict may give rise to haphazard reactions such as impeded flight.
Simmons (1952)	Emphasized the importance of displacement activities, which are low-intensity responses, while DD are high-intensity responses to intruders. DD results from simultaneous activation of attack and escape (following Tinbergen) but disagrees with Lack about the role of conflict, believing that it is important proximately. Disagrees with Williamson and suggests pseudo-sleeping is indeed a displacement activity.
Skutch (1955)	Objected to conflict theory; notes lack of "anguish." Birds are always in full command of their faculties. DD are easily acquired or lost.
Olney (1970)	Intruder near nest causes conflict in Eurasian Avocet between escaping and tending the nest. The compromise has become a ritualized threat and DD.
Griffin (1976)	Emphasized that displaying bird may indeed be aware of the effect of its behavior on the intruder.
Gochfeld (herein)	DD includes some stereotyped, heritable motor patterns (e.g., broken wing, rodent run, impeded flight, threat behaviors). Displacement activities may arise *de novo* or may have become ritualized in certain species, but little evidence for "conflict" in the proximate sense. An individual's selection among different available behavioral options depends on many factors, including the nature and behavior of the intruder.

chronologically ideas regarding the origin of avian distraction displays. Important questions include (1) to what extent is antipredator behavior innate?, (2) to what extent are the displays stereotyped?, (3) to what extent are the displays a result of two conflicting motivations, namely the tendency to escape and the tendency for parental protection? The following discussion examines some of the interpretations offered.

Innateness. Eibl-Eibesfeldt (1964) emphasizes that in some species there is strong evidence for innateness of distraction displays. The Galapagos Barred Dove (*Geopelia lineata*), a species residing in a land free of terrestrial predators, gives a well-developed distraction display that he considered a residual of a behavioral trait evolved by its mainland ancestors to thwart mammalian predators. Pitman (1965) suggests that egg-covering behavior of St. Helena Plover is a residual of behavior evolved in its mainland progenitor, Kittlitz's Plover, exposed to intense risk of predation.

Williamson (1952) was impressed by the apparent innateness and ritualization of displays as indicated by the similarity of injury-feigning displays among the small *Charadrius* plovers, among the rodent run displays of the Calidridine sandpipers, and between displays of true snipes (*Gallinago* spp.) and Greater Painted Snipes.

Fixed Action Patterns. To some authors distraction displays appear highly stereotyped, while others are impressed with the seeming convulsiveness or chaotic nature and variability of the behaviors. Most injury-feigning behavior incorporates certain more or less stereotyped components into a complex and flexible train of displays. One may ask whether the stereotyped components can be interpreted as fixed action patterns (see Gould, 1982, p. 37). Fixed action patterns once released may proceed to completion until terminated, even though the original stimulus is withdrawn (Gould, 1982). Only in the full expression of injury-feigning displays could one find evidence of sufficient stereotypy to consider these fixed action patterns. Instead one finds graded displays that are highly responsive not only to the stimulus presented by the intruder, but to changes in the position and behavior of the intruder.

Fixed action patterns, although modifiable by learning, are considered to have a strong innate component. Conversely, although a fixed action pattern is innate as well as stereotyped, learned behavior can also possess a high degree of stereotypy (Eibl-Eibesfeldt, 1970). There is little evidence regarding the consistency of distraction display or its stereotypy. The most impressive anecdote is from Haverschmidt (1949) who watched a Eurasian Oystercatcher leave its nest and run until its path was blocked by a low wire fence. The bird ran against the fence and struggled continually to run through it, though it could easily have turned aside or flown over it. The author interpreted this as evidence that the nest departure was part of a predetermined, stereotyped behavior pattern, released by an intruder, requiring the individual to run a minimum distance before flying.

Possible Derivation from Other Displays. Aggressive or distraction displays may be derived from modifications of displays used in contexts of courtship or intraspecific aggression. Thus, ungulate displays may derive from generalized threat displays (Fig. 5), or from such conspicuous displays as the Buff-breasted Sandpiper's double-wing embrace (Fig. 5m; Myers, 1978), which serves to attract females and establish the male's copulatory station on the lek. Fluffing of plumage seen in both aggressive and distraction display (Figs. 5c, f) is a widespread component of avian threat (giving the illusion of large size), but also commonly serves for heat dissipation in hot climates (e.g., North, 1937; Little, 1967; Ade, 1979; Howell, 1979). Jenni and Betts (1978) note that the antipredator displays of the Northern Jacana (*Jacana spinosa*) resemble those used in intraspecific fighting. I find that the crouched run of the Piping Plover resembles the horizontal threat display (Fig. 5f) described by Cairns (1982).

Origin in Conflict. Selous (1927), Friedmann (1934), and later Armstrong (1954) invoked conflicting motivations as a basis for the generation of distraction displays. It is important to separate two possible roles of conflict: (1) proximate conflict wherein each instance of a display arises due to the existence of a motivational conflict, or (2) conflict as an ultimate cause in which ritualized displays evolved in situations of conflict are given in situation wherein conflict is not evident.

A. Decision Making

Krebs and Davies (1978, p. 378) describe decision making in animals. They argue that "rational" decisions are made, in the sense that the options available to the animal can be ordered in terms of costs and benefits, and that the animal's choice serves to maximize the payoff or minimize the cost. To these assumptions one must add that the animal has a way of obtaining and processing information regarding the costs and payoffs (i.e., the order of the options must be available to the animal, not just to the ethologist).

Whether one accepts that a parent shorebird makes a "rational decision" as opposed to an "instinctive act," it is apparent that the consequence of an intruder's approach may be variable between individuals (phenotypic variation), or even for the same individual at different times. Cooper and Kaplan (1982) present a challenging suggestion, that the observed variability in phenotypes (in this case antipredator behavior) may not be the direct product of selection, rather organisms are selected for what they call "coin-flipping" decision making, which operates randomly, under rather fixed guidelines to produce one of a choice of behaviors.

B. Parental Investment

Although the typical distraction display appears to put the displaying parent at substantial risk of predation, only rarely does a bird performing distraction displays actually misjudge its target and succumb. The performance of distraction display may be associated with a hyperalert state in which the performer is constantly monitoring the intruder, the threat to its young, and the threat to itself. To what extent can one ascribe the changes in display intensity during the breeding cycles to the bird's decisions regarding its parental investment, and how can the birds appraise their investment at any point in the cycle or the cost of abandoning a breeding effort and reinvesting in a new one (Barash, 1975; Dawkins and Carlisle, 1976)? Finally, one must determine whether a bird is in a position to judge whether it has a chance at successfully distracting a predator or whether it can accurately estimate the immediate risk to itself.

Table II clearly indicates that there are many species that near the time of hatching show a marked increase in the tendency to mob or give distraction displays and in the intensity of such displays (see also Fig. 9). It is a reasonable generalization that adult shorebirds are more active in protecting chicks than eggs. Does this reflect an evaluation of their investment? Two alternative physiologic mechanisms appear more plausible: (1) such behavior could be triggered by a stimulus change such as the appearance of a pipping egg or chick or (2) the fluctuations in hormonal level during the course of the breeding cycle could program the bird for different levels of nest protection behavior.

C. The Role of Fear

An approaching predator is likely to represent a fearsome stimulus to a nesting bird, and many writers have been impressed with the importance of fear as a factor leading to distraction behavior. Impressed by some of the more erratic, agitated behavior or exaggerated injury feigning, early writers believed that birds incapacitated by fear, experienced actual convulsions or inability to coordinate their movements. Selous (1927), Dewar (1928), and Friedmann (1934) believed that injury-feigning movements were actually uncoordinated and spasmodic. Dewar (1928) was impressed that when threatened, many birds behave in a fashion that cannot be construed as injury feigning, yet could represent temporary locomotory incompetence (my terms). Lack (1932) also saw elements of paralysis in distraction display.

D. Consciousness

In contrast to the fear and convulsion theory were suggestions that distraction displays represented conscious attempts to deceive intruders, and that birds feigning injury were actually aware that it was more fruitful to feign injury than to perform other acts. These suggestions, prevalent before 1930 at a time prior to the new evolutionary synthesis, resulted from the apparent responsiveness of displaying shorebirds to human intruders.

Jourdain (1936, 1937) examined a variety of antipredator behavior. He criticized suggestions that injury feigning was conscious, as well as suggestions that birds were paralyzed or incapacitated by fear. He emphasized that it was not necessary to invoke fear as a mechanism leading to distraction display. Jourdain was by no means alone in his conclusion. Allen (1936) argued that however behavior may have evolved, the specific performance of an injury-feigning display does not arise out of fear. A bird giving injury feigning continually assesses the responses of the intruder and circles back to repeat the act, which is hardly consistent with intense fear. Deane (1944) proposed that a given distraction display in an individual was shaped by training and "intelligence" and indicated that the variability in response was in some way matched to the nature of the intruder and the context. Skutch (1955, 1976) most strongly rejected the role of fear in the origin or cause of distraction displays.

E. Conflict Hypotheses

The hypothesis that has received most attention and favor over the years can be called the "conflict hypothesis." Although Selous (1927) may have been the earliest proponent of the idea that distraction displays represented a clash of conflicting emotional drives each inhibiting the normal expression of the other and producing distorted action, Friedmann (1934) was its champion calling attention to a conflict between parental instincts and fear. Armstrong (1949, 1954) attributed distraction display to the outcome of conflicting drives (fleeing and protection), leading to a ritualized compromise activity. Tinbergen (1952) explained it as simultaneous activation of escape and aggression.

It is necessary to realize that some authors invoked this conflict to account for behaviors at the time they observed it [proximate cause (e.g., Simmons, 1952)], while others suggested that conflict may have operated in the past to shape and ritualize certain behaviors which can be seen today (Lack, 1941).

Ultimate Role. If we assume that under the anxiety-provoking experience of having one's nest threatened by an intruder, a naive adult can employ a variety of motor patterns, then those that are effective in thwarting predation can be readily fixed in a population. If such a bird does not "know what to do," it may perform one type of behavior or another, either with consistency or rapid alternations. In some cases the behavior will be unsuccessful and the bird will lose its current eggs or young. In other cases the behavior may expose the adult to predation, a more serious consequence for its genotype. In still other cases the behavior may be successful. In the latter case, the bird may be in a position to learn that such behavior is successful. Alternatively, the bird may have no measure or recollection of what behavior it performed, and on subsequent exposure to an intruder it may behave as a *tabula rasa*. In either case successful behavior may be selected, either quickly or slowly, and the tendency of members of the population to employ successful antipredator behavior will tend to be fixed. As ineffective behavior may also be dangerous, the pressure to perfect safe and effective displays will be strong.

In a situation where adult escape behavior jeopardizes the young and close approach or aggressive behavior jeopardizes itself, a situation of conflict may, as Friedmann (1934) and Armstrong (1954) point out, result in incomplete expression of either behavior. The resultant behavior patterns could be either a stereotyped display such as ineffective locomotion (seen in injury feigning, for example), or aborted attacks (seen in agitated circling). In addition, I suggest that one historical outcome of conflict could be selection for rapid changes in behavior that initially might have been the product of anxiety and "indecision" but that might prove effective antipredator behavior. It is possible that conflict and incomplete expression of motor patterns could have provided the evolutionary substrate for the antipredator behavior manifest today. One weakness lies in the fact that antipredator displays are much less stereotyped than some earlier authors would have believed. There is a strong tendency to see and remember the peak of displays, while closer examination (see for example Figs. 2, 3) reveals much more variable behavior.

Proximate Role. Alternatively, some writers emphasize the proximate role of conflicting motivations. Lind (1961) speaks of the relative strengths of the two tendencies in the Black-tailed Godwit. Attack behavior reflects a strong attack tendency, flight a strong fear or escape tendency, while hovering flight results when the two tendencies are equally strong. It is clear that Lind (1961) is discussing conflict as a proximate factor shaping a particular display. Dewar (1928) argued that broken wing displays arose, not out of a pretense at being wounded, but from a true mental disturbance arising out of a "clash of conflicting instincts—a parental instinct in conflict with the instinct of self-preservation," resulting in a temporary inability to properly control its movements. Dewar (1928) further (1) argued that birds cannot "know" how a wounded bird is supposed to behave, (2) doubted that species of many families and regions would converge on the same behavior, (3) observed that birds often show broken wing behavior in circumstances where there is no benefit to the young, and (4) emphasized that if birds were "feigning" they would attempt to keep in view of the intruder at all times, contrary to their actual behavior. Moreover, he predicted that if birds were pretending anything, the same bird should show comparable levels of pretense on repeated exposure.

Displacement Activities. Displacement activities arise when a particular sequence of behavior has been thwarted (e.g., by the removal of the target or the presentation of conflict), yet the individual remains motivated to perform some act. Ethologists have discussed this at some length providing models such as the hydrodynamic model of Lorenz [see discussion in Gould (1982, p. 196)] or the electricity model of Tinbergen (1952). Either the resulting action or its target is inappropriate or irrelevant to the situation in which the original action arose [i.e., grass-pulling during gull aggression (Tinbergen, 1952)]. Williamson (1952), for example, ascribes false brooding to the thwarting of normal broodiness by the forced departure of the bird from the nest, leading to a tendency to brood in an inappropriate location, and other activities such as false feeding and pseudo-sleeping can be similarly interpreted. As with injury feigning, one must distinguish whether the behavior represents displacement due to thwarting at the present time, or represents a ritualized distraction display, which may have arisen as a displacement behavior.

Arguments Pro and Con. Hinde (1966, p. 250), discussing various behaviors in terms of conflicting motivations, invoked a model showing two tendencies (avoidance and approach) as a function of distance, such that at close distance the avoidance tendency exceeds approach, while the converse occurs when the stimulus is far. I would add the dimension of habituation or previous exposure to intruders, which can serve to reduce fearfulness (Burger and Gochfeld, 1983). Lind's (1961) discussion of mobbing in the Black-tailed Godwit employs this line of reasoning. However, in my observations of injury feigning there are many cases of a sudden, categorical rather than gradual change in behavior from a full injury-feigning display to suddenly normal flight. This suggests that a bird is monitoring its situation and choosing from a menu of available options that at any moment includes some stereotyped displays, certain aggressive activities, and normal locomotion. An alternative explanation of the categorical shift in behavior would be that the behavior does not directly track the motivational balance, but shifts only after a certain lag period.

The conflict hypothesis has certainly had the greatest number of proponents (e.g., Selous, 1927; Friedmann, 1934; Armstrong, 1949; Tinbergen, 1952; Simmons, 1952; *inter alia*). However, in the proximate sense it does not stand up well to a consistency test. For example, under conditions where the tendency to incubate is strong, one would anticipate that an incubating bird will tend to sit tight, and once flushed will show a strong inclination to return. When "fearfulness" or tendency to avoid a predator is strong, the incubating bird would leave the nest early and tend not to return to it until the predator is far away. However, birds that sit tight may flush explosively and fly far away, while birds that leave early may circle around and actually approach the intruder, in some cases returning to the nest while the intruder is within 1 m (e.g., Austin, 1932). The outcome or observed behavior, as the intruder approaches the nest, varies in a manner not predicted by the conflict hypothesis.

Brown (1962), using the alternation of approach and retreat as evidence of motivational conflict, tested whether such conflict predisposed to distraction displays. He found that Western Sandpipers that did show alternation were significantly more likely to also show distraction displays than were birds that only retreated or approached.

Figure 8 reveals that the response distance (which may be one measure of the relative strengths of conflicting tendencies) and the intensity of subsequent distraction displays, are independent, some taxa performing intense displays after early departure, others after late departure. Moreover, even assuming that at some intermediate distance the tendencies to flee and remain are balanced, one would be left with very variable response distance.

I conclude that whatever role conflict may have played in the evolution of distraction behavior, it is not often useful as a predictor of immediate responses. My interpretation coincides with Jourdain's (1936) that in a proximate sense conflict plays little role in determining the resultant behavior.

F. Conclusion

The literature on avian distraction displays reflects the evolution of ethological thought. The literature is replete with discussions of innateness, fixed action patterns, derivations of displays, and conflicts among motivations. Much of the literature of the 1930s and 1940s interpreted distraction behavior quite rigorously as a result of conflict between the tendency to incubate or brood, the tendency to escape (e.g., Friedmann, 1934; Williamson, 1950), or the tendency for aggression. Alternative views held that the distraction display represented some kind of seizure resulting from uncontrolled panic (Dewar, 1928).

It is now clear that no simple single view can account for the large amount of inter- and intraspecific variation or the variability in behavior manifested by the same individual at different times. It is difficult to think of another problem where the ratio of anecdotal accounts to scientific studies so nearly approaches infinity. This section should make it clear that the origin of distraction display involves factors not yet explored. It is essential that careful descriptive and experimental studies of antipredator behavior, aggressive and distractive, be conducted.

Although scientific progress clearly requires the formulation of predictive and testable hypotheses, one must not be afraid to consider questions simply because tests are not immediately apparent, because the behavior is complex, or because of prevailing dogma. I am impressed with this quote from Griffin (1976, p. 79), who raised the challenging question of how "aware" a displaying bird might be of its own circumstances and the possible response of the predator: these are "questions that neither we nor our descendants may be able to answer in the foreseeable future." The question of "intentionality" is nonetheless challenging.

VIII. PERSPECTIVES FOR FUTURE RESEARCH

Despite the difficulties encountered in describing distraction behavior, and the general lack of detailed studies of either distraction or mobbing behavior, I am not nihilistic about our ability to bring detailed field descriptions and experimental studies to bear on the problem. Quantitative studies of distraction behavior (Brown, 1962), nest departure (Burger and Gochfeld, 1981), and mobbing (Lemmetyinen, 1971) indicate that responses of nesting birds to intrusion can be studied rigorously. Although traditional ethological studies have attempted to reduce or eliminate the impact of the observer, this impact is an important part of the stimulus for antipredator behavior, and if it is not possible to eliminate it, standardized presentation of human intruders must be exploited by the researcher.

I pose a series of questions for future study. I have divided these into (1) questions relating to the bird's state (e.g., individual and species differences, hormonal levels, experience), (2) questions relating to the intruder (e.g., species, response to display), (3) questions relating to the context, and (4) methodologic problems.

A. The Displaying Bird

1. Species

Comparative studies of antipredator behavior among closely related species, and among distantly related species with otherwise similar behavior or habitats are desirable. Detailed analysis of motor patterns and responses to intrusion will answer such questions, for example, as: how similar are the apparently uniform distraction displays of the Calidridini sandpipers or small *Charadrius* plovers (see Williamson, 1952)?

2. Individual Variation

Which aspects of antipredator behavior show substantial individual variation within a species, and what factors contribute to the variation?

Sex. In many shorebird species both sexes take an active part in incubation and parental care. The early literature, assuming that females incubated and males either "stood guard" or departed, posed well-defined roles or stereotypes based on sex. Some better defined sex role differences are given in Tables XIVa and b. More rigorous studies of sex differences in the likelihood of standing guard, performing mobbing and/or distraction displays are needed. In species such as the Piping Plover or Killdeer in which both sexes engage in distraction displays, are there consistent differences in frequency, form, or intensity of the displays? Are the displays of one bird modified by those of its mate?

Hormonal Levels. In view of the clear potential role of hormone-mediated aggression in mobbing behavior and of the role ascribed to conflict between fear and defense in producing distraction behavior, it is essential to test the hypothesis that the immediate circulating hormone levels influence the bird's response to intruders. Does the perception of an approaching intruder result in an immediate and short-term change in circulating hormone levels that results in predictable antipredator behavior?

Reproductive Cycle. The stage of the cycle, per se, is an important determinant of antipredator responses. Important questions relate to gradual versus categorical changes in behavior prior to and during incubation, at hatching, and while the young are dependent. I discussed on p. 336 how the intensity and response distance may vary during the cycle. Figure 9 provides alternative models relating the intensity of antipredator behavior to the reproductive cycle.

I am strongly impressed by certain long-term categorical changes in nest protection behavior. For example, in many species (see Table II) there is described a sudden change in parental behavior when the young have hatched (Fig. 9f). An alternative prediction, from theories of parental investment, would have a gradual increase in attentiveness throughout the cycle, associated with increased prior investment (Fig. 9b–e). Can one identify certain stimuli that lead to categorical shifts during the cycle (e.g., the pipping egg)? Can parental responses be altered by substituting dummy eggs and prolonging incubation? Unfortunately, shorebirds are precocial and one cannot readily substitute eggs for chicks once the young have left the nest site, but the converse can be accomplished.

Categorical versus Intensity Shifts. I view detailed study of the factors influencing categorical shifts versus intensity changes in behavior as an important area for study. Categorical rather than intensity shifts can occur in the short term (transition between behavior patterns) as well as in the long term (at different phases of the cycle). Transitions among motor patterns may be either quantitative (changes in intensity shown in Figs. 2 and 3) or qualitative (e.g., injury feigning to crouched run to false brooding to circling flight). The latter categorical changes in behavior are not necessarily predicted by measuring the relative strength of so-called conflicting motivations (e.g., fear and aggression or fear and parental attentiveness). What stimuli or changes in stimuli account for such short-term categorical changes in behavior?

Habituation and Experience. Habituation is one of the most prominent confounding variables influencing a bird's response to intruders. In some cases there are good accounts suggesting the lack of habituation, for instance in Bryant's (1948) description of avocets responding to a camera near the nest. In other cases [e.g., Portenko (1981) and my own studies of Piping Plover], apparent habituation occurs rapidly. By returning to a nest after an interval of several days one can examine the waning of the habituation (it is essential for species with biparental care that the birds be individually marked in order to distinguish them). While habituation is a psychophysiologic process of decreased response to repeated stimuli, prior experience with certain intruders may result in a change of behavior that is not a result of habituation. Birds may learn that certain intruders are either too much of a threat to themselves or are too unresponsive for certain displays. Unlike habituation, changes in behavior due to experience or learning may wane very slowly.

Many shorebirds, particularly at high latitudes, seldom encounter humans on the breeding ground. Behavior when first confronted by a human intruder may be quite different from responses given after birds have experience with humans. Moreover, there may be strong intraspecific differences in how breeding birds in different parts of their range respond to humans. Thus, the dramatic tameness of certain shorebirds, which may return and brood eggs held in the intruder's hand (e.g., Portenko, 1981), suggests that a comparative study across populations of a given species would be useful. The extent to which such experience with humans may account for some of the observed geographic differences remains to be examined. The Greater (see Ratcliffe, 1977) and American (see McCaffery, 1982) Golden Plovers appear to be good candidates for such study.

B. The Nature and Response of the Intruder

1. Influence of Species

A number of studies have been made of responses of birds to models, stuffed birds, or live tethered birds or mammals (e.g., Kruuk, 1964; Lemmetyinen, 1971, 1972). These are readily adaptable to studies of distraction displays (C. Ristau, personal communication) There are numerous anecdotes that birds respond differently to different intruders, so that experimental studies involving presentation of different organisms in different ways will be fruitful. Lemmetyinen (1971) was able to study the effects of mobbing on naturally occurring predators, but this requires a situation in which such predators are of frequent occurrence. The advantage of such situations is that one can then study the efficacy of the antipredator behavior on a live predator, but the impact of the predator behavior itself introduces new variables.

2. Influence of Intruder's Response

The response of breeding shorebirds to the human intruder is dramatic. Many anecdotes report that if the displaying bird is not followed, it returns to the intruder and "tries again." A study of reentrapment is valuable as it can address the questions of how much and under what circumstances the bird is responding to changes in the intruder's behavior.

Cognitive Features. Can shorebirds form concepts about intruders? It is well documented for many species that responses to ungulates, carnivores, humans, and avian predators differ. How do particular individuals or species classify intruders? Are classifications consistant, and are they innate or learned? Are the concepts dependent on the static appearance of the intruder or on its behavior? Although central to developmental psychology, ornithologists have tended to avoid dealing with this area.

Intentionality. Even more unpopular than studies of cognitive behavior are considerations of animal consciousness or awareness (see Griffin, 1976). C. Ristau and D. Griffin (personal communication) invoke Dennett's (1978) proposal of "intentionality" in their studies of distraction behavior. Can one better understand antipredator behavior by invoking the mental state of the bird? Does a displaying bird "want" the intruder to follow it rather than approach the nest? Does the bird have a concept that the intruder is a threat to its eggs or young? Such analysis stands in contrast to traditional stimulus–response interpretations of antipredator behavior, but despite its apparent anthropomorphism, offers new opportunities to investigate behavior as highly variable as antipredator behavior. In fact, although ethologists eschew anthropomorphism, there is no evidence that birds are incapable of consciousness or belief-formation.

C. Nature of the Context

The influence of habitat on nest protection behavior is important for several reasons. The ability of an incubating bird to respond early to an intruder depends in part on its field of visibility (or the presence of a mate on guard). In some circumstances, individuals (of a species that normally leaves the nest early, e.g., the Piping Plover) may be surprised at the nest and may give quite different responses. In addition, birds make different use of the habitat in their distraction displays—some false brooding in appropriate depressions, others momentarily disappearing in vegetation and then reappearing. Habitat modifications, particularly on flat unbroken landscape, can be used to assess the plasticity of displays (i.e., will a bird take advantage of an object suddenly placed in its territory?). Reported cases of birds brooding at neighboring nests suggest another line of experimentation aimed at inducing such behavior.

If, on the other hand, a warning by a mate on guard is crucial, what changes in nest-departure and subsequent antipredator behavior will be occasioned by temporarily or permanently removing the mate?

D. Vocal Components

Although most published reports mention certain calls that characterize aggressive or distractive antipredator behavior, I have found it desirable to ignore the vocal components of distraction displays in this review because of the lack of published objective descriptions or sonograms of the sounds accompanying distraction displays in shorebirds. Published spectrograms of mobbing calls indicate at least a superficial similarity across species, but this warrants more detailed examination. Sound playback techniques offer an excellent opportunity for examining certain aspects of antipredator behavior. For example, how do different individuals or species respond to playbacks of the mobbing calls or distraction display calls of conspecifics and heterospecifics? How do predators respond to the vocalizations? Anecdotal accounts of the dramatic approach responses to distress calls of chicks (e.g., Austin, 1932) suggest an additional line of fruitful research with playbacks. Can such behavioral changes be achieved by playback of chick distress calls during incubation?

E. Methodologic Considerations

More detailed objective descriptions of the motor patterns involved in the antipredator displays are certainly desirable. As indicated in Figs. 2 and 3 and Table IX, there is great variability in how birds hold their body, wings, and tail, and these may represent qualitative or quantitative differences. A simple means of describing the displays is to identify a particular act, and quantitatively describe the position of the head and neck, body, wings, and tail, in terms of the degree of elevation or depression and percent abduction, adduction, extension, or flexion of the extremities and the percent spreading or fanning of the tail.

The transitions among various behavioral components and the correlation of such transitions with the intruder's behavior, will also yield to detailed description. Such description benefits from documentation with portable video recorders.

Captive Experimentation. Ethology has been greatly advanced by studies of captive and captive-reared birds under controlled laboratory conditions. Although shorebirds are not, as a group, easily reared, they are not impossible to breed, or at least raise in captivity. Moreover, certain other taxa such as the tinamous (Tinamidae), grouse (Tetraonidae), and doves (Columbidae), which include species with well-developed distraction displays, are easily raised in captivity.

F. The Future

Why ethological study of this particularly fascinating type of behavior should have lagged so far behind studies of social bonding, courtship behavior, incubation, parent–young recognition, etc., is not immediately clear. Perhaps most field students who have encountered distraction displays have been more impressed by their variability and inexplicability than by the possibility of fruitful examination. Griffin's warning (1976) that one must examine the bird's awareness of itself, its surroundings, and the effect of its behavior, would probably have made a whole generation of ethologists quite uncomfortable, yet appraisal of nest protection behavior leads me inexorably to this conclusion. Definitions of behavior patterns and questions raised in this chapter will hopefully provoke and facilitate detailed descriptive and experimental studies.

ACKNOWLEDGMENTS

I wish to thank Dr. J. Burger for her encouragement during preparation of this chapter, and for providing photographs of displays. For company and assistance during field studies of Willets, Piping Plovers, and Killdeer, I thank J. Burger, B. Jones, and J. Jones. A. Gochfeld kindly translated selected Russian literature. Dr. R. Paxton generously provided accommodations close to field study sites. Dr. C. Ristau shared with me her ideas on distraction displays and brought Dennett's work to my attention. Dr. D. Griffin provided important encouragement at an early stage of my study, confirming my belief that this neglected area of behavior warranted a new look, of which this review is only the first step.

REFERENCES

Ade, B., 1979, Some observations on the breeding of Crowned Plovers (1977), *Bokmakierie* **31**:9–16.

Ali, S., and Ripley, S. D., 1969, *Handbook of the Birds of India and Pakistan*, Vol. II, Oxford University Press, London.

Allen, F. H., 1936, "Injury-feigning" by birds, *Auk* **53**:125–127.

Armstrong, E. A., 1949, Diversionary displays—The nature and origin of distraction display, Parts 1 and 2, *Ibis* **91**:88–97, 179–188.

Armstrong, E. A., 1952, The distraction displays of the Little Ringed Plover and territorial competition with the Ringed Plover, *Br. Birds* **45**:55–59.

Armstrong, E. A., 1954, The ecology of distraction display, *Br. Anim. Behav.* **2**:121–135.
Armstrong, E. A., 1956, Distraction display and the human predator, *Ibis* **98**:641–654.
Arnold, P., and Ferguson, W., 1962, *Birds of Israel*, Shalit Publishers, Haifa.
Ashkenazie, S., and Safriel, U. N., 1979, Breeding cycle and behavior of the Semipalmated Sandpiper at Barrow, Alaska, *Auk* **96**:56–67.
Audubon, J. J., 1839, *Ornithological Biography*, Edinburgh.
Austin, O. L., Jr., 1932, The birds of Newfoundland Labrador, *Mem. Nuttall Ornithol. Club* **7**:1–229.
Bailey, A. M., 1943, The birds of Cape Prince of Wales, *Proc. Color. Mus. Nat. Hist.* **18**:1–113.
Bailey, R. F., 1935, Field notes on the Australian Pratincole, *Emu* **35**:1–10.
Barash, D. P., 1975, Evolutionary aspects of parental behavior: Distraction behavior of the Alpine Accentor, *Wilson Bull.* **87**:367–373.
Barlow, M. L., Muller, P. M., and Sutton, R. R., 1972, Breeding data on the Spur-winged Plover in Southland, New Zealand, *Notornis* **19**:212–249.
Bates, R. S., and Lowther, E. H. N., 1952, *Breeding Birds of Kashmir*, Oxford University Press, London.
Bengtson, S.-A., 1968, The breeding behavior of the Gray Phalarope in west Spitsbergen, *Var Fagelvarld* **27**:1–13.
Bengtson, S.-A., 1970, Breeding behavior of the Purple Sandpiper, *Calidris maritima* on west Spitsbergen, *Ornis Scand.* **1**:17–26.
Benson, S. V., 1948, "Injury feigning" of oystercatcher, *Br. Birds* **41**:222.
Bent, A. C., 1927, Life histories of North American shorebirds, Part I, *U.S. Natl. Mus. Bull.* **142**:1–420.
Bent, A. C., 1929, Life histories of North American shorebirds, Part II, *U.S. Natl. Mus. Bull.* **146**:1–412.
Brown, P. E., and Lynn-Allen, E., 1979, The breeding of avocets in England in 1947, *Br. Birds* **41**:14–17.
Brown, R. G. B., 1962, The aggressive and distraction behaviour of the Western Sandpiper *Ereunetes mauri*, *Ibis* **104**:1–12.
Brown, R. G. B., Blurton Jones, N. G., and Hussell, D. J. T., 1967, The breeding behaviour of Sabine's Gull, *Xema sabinii*, *Behaviour* **28**:110–140.
Bryant, C. E., 1947, Notes on avocets breeding near Melbourne, *Emu* **46**:241–245.
Bryant, C. E., 1948, More observations on nesting avocets, *Emu* **48**:89–92.
Burger, J., 1963, Comparative behavior of the Killdeer and the Spotted Sandpiper, *Kingbird* **13**:14–17.
Burger, J., 1974a, Breeding adaptations of Franklin's Gull (*Larus pipixcan*) to a marsh habitat, *Anim. Behav.* **22**:521–567.
Burger, J., 1974b, Breeding biology and ecology of the Brown-hooded Gull in Argentina, *Auk* **91**:601–613.
Burger, J., and Gochfeld, M., 1981, Discrimination of the threat of direct versus tangential approach to the nest by incubating Herring and Great Black-backed Gulls, *J. Comp. Physiol. Psychol.* **95**:676–684.
Burger, J., and Gochfeld, M., 1983, Behavioral responses to human intruders of Herring Gulls *Larus argentatus* and Great Black-backed *L. marinus* with varying exposure to human disturbance, *Behav. Processes* **8**:327–344.
Cairns, W. E., 1982, Biology and behavior of breeding Piping Plovers, *Wilson Bull.* **94**:531–545.

Carpenter, C. C., and Ferguson, W., 1977, A survey of stereotyped reptilian behavioral patterns, in: *Biology of the Reptilia*, Vol. 7 (C. Gans and D. W. Tinkle, eds.), pp. 335–404, Academic Press, New York.
Cawkell, E. M., 1948, "Injury feigning" of oyster-catcher, *Br. Birds* **41:**125.
Chisholm, A. H., 1936, Injury feigning in birds, *Auk* **53:**251–253.
Clay, A., 1967, Intraspecific distraction-display of Ringed Plover, *Br. Birds* **60:**369.
Cooper, W. S., and Kaplan, R. H., 1982, Adaptive "coin-flipping": A decision-theoretic examination of natural selection for random individual variation, *J. Theor. Biol.* **94:**135–151.
Cruickshank, A. D., 1936, Injury feigning in birds, *Auk* **53:**480.
Cunningham, J. M., 1973, The banded dotterel, *Charadrius bicinctus*: Pohowera or tuturiwhatu? Call notes and behavior, *Notornis* **20:**21–27.
Curio, E., 1976, *The Ethology of Predation*, Springer-Verlag, Berlin.
Davidson, A., 1954, *A Birdwatcher in Scandinavia*, Chapman & Hall, London.
Dawkins, R., and Carlisle, T. R., 1976, Parental investment, mate desertion and a fallacy, *Nature (London)* **262:**131–133.
Deane, C. D., 1944, The broken-wing behavior of the Killdeer, *Auk* **61:**243–247.
Dement'ev, G. P., Gladkov, N. A., and Spangenberg, E. P., 1969, *The Birds of the Soviet Union*, Vol. 5, Israel Program for Scientific Translations, Jerusalem.
Dennett, D. C., 1978, *Brainstorms*, Montgomery, Bradford.
Dewar, D., 1928, *Birds at the Nest*, London.
Dixon, J., 1927, The Surf-bird's secret, *Condor* **29:**3–16.
Dixon, J. S., 1933, Nesting of the Wandering Tattler, *Condor* **35:**173–179.
Drury, W. H., Jr., 1961, The breeding biology of shorebirds on Bylot Island, Northwest Territories, Canada, *Auk* **78:**176–219.
Drycz, A., Witkowski, J., and Okulewiz, J., 1981, Nesting of 'timid' waders in the vicinity of 'bold' ones as an anti-predator adaptation, *Ibis* **123:**542–545.
Duffey, E., Creasey, N., and Williamson, K., 1950, The rodent-run distraction behaviour of certain waders, *Ibis* **92:**27–33.
Eaton, E. H., 1910, *Birds of New York State*, New York State Museum, Albany.
Edwards, G., Hosking, E., and Smith, S., 1948, Aggressive display of the oystercatcher, *Br. Birds* **42:**236–243.
Eibl-Eibesfeldt, I., 1964, *Galapagos, die Arche Noah im Pazifik*, Piper, Munich.
Eibl-Eibesfeldt, I., 1970, *Ethology: The Biology of Behavior*, Holt, Rinehart & Winston, New York.
Ellison, N. F., and Johnston, T. L., 1946, "Injury feigning" of the oystercatcher, *Br. Birds* **39:**60.
Ennion, E. A. R., 1949, *The Lapwing*, Methuen, London.
Erard, C., 1963, Sur le comportement de diversion du Gravelot a Collier Interrompu, *Charadrius alexandrinus* L., a l'egard de l'homme, *Alauda* **31:**262–284.
Faris, R. C., 1942, Snipe feigning injury, *Ir. Nat. J.* **8:**78.
ffrench, R. P., 1973, *A Guide to the Birds of Trinidad and Tobago*. Livingston Publishers, Wynnewood, Pa.
Ficken, M. S., and Ficken, R. W., 1962, The comparative ethology of Wood Warblers: A review, *Living Bird* **1:**103–122.
Flint, V. E., 1973a, Data on the biology of the Broad-billed Sandpiper, *Fauna Ekolo. Kulikov* **1:**98–100.
Flint, V. E., 1973b, Chick-feign as a peculiar type of distraction display of waders, *Fauna Ekol. Kulikov* **1:**23–24.

Flint, V. E., and Kistchinski, A. A., 1972, Data on the biology of the Siberian Pectoral Sandpiper, *Fauna Ekol. Kulikov* **1**:100–105.

Freese, C. H., 1975, Notes on nesting in the Double-striped Thick-knee *Burhinus bistriatus* in Cost Rica, *Condor* **77**:353–355.

Friedmann, H., 1934, The instinctive emotional life of birds, *Psychoanal. Rev.* **21,** Nos. 3 & 4.

Gabrielson, I. N., 1922, Short notes on the life histories of various species of birds, *Wilson Bull.* **34**:193–198.

Gabrielson, I. N., and Lincoln, F. C., 1959, *Birds of Alaska*, Wildlife Management Institute and Stackpole Company, Washington, D.C.

Gibson, F., 1971, The breeding biology of the American Avocet *Recurvirostra americana* in central Oregon, *Condor* **73**:444–454.

Gochfeld, M., 1974, An incipient distraction display of the Least Tern, *Proc. Linn. Soc. N.Y.* **73**:80–82.

Gochfeld, M., 1979, Brood parasite and host co-evolution: Interactions between Shiny Cowbirds and two species of Meadowlarks, *Am. Nat.* **113**:855–870.

Gould, J. L., 1982, *Ethology: The Mechanisms and Evolution of Behavior*, Norton, New York.

Graul, W. D., 1971, Observations at a Long-billed Curlew nest, *Auk* **88**:182–184.

Graul, W. D., 1973, Adaptive aspects of the Mountain Plover social system, *Living Bird* **12**:69–94.

Graul, W. D., 1975, Breeding biology of the Mountain Plover, *Wilson Bull.* **87**:6–31.

Griffin, D. R., 1976, *The Question of Animal Awareness*, Rockefeller University Press, New York.

Hagar, J. A., 1966, Nesting of the Hudsonian Godwit at Churchill, Manitoba, *Living Bird* **5**:5–44.

Hall, K. R. L., 1959, Nest records and additional behavior for Kittlitz's Sand Plover *Charadrius pecuarius* in the S. W. Cape Province, *Ostrich* **30**:33–38.

Hamilton, R. B., 1975, Comparative behavior of the American Avocet and the Black-necked Stilt (Recurvirostridae), *A.O.U. Ornithol. Monogr.* **17**:1–98.

Hanna, G. D., 1921, The Pribilof Sandpiper, *Condor* **23**:50–57.

Haverschmidt, F., 1946, "Injury feigning" of the oystercatcher, *Br. Birds* **39**:59–60.

Haverschmidt, F., 1949, Behavior of an incubating oystercatcher, *Br. Birds* **42**:63.

Helversen, O., 1963, Beobachtungen zum Verhalten und zer Brutbiologie des Spornkeibitzes (*Hoplopterus spinosus*), *J. Ornithol.* **104**:89–96.

Hinde, R. A., 1966, *Animal Behaviour: A Synthesis of Ethology and Comparative Psychology*, McGraw–Hill, New York.

Hobbs, J. N., 1972, Breeding of Red-capped Dotterel at Fletcher's Lake Dareton, NSW, *Emu* **72**:121–125.

Hohn, E. O., 1958, Observations on certain Arctic birds, *Arctic* **11**:93–100.

Hohn, E. O., 1967, Observations on the breeding biology of the Wilson's Phalarope (*Steganopus tricolor*) in central Alberta, *Auk* **84**:220–244.

Hohn, E. O., 1968, Some observations on the breeding of Northern Phalaropes at Scammon Bay, Alaska, *Auk* **85**:316–317.

Holmes, R. T., 1971, Density, habitat and the mating systems of the Western Sandpiper (*Calidris mauri*), *Oecologia (Berlin)* **7**:191–209.

Holmes, R. T., 1973, Social behavior of breeding Western Sandpipers (*Calidris mauri*), *Ibis* **115**:107–123.

Hopcraft, J. B. D., 1968, Some notes on the chick-carrying behavior in the African Jacana, *Living Bird* **7**:85–88.

Hortling, I., and Baker, E. C. S., 1932, Bird notes on a trip to Lapland, *Ibis* **1932:**100–127.

Howe, M. A., 1982, Social organization in a nesting population of eastern Willets (*Catoptrophorus semipalmatus*), *Condor* **84:**88–102.

Howell, T. R., 1979, Breeding biology of the Egyptian Plover (*Pluvianus aegyptius*), *Univ. Calif. Berkeley Publ. Zool.* **118:**1–76.

Hudson, W. H., 1920, *Birds of La Plata*, Vol. 1, Drent, London.

Hussell, D., and Page, G. W., 1976, Observations on the breeding biology of the Black-bellied Plovers on Devon Island, N.W.T., Canada, *Wilson Bull.* **88:**632–653.

Hutson, H. P. W., and Bannerman, D. A., 1931, The birds of northern Nigeria, II, *Ibis* **1931:**18–43.

Jehl, J. R., Jr., 1968, The systematic position of the Surfbird (*Aphriza virgata*), *Condor* **70:**200–210.

Jehl J. R., Jr., 1973, Breeding biology and systematic relationships of the Stilt Sandpiper, *Wilson Bull.* **85:**115–147.

Jenni, D. A., and Betts, B. J., 1978, Sex differences in nest construction, incubation and parental behaviour in the polyandrous American Jacana (*Jacana spinosa*), *Anim. Behav.* **26:**207–218.

Johnsgard, P. A., 1981, *The Plovers, Sandpipers, and Snipes of the World* University of Nebraska Press, Lincoln.

Johnson, A. W., and Goodall, J. D., 1965, *The Birds of Chile*, Vol. I, Platt Estableimientos Graficos SA, Buenos Aires.

Jones, A., 1979, Notes on the behaviour of the Variable Oystercatchers, *Notornis* **26:**47–52.

Jourdain, F. C. R., 1936, The so-called injury feigning in birds, *Ool. Rec.* **16:**25–37.

Jourdain, F. C. R., 1937, The so-called injury-feigning in birds, Part III, *Ool. Rec.* **17:**14–16.

Kapitonov, V. I., 1962, Ornithological observations on the Lena lowlands, *Ornitologiya* **4:**40–63.

Kistchinski, A. A., and Flint, V. E., 1973, Materials on the biology of the dowitcher in east Siberian tundras, *Fauna Ekol. Kulikov* **1:**52–55.

Komeda, S., 1983, Nest attendance of parent birds in the Painted Snipe, *Auk* **100:**48–55.

Krebs, J. R., and Davies, N. B., 1978, *Behavioural Ecology: An Evolutionary Approach*, Blackwell, Oxford.

Kruuk, H., 1964, Predators and anti-predator behaviour of the Black-headed Gull *Larus ridibundus*, *Behav. Suppl.* **11:**1–129.

Kuzyakin, A. P., 1959, The Semipalmated Sandpiper on the East Chukotsk peninsula, *Ornitologiya* **2:**130–134.

Labutin, Y. V., Leonovitch, V. V., and Veprintsev, B. N., 1982, The Little Curlew *Numenius minutus* in Siberia, *Ibis* **124:**302–309.

Lack, D., 1932, Some breeding-habits of the European Nightjar, *Ibis* **1932:**266–284.

Lack, D., 1941, Some aspects of instinctive behaviour and display in birds, *Ibis* **1941:**407–441.

Larson, S., 1960, On the influence of the Arctic fox (*Alopex lagopus*) on the distribution of Arctic birds, *Oikos* **11:**277–305.

Lemmetyinen, R., 1971, Nest defence behaviour of Common and Arctic Terns and its effects on the success achieved by predators, *Ornis Fenn.* **48:**13–24.

Lemmetyinen, R., 1972, Nest defence behaviour in the Arctic Tern, *Sterna paradisaea* towards stuffed nest predators on Spitsbergen, *Rep. Kevo Subarctic Res. Stn.* **9:**28–31.

Lind, H., 1961, *Studies on the Behaviour of the Black-tailed Godwit*, (*Limosa limosa* (*L.*)), Munksgaard, Copenhagen.
Little, J. de V., 1967, Some aspects of the behaviour of the Wattled Plover (*Afribyx senegallus*) (Linnaeus), *Ostrich* **38**:259–280.
Lowe, V. P. W., 1972, Distraction display by a woodcock with chicks, *Ibis* **114**:106–107.
Lye, R. J., 1948, Two types of "injury-feigning" by snipe, *Br. Birds* **41**:90.
Lynes, H., 1910, Manoevres of lapwing in defence of young, *Br. Birds* **4**:157.
McCaffery, B. J., 1982, Geographic variation in predator-mobbing behavior—A hypothesis, *Wader Study Group Bull.* **35**:30.
MacGillivray, W., 1924, A contribution to the life-story of the Australian Pratincole, *Emu* **24**:81–85.
McKenzie, H. R., 1953, Nesting of the New Zealand dotterel, 1951, *Notornis* **5**:121–122.
Maclean, G. L., 1967, The breeding biology and behaviour of the Double-banded Courser *Rhinoptilus africanus* (Temminck), *Ibis* **109**:556–569.
Maclean, G. L., 1969, A study of seedsnipe in southern South America, *Living Bird* **8**:33–80.
Maclean, G. L., 1972, Problems of display postures in the Charadrii (Aves: Charadriiformes), *Zool. Afr.* **7**:57–74.
Maclean, G. L., 1973, A review of the biology of the Australian desert waders, *Stilta* and *Peltohyas*, *Emu* **73**:61–70.
Maclean, G. L., 1974, Egg-covering in the Charadrii, *Ostrich* **45**:167–174.
Maclean, G. L., 1976, A field study of the Australian Dotterel, *Emu* **76**:207–215.
McNicholl, M., 1973, Habituation of aggressive responses to avian predators by terns, *Auk* **90**:902–904.
Makkink, G. F., 1936, An attempt at an ethogram of the European avocet, *Ardea* **25**:1–74.
Male, A. E., 1950, Whitethroat attacking weasel, *Br. Birds* **43**:155.
Manniche, A. L. V., 1910, The terrestrial mammals and birds of north-east Greenland: Biological observations, *Dan. Exped. Gronlands Nordostkyst.* **5**:93–199.
Martin, J., 1972, Nesting habits of our three resident sandplovers, *Bokmakierie* **24**:40–41.
Mason, C. F., and MacDonald, S. M., 1976, Aspects of the breeding biology of the snipe, *Bird Study* **23**:33–38.
Mathew, D. N., 1964, Observations on the breeding habits of the Bronze-winged Jacana (*Metopidius indicus* (Latham)), *J. Bombay Nat. Hist. Soc.* **61**:295–302.
Matthiessen, P., 1967, *The Shorebirds of North America* (Chapter 9) (G. Stout, ed.), Viking Press, New York.
Mayfield, H. F., 1979, Red Phalaropes breeding on Bathurst Island, *Living Bird* **17**:7–40.
Meltofte, H., 1977, "Fly-away trick" of some Arctic waders, *Br. Birds* **70**:333–336.
Meltofte, H., 1979, The population of waders Charadriidae at Danmarks Haven, northeast Greenland, 1975, *Dan. Ornithol. Foren. Tidsskr.* **73**:69–94.
Miller, A. H., 1931, Observations on the incubation and the care of the young in the jacana, *Condor* **33**:32–33.
Miller, E. H., and Baker, A. J., 1980, Displays of the Magellanic Oystercatcher (*Haematopus leucopodus*), *Wilson Bull.* **92**:149–168.
Miller, J. R., and Miller, J. T., 1948, Nesting of the Spotted Sandpiper at Detroit, Michigan, *Auk* **65**:558–567.
Mousley, H., 1937, Nesting habits of the Spotted Sandpiper, *Auk* **54**:445–451.
Mousley, H., 1939, Nesting behavior of Wilson's Snipe and Spotted Sandpiper, *Auk* **56**:129–133.
Mundahl, J. T., 1982, Role specialization in the parental and territorial behavior of the Killdeer, *Wilson Bull.* **94**:515–530.

Murie, O. J., 1924, Nesting records of the Wandering Tattler and Surf-bird in Alaska, *Auk* **41**:231–237.
Murphy, R. C., 1926, Nest protecting display of the Woodcock, *Bird-lore* **28**:265-266.
Murphy, R. C., 1936, *Oceanic Birds of South America*, American Museum of Natural History, New York.
Myers, J. P., 1978, One deleterious effect of mobbing in the Southern Lapwing (*Vanellus chilensis*), *Auk* **95**:419.
Nechaev, V. A., 1980, Okhotskii ulit, *Priroda (Sofia)* **4**:102–106.
Nethersole-Thompson, D., 1940, Threat and "injury feigning" display of lapwing, *Br. Birds* **23**:138–139.
Nethersole-Thompson, D., 1973, *The Dotterel*, Collins, Glasgow.
Nice, M. M., 1943, Studies in the life history of the Song Sparrow. II. The behavior of the Song Sparrow and other passerines, *Trans. Linn. Soc. N.Y.* **6**:1–328.
North, M. E. W., 1937, Breeding habitats of the Crested Wattled Plover, *J. East Afr. Nat. Hist. Soc.* **13**:132–145.
Nowicki, T., 1973, A behavioral study of the Marbled Godwit in North Dakota, M.S. thesis, Central Michigan University.
Oberholser, H. C., 1938, *The Bird Life of Louisiana*, Bull. No. 28, Department of Conservation, Louisiana.
Olney, P. J. S., 1970, Studies of avocet behaviour, *Br. Birds* **63**:206–209.
Osborne, D. R., and Bourne, G. R., 1977, Breeding behavior and food habits of the Wattled Jacana, *Condor* **79**:98–105.
Page, G. W., Stenzel, L. E., Winkler, D. W., and Swarth, C. W., 1983, Spacing out at Mono Lake: Breeding success, nest density and predation in the Snowy Plover, *Auk* **100**:13–24.
Parmelee, D. F., 1970, Breeding behavior of the Sanderling in the Canadian high Arctic, *Living Bird* **9**:97–146.
Parmelee, D. F., Stephens, H. A., and Schmidt, R. H., 1967, The birds of southeastern Victoria Island and adjacent small islands, *Nat. Mus. Can. Bull.* **222**:1–229.
Parmelee, D. F., Greiner, D. W., and Graul, W. D., 1968, Summer schedule and breeding biology of the White-rumped Sandpiper in the central Canadian Arctic, *Wilson Bull.* **80**:5–29.
Parrinder, E. R., 1948, The Little Ringed Plover in southern England in 1948, *Br. Birds* **41**:377–381.
Peters, H. S., and Burleigh, T. D., 1951, *The Birds of Newfoundland*, Department of Natural Resources, St. John's.
Pitman, C. R. S., 1965, The eggs and nesting habits of the St. Helena Sand-plover or Wirebird, *Charadrius pecuarius sanctae-helenae* (Harting), *Bull. Br. Ornithol. Club* **85**:121–129.
Portenko, L. A., 1981, *Birds of the Chukotsk Peninsula and Wrangel Island*, Vol. I, Nauka Press, Moscow.
Ratcliffe, D. A., 1977, Observations on the breeding of the Golden Plover in Great Britain, *Bird Study* **23**:63–116.
Ridley, M., 1980, The breeding behaviour and feeding ecology of Gray Phalaropes *Phalaropus fulicarius* in Svalbard, *Ibis* **122**:210–226.
Rockwell, R. R., 1912, Notes on the wading birds of the Barr Lake region Colorado, *Condor* **14**:117–131.
Ruttledge, R. F., 1944, Distraction displays of Common Gulls, *Br. Birds* **55**:133.
Sargeant, A. B., and Eberhardt, L. E., 1975, Death feigning by ducks in response to predation by Red Foxes (*Vulpes fulva*), *Am. Midl. Nat.* **94**:108–119.

Sauer, E. G. F., 1962, Ethology and ecology of Golden Plovers on St. Lawrence Island, Bering Sea, *Psychol. Forsch.* **26**:399–470.
Seebohm, H., 1901, *The Birds of Siberia*, Murray, London.
Selous, E., 1927, *Realities of Bird-life*, London.
Serle, W., 1956, Notes on *Anomalophrys superciliosus* (Reichnow) in west Africa with special reference to its nidification, *Bull. Br. Ornithol. Club* **76**:101–104.
Simmons, K. E. L., 1951, Distraction display in the Kentish Plover, *Br. Birds* **45**:181–187.
Simmons, K. E. L., 1952, The nature of the predator-reactions of breeding birds, *Behaviour* **4**:161–171.
Simmons, K. E. L., 1955, The nature of the predator-reactions of waders towards humans; with special reference to the role of the aggressive escape, and brooding drives, *Behaviour* **8**:130–173.
Skead, C. J., 1955, A study of the Crowned Plover *Stephanibyx coronatus coronatus* (Boddaert), *Ostrich* **26**:88–98.
Skutch, A. F., 1955, The parental stratagems of birds, *Ibis* **97**:118–142.
Skutch, A. F., 1976, *Parent Birds and Their Young*, University of Texas Press, Austin.
Smith, J., and Smith, M., 1974, Behaviour of Ringed Plover in defense of nest, *Br. Birds* **67**:78.
Sordahl, T. A., 1979, Vocalizations and behavior of the willet, *Wilson Bull.* **91**:551–574.
Sordahl, T. A., 1981, Predator mobbing in the shorebirds of North America, *Wader Study Group Bull.* **31**:41–44.
Spencer, J. F., 1948, "Injury feigning" of Common Snipe, *Br. Birds* **41**:27–28.
Stone, W., 1937, *Bird Studies at Old Cape May*, Vol. 1, Dover Reprint, New York.
Street, J. F., 1923, On the nesting grounds of the Solitary Sandpiper and the Lesser Yellowlegs, *Auk* **40**:577–583.
Summers, R. W., and Hockey, P. A. R., 1981, Egg-covering behaviour of the White-fronted Plover, *Charadrius marginatus*, *Ornis Scand.* **12**:240–243.
Sutton, G. M., 1967, Behaviour of the Buff-breasted Sandpiper at the nest, *Arctic* **20**:3–7.
Sutton, G. M., and Parmelee, D. F., 1956, On certain charadriiform birds of Baffin Island, *Wilson Bull.* **68**:210–223.
Swarth, H., 1935, "Injury-feigning" in nesting birds, *Auk* **52**:352–354.
Taka-Tsukasa, N., 1967, *The Birds of Nippon*, Methuen, London.
Tarboton, W. R., and Nel, F., 1980, On the occurrence of the White-crowned Plover in the Kruger National Park, *Bokmakierie* **32**:19–21.
Taverner, P. A., 1936, Injury feigning by birds, *Auk* **53**:366.
Thomas, D. G., 1969, Breeding biology of the Australian Spur-winged Plover *Emu* **69**:81–102.
Thompson, A. L., 1964, *A New Dictionary of Birds*, McGraw–Hill, New York.
Tinbergen, N., 1952, Derived activities: Their causation, biological significance, origin and emancipation during evolution, *Q. Rev. Biol.* **27**:1–32.
Tomkins, I. R., 1944, Wilson's Plover in its summer home, *Auk* **61**:259–269.
Towsend, C. W., and Allen, G. M., 1907, Birds of Labrador, *Proc. Boston Soc. Nat. Hist.* **33**:277–428.
Tyler, S., 1978, Observations on the nesting of the Three-banded Plover, *Charadrius tricollaris*, *Scopus* **2**:39–41.
Vogt, W., 1938, Preliminary notes on the behavior and the ecology of the eastern Willet, *Proc. Linn. Soc. N.Y.* **49**:8–42.
Walker, J., 1955, Mountain Plover, *Audubon Mag.* **55**:210–212.
Walters, J., 1979, Interspecific aggressive behaviour by Long-toed Lapwings (*Vanellus crassirostris*), *Anim. Behav.* **27**:969–981.

Walters, J., and Walters, B. F., 1980, Cooperative breeding by Southern Lapwings *Vanellus chilensis*, *Ibis* **122:**505–508.

Webster, J. D., 1941, The breeding of the Black Oystercatcher, *Wilson Bull.* **53:**141–156.

Williamson, K., 1943, The behaviour patterns of the western oystercatcher (*Haematopus ostralegus occidentalis* Neumann) in defence of nests and young, *Ibis* **1943:**486–490.

Williamson, K., 1946, Field-notes on the breeding-biology of the Whimbrel, *North West. Nat.* **1946:**147–184.

Williamson, K., 1947, Distraction display in the Ringed Plover, *Ibis* **1947:**511–513.

Williamson, K., 1948, Field notes on nidification and distraction display in the Golden Plover, *Ibis* **90:**90–98.

Williamson, K., 1949, The distraction display of the Arctic skua, *Ibis* **91:**307–313.

Williamson, K., 1950a, The "rodent-run" distraction display of certain waders. Part II. Interpretation of "rodent-run" display, *Ibis* **92:**28–33.

Williamson, K., 1950b, The distraction behaviour of the Faeroe Snipe, *Ibis* **92:**66–74.

Williamson, K., 1950c, The pseudo-sleeping attitude of the oystercatcher, *Br. Birds* **43:**1–4.

Williamson, K., 1951, The distraction behaviour of the Faeroe snipe, *Ibis* **93:**306.

Williamson, K., 1952, Regional variation in the distraction displays of the oyster-catcher, *Ibis* **94:**85–96.

Wilson, A., and Bonaparte, C. L., 1832, *American Ornithology*, Porter & Coates, Philadelphia.

Witherby, H. F., Jourdain, F. C. R., Ticehurst, N. F., and Tucker, B. W., 1940, *The Handbook of British Birds*, Vol. 4, Witherby, London.

Yeates, G. K., 1948, Some notes on the nesting habits of the pratincole, *Br. Birds* **41:**301–303.

Chapter 9

CONSERVATION OF NEARCTIC SHOREBIRDS

Stanley E. Senner

Hawk Mountain Sanctuary Association
Kempton, Pennsylvania 19529

and

Marshall A. Howe

U.S. Fish and Wildlife Service
Patuxent Wildlife Research Center
Laurel, Maryland 20708

The time to save a species is while it is still common
Rosalie Edge, 1907–1962

I. INTRODUCTION

The public, and the government agencies that they influence, are often aroused to the cause of wildlife conservation only after species are seriously endangered. Unfortunately, the protection and restoration of already endangered species is at best an expensive and difficult endeavor; at worst, it can be a futile one. Thus, there is no question that preventive measures are the best ones. This is certainly the case in conservation efforts directed toward shorebirds.

Although some shorebird (Charadrii) species are extinct (e.g., White-winged Sandpiper, *Prosobonia leucoptera*) and others are nearly so (e.g., Eskimo Curlew, *Numenius borealis*), most are still common and even abundant. One of the reasons for this is that the breeding grounds of many species are at high latitudes remote from human settlement. However, some of the very qualities that make shorebirds attractive subjects for ornithologists and birdwatchers—particularly the tendency to congregate in migration and to winter in geographically limited, accessible habitats—also expose a large proportion of shorebird species to environmental modification, pollution, and disturbance.

The purposes of this chapter are to give an overview of activities and developments that may adversely affect shorebird populations and to describe current efforts in, and legal tools for, the management and conservation of shorebirds in the Western Hemisphere. Formulation of effective strategies for conserving shorebird populations requires knowledge of distribution patterns and ecological requirements at different times of the year. Our description of problems and opportunities in shorebird conservation is cast in the context of those periods and areas that are especially sensitive and critical in the annual cycles of shorebird populations.

II. SENSITIVE PERIODS IN THE ANNUAL CYCLE

In the past two decades, the proliferation of shorebird research in both Northern and Southern Hemispheres has provided a sound basis for identifying and interpreting sensitive periods in the annual cycle. Although the majority of American studies have dealt with aspects of breeding biology, the focus of recent research has shifted to migrating shorebirds. Because of these studies (e.g., Burger *et al.*, 1977; McNeil and Cadieux, 1972; Morrison and Harrington, 1979; Pitelka, 1979; Schneider and Harrington, 1981; Senner *et al.*, 1981), we now have a better understanding of the importance of staging areas and a clearer perspective on the requirements of Western Hemisphere shorebirds in migration. However, only scattered studies have concerned wintering shorebirds (e.g., Ashmole, 1970; Baker and Baker, 1973; Spaans, 1978; Myers *et al.*, 1979; Myers, 1980). A major gap of knowledge continues to be wintering distribution and ecology of New World species.

The purpose of this section is to identify potentially sensitive periods in the annual cycle of shorebirds and to illustrate with specific examples from the Western Hemisphere. For convenience, we divide the annual cycle of migratory species into three periods: breeding, migration, and wintering.

A. Breeding Season

Most species of North American shorebirds breed on Arctic and subarctic tundra and in the tundra-boreal forest ecotone remote from direct human influence. Most of these species have rather broad geographic distributions and are not presently vulnerable during this season. An exception may be those segments of populations that occur coastally near offshore oil operations such as at Barrow, Alaska, but even here most species are largely restricted to the tundra during nesting and are unlikely to contact oil. A few species, however, such as Western Sandpipers (*Calidris mauri*), Semipalmated Sandpipers (*C. pusilla*), and Baird's Sandpipers (*C. bairdii*), make extensive use of littoral areas when breeding (Connors *et al.*, 1979; see Burger, this volume).

Geographically restricted breeding grounds are mainly a problem for species that nest at lower latitudes where human pressures are most evident. Such species should be monitored carefully. Potentially the most vulnerable are species that nest on the sandy beaches typical of the Atlantic and Gulf coasts of the United States and certain coastal stretches of the Pacific and Great Lakes. For example, the Piping Plover (*Charadrius melodus*) is endemic to this habitat in eastern North America and to widely scattered alkaline lakes in the interior. Declines in its numbers and reproductive success have been documented through most of its range (Lambert and Nol, 1978; Cairns and McLaren, 1980; Lambert and Ratcliff, 1981). The species is now considered endangered in Ontario (McCracken *et al.*, 1981). This species does not appear capable of exploiting habitats other than sandy beaches or alkaline flats for nesting. The same is true for the Snowy Plover (*C. alexandrinus*).

Several species of shorebirds that breed in the grasslands of the central United States and southern Canada strongly prefer uncultivated sites and have undergone range reductions due to agricultural expansion. Examples are the Long-billed Curlew (*Numenius americanus*) (Pampush, 1980), Upland Sandpiper (*Bartramia longicauda*) (Kirsch and Higgins, 1976), and Mountain Plover (*Charadrius montanus*) (Graul and Webster, 1976). Of these species, a large proportion of the curlew population inhabits littoral sites at some times of the year. The near disappearance of the curlew as a wintering species on southeastern coasts of the United States (Sugden, 1933) is probably related to habitat loss in the eastern Great Plains and corollary impacts on breeding success. Recent evidence suggests a possible resurgence of curlew numbers on undisturbed rangeland in the West (R. Redmond, personal communication).

In general, the breeding season does not appear to be a time of extraordinary energy stress for most shorebirds, despite the burden on females of egg production. Normal summer conditions in the subarctic produce a great abundance of insect larvae that tend to peak during the chick growth period (Holmes, 1966a; Baker and Baker, 1973). Moreover, the nearly continuous daylight of the summer months at high latitudes affords extended feeding opportunities. In spite of the prevailing situation, however, food shortages can occur in periods of inclement weather, a possibility enhanced by the low diversity of prey species (Pitelka *et al.*, 1974). The most critical time is shortly after arrival on the breeding grounds. Late spring cold and snow can in some years delay breeding (Green *et al.*, 1977) and, in extreme cases, cause widespread mortality of adults (Marcström and Mascher, 1979). The susceptibility of breeding shorebirds to weather and corresponding food shortages probably increases with latitude. Holmes (1970), for example, showed that Dunlin (*Calidris alpina*) territories (from which food is procured) on Arctic tundra are five times as large as those on subarctic tundra. This suggests a lower or less predictable food supply higher in the Arctic. High Arctic species are occasionally unable to breed at all because of stressful conditions (Bird and Bird, 1940). At lower latitudes, food stress appears to be a regular phenomenon in harsh environments like the arid prairies. For example, rainfall patterns strongly influence insect production and, secondarily, may affect the ability of Mountain Plovers to produce multiple clutches (Graul, 1974).

B. Migration Period

Postbreeding and migratory movements can result in extremely large local aggregations of shorebirds. Although flocks of many thousands can occur in almost any suitable habitat along major migration routes, many such sites seem to be traditional.

Connors *et al.* (1979) have shown that several summer resident shorebirds at Barrow, Alaska, shift their activity centers from tundra to littoral habitats in the postbreeding period. Juvenile Red Phalaropes (*Phalaropus fulicarius*) are most striking in this respect, but adult and juvenile Dunlin and juvenile Long-billed Dowitchers (*Limnodromus scolopaceus*) show a similar pattern. Gill and Handel (1981) have observed large numbers of many species moving into the Yukon River Delta, Alaska, late in the breeding season. Postbreeding movements to nearby littoral areas are probably a normal phenomenon in many species and might be considered the beginning of migration.

During migration proper, many coastal and interior areas are well known as sites where shorebirds gather for varying periods of time, accumulating lipid reserves and sometimes molting before moving on. Traditional staging areas may be different for different species and may vary for a given species between the southbound and northbound flights. The locations of staging areas are best known for North America and more

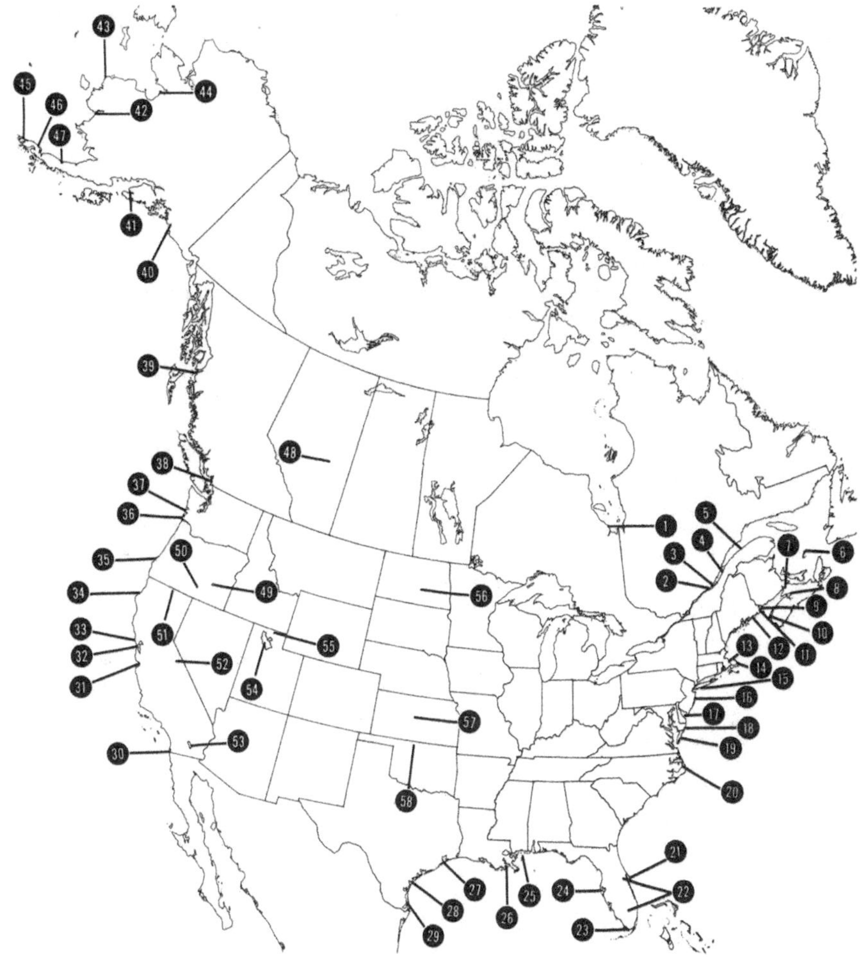

Fig. 1. Important, traditional stopover sites for migrating shorebirds in North America. Sites shown are those believed to support 20,000 or more shorebirds of all species combined during the peak of migration. See Table I for site identification and information on seasonality and species composition.

poorly known for Latin America. In certain cases a very significant proportion of an entire North American species population or a major regional subpopulation stop over at one or a series of local sites in migration. Figure 1 and Table I identify some of the stopovers in North America where at least 20,000 shorebirds are believed to occur at any one time during spring and/or fall migration.

Table I. Description of the Stopover Sites of Migrating Shorebirds Shown in Fig. 1[a]

Site No.	Site	Comments	Source
1	North Point, Ontario	S; especially Semipalmated Sandpiper	R. I. G. Morrison (personal communication)
2	Beauport, Quebec	S; especially Semipalmated Sandpiper and Black-bellied Plover. 50,000 typical	S. Lemieux (personal communication)
3	Batture aux loups Marins, Quebec	S; especially Semipalmated Sandpiper	S. Lemieux (personal communication)
4	Montmagny, Quebec	S; especially Semipalmated Sandpiper	S. Lemieux (personal communication)
5	Baie Ste. Catherine, Quebec	S; especially Semipalmated Sandpiper	S. Lemieux (personal communication)
6	Lagune Havre aux Basques, Magdalen Islands, Quebec	S; especially Semipalmated Sandpiper	S. Lemieux (personal communication)
7	Chignecto Bay, New Brunswick	S; 98% Semipalmated Sandpiper. Up to 300,000	P. Hicklin (personal communication), Morrison (1976)
8	Cobequid Bay and Minas Basin, Nova Scotia	S; 98% Semipalmated Sandpiper. Up to 250,000	P. Hicklin (personal communication), Morrison (1976)
9	Deer Island, New Brunswick, and southeast of Grand Manan Island	S; Northern Phalarope, pelagic waters. Many hundreds of thousands	N. Famous (personal communication)
10	Brier Island, Nova Scotia	N; Northern Phalarope, pelagic waters S; Red Phalarope, pelagic waters	R. G. B. Brown (personal communication)
11	Cobscook Bay, Maine	S; especially Semipalmated Sandpiper	McCollough and May (1980)
12	Harrington, Maine	S; especially Semipalmated Sandpiper and Semipalmated Plover	N. Famous (personal communication)
13	Plymouth and Scituate, Mass.	S	B. A. Harrington (personal communication)
14	Monomoy Island, Mass.	S	B. A. Harrington (personal communication)
15	Jamaica Bay to Captree, Long Island, N.Y.	N, S; 100,000 estimated, both seasons	P. A. Buckley (personal communication)
16	Brigantine National Wildlife Refuge, N.J.	N, S	B. A. Harrington (personal communication)

[a] Comments indicate whether the site is important for northbound (N) or southbound (S) migrants and include information on species composition and peak numbers when of exceptional interest.

Table I. (*Continued*)

Site No.	Site	Comments	Source
17	Delaware Bay—New Jersey and Delaware shorelines	N; especially Semipalmated Sandpiper, Sanderling, Red Knot, and Ruddy Turnstone. Up to 300,000	Wander and Dunne (1981)
18	Chincoteague National Wildlife Refuge, Assateague Island, Va.	N, S	C. Wilds (personal communication)
19	Virginia barrier island system, Va.	N, S	M. Howe (personal observation)
20	Outer Banks barrier island system, N.C.	N, S	M. Howe (personal observation)
21	Merritt Island, Fla.	N, S	P. Sykes (personal communication)
22	Lake Apopka and Zellwood, Fla.	S; agricultural fields flooded in summer; especially Lesser Yellowlegs (*Tringa flavipes*)	P. Sykes (personal communication), Sykes and Hunter (1978)
23	Northern Florida Bay, Fla.	N, S	P. Sykes (personal communication)
24	McKay's Bay in Tampa Bay, Fla.	N, S	P. Sykes and B. A. Harrington (personal communication)
25	Chandeleur Islands, La.	N, S	B. Ortego (personal communication)
26	Grand Terre Island, La.	N, S	B. Ortego (personal communication)
27	Bolivar flats, Galveston Bay, Tex.	N, S	G. Blacklock (personal communication)
28	Oso Bay and Nueces Bay, Tex.	N, S	G. Blacklock (personal communication)
29	Laguna Madre barrier island system, Tex.	N, S	G. Blacklock (personal communication)
30	San Diego Bay, Calif.	N, S	G. McCaskie (personal communication)
31	Elkhorn Slough, Calif.	N; mainly Western Sandpiper	B. Allen (personal communication)
32	San Francisco Bay, Calif.	N, S; at least 100,000	G. Page (personal communication)
33	Bolinas Lagoon, Calif.	N	G. Page (personal communication)

(*Continued*)

Table I. (*Continued*)

Site No.	Site	Comments	Source
34	Humbolt Bay, Calif.	N, S; over 100,000 in spring	R. Gerstenberg (personal communication)
35	Coos Bay and Coquille Bay, Oreg.	N, S	A. McGie (personal communication)
36	Willapa Bay, Wash.	N, S	S. Herman (personal communication)
37	Grays Harbor, Wash.	N, S; up to 1 million in spring	S. Herman (personal communication), Herman and Bulger (1981)
38	Fraser River Delta, British Columbia	N, S; up to 250,000	G. Kaiser (personal communication)
39	Big Bay, Prince Rupert, British Columbia	N, S	G. Kaiser (personal communication)
40	Copper–Bering River Delta, Alaska	N, S; especially Western Sandpiper (most of world population) and Dunlin in spring. Million or more. See Section V	Senner *et al.* (1981), Isleib and Kessel (1973)
41	Kachemak Bay, Alaska	N, S; mainly Western Sandpiper	R. Gill (personal communication), Senner *et al.* (1981)
42	Kuskokwim Bay, Alaska	S	R. Gill (personal communication)
43	Yukon River Delta, Alaska	S; mainly Western Sandpiper and Dunlin. Several hundred thousand	R. Gill (personal communication)
44	Norton Sound, Alaska	S	R. Gill (personal communication)
45	Izembek Lagoon, Alaska	S; mainly Dunlin and Rock Sandpiper (*Calidris ptilocnemis*)	R. Gill (personal communication)
46	Nelson Lagoon, Alaska	S; mainly Dunlin, Western Sandpiper, and Bar-tailed Godwit (*Limosa lapponica*). Up to 100,000	Gill *et al.* (1981)
47	Port Heiden, Alaska	S; same species as Nelson Lagoon	R. Gill (personal communication)
48	Beaver Hill Lake, Alberta	N, S	P. Stepney (personal communication)
49	Harney Lake, Oreg.	S; mainly Western Sandpiper and Dunlin	S. Herman (personal communication)
50	Abert Lake, Oreg.	S; up to 50,000 Wilson's Phalaropes (*Phalaropus tricolor*)	J. R. Jehl, Jr. (personal communication)

(*Continued*)

Table I. (*Continued*)

Site No.	Site	Comments	Source
51	Lower Klamath, Tule, and Clear Lakes, Calif.	N, S; 20,000+ at each lake	Jurek (1974)
52	Mono Lake, Calif.	S; 50,000 Northern Phalaropes and 100,000 Wilson's Phalaropes	Jehl (1981)
53	Salton Sea, Calif.	N, S	McCaskie (1970)
54	Great Salt Lake, Utah	S; hundreds of thousands of Wilson's Phalaropes	J. R. Jehl, Jr. (personal communication)
55	Bear River Marsh, Utah	N, S	T. Sordahl (personal communication)
56	Horsehead and Sibley Lakes, N.D.	S	C. Faanes (personal communication)
57	Cheyenne Bottoms, Kans.	N, S; up to 200,000	E. Martinez (personal communication)
58	Great Salt Plains National Wildlife Refuge, Oklahoma	S	B. A. Harrington (personal communication)

Although the bases for selection of traditional migratory stopovers have not been convincingly demonstrated, it is logical to assume that these sites provide more predictable feeding and roosting areas than other sites along the migration routes. Given the high densities of intertidal invertebrates in the Bay of Fundy (Morrison, 1976) and the annual concentrations of horseshoe crab (*Limulus polyphemus*) eggs in spring on Delaware Bay in Delaware and New Jersey (Wander and Dunne, 1981), it is not a chance event that large numbers of shorebirds stage in those areas. On a local basis, Schneider and Harrington (1981) have shown shorebird distribution to correlate strongly with total quantity of prey available.

The availability of a predictably abundant food source is especially critical for species that must undertake long, sustained flights to complete migration. This applies to many North American breeders. Many of these species require an energy source sufficient to permit accumulation of substantial lipid reserves in a short period of time. For northbound migrants, fat reserves accumulated at final staging areas like the Copper River Delta in Alaska (Senner, 1979) and Delaware Bay in Delaware and New Jersey (Wander and Dunne, 1981) may be essential, not only for long overland flights, but also to provide insurance against adverse early weather conditions on the breeding grounds. For southbound migrants, lipid reserves are essential to fuel remarkably long transoceanic flights from the Atlantic coast of Canada and the United States to northern South America (McNeil and Burton, 1977) and from Alaska to the Pacific coast of the United States (Gill and Jorgensen, 1979) or to Pacific islands (Thompson, 1974). Most migrating White-rumped Sandpipers (*C. fuscicollis*) and Hudsonian Godwits (*Limosa haemastica*) have been shown to store fat levels sufficient for a direct flight from the Canadian Maritime Provinces to South America (McNeil and Cadieux, 1972). The same capability exists in many individuals of other species such as Semipalmated Sandpiper, Least Sandpiper (*Calidris minutilla*), and Short-billed Dowitcher (*Limnodromus griseus*) (McNeil and Cadieux, 1972; T. May, personal communication).

C. Winter Season

The locations of major winter concentrations of shorebirds in the Western Hemisphere are not as well known as in Europe and Africa, where large numbers winter near human population centers (e.g., Prater, 1981; Summers *et al.*, 1977). With the exception of well-known sites on the Pacific coast between southern British Columbia and Baja California (e.g., Jurek, 1974; Page *et al.*, 1979) and scattered locations in the southeastern United States, particularly Tampa Bay (B. A. Harrington, personal communication), the great majority of Western Hemisphere shorebirds winter in tropical and south temperate portions of the Americas seldom visited by ornithologists and birdwatchers. Some of the most important known sites in this extensive region are the coast of Surinam (Spaans, 1978), eastern Buenos Aires Province, Argentina (Myers, 1980), and locally on the Peruvian coast (Hughes, 1979). Other sites will be revealed by aerial surveys currently being conducted by the Canadian Wildlife Service (R. I. G. Morrison, personal communication). Wintering shorebird distribution in Central America is poorly known, although very large concentrations reportedly occur in Panama (B. A. Harrington, personal communication). Because most shorebirds spend nearly two-thirds of the year on the wintering grounds, we need much better knowledge of winter distribution and the factors contributing to mortality.

Energy stress faced by wintering shorebirds varies greatly as a function of the severity of winter conditions encountered. Extensive study of the wintering ecology of shorebirds in the British Isles has revealed many of the adaptations and acclimations of shorebirds to high-latitude winters. In this area, widespread kills rarely occur (Pilcher *et al.*, 1974). Winter mortality is higher than at other seasons (Table II) and sublethal effects

Table II. The Probability (× 5000) That Waders Counted on the Wash Would Be Found Dead in Winter (November–March) and at Other Times of the Year (August–October and April–May). Data from 1969 to 1974[a]

Species	Winter	Nonwinter
Common Redshank	10.717	0.215
Red Knot	0.290	0.093
Dunlin	0.247	0.046
Eurasian Oystercatcher	0.518	0.258
Eurasian Curlew	1.583	0.326
Ruddy Turnstone	2.697	0.313
Black-bellied Plover	0.284	0.372
Bar-tailed Godwit	0.451	0.063
Mean[b]	2.098	0.211

[a] Adapted from Goss-Custard *et al.* (1977).
[b] Difference significant ($p < 0.05$, Wilcoxon matched-pairs signed-ranks test, two-tailed).

on body condition and foraging effort are routinely found. For most species, food availability decreases between late autumn and late winter (Goss-Custard *et al.*, 1977). Shorebirds respond to this change in several ways. Fat levels tend to increase during the autumn, presumably as insurance against winter weather and adverse feeding conditions (Dugan *et al.*, 1981). In the Black-tailed Godwit (*Limosa limosa*), fat levels peak in December when daylight hours are shortest (Evans and Smith, 1975). In severe winters, Black-bellied Plovers (*Pluvialis squatarola*) have been shown to lose up to 35% of their lean weight and still survive (Davidson, 1981). Many species feed over a longer portion of the tide cycle than at more favorable times of the year (Goss-Custard *et al.*, 1977). Some species respond to adverse conditions by switching to alternative foods such as grain provided for wintering ducks (Pienkowski, 1979) or by feeding at night (Goss-Custard, 1969). Movement between estuaries in Britain appears to be another mechanism by which shorebirds react to food stress.

In the Western Hemisphere it is likely that shorebirds wintering at high latitudes respond to cold conditions in similar ways, though few data are available. A study in British Columbia suggests that fats are deposited as insurance against adverse weather and feeding conditions (Kaiser and Gillingham, 1981). Data on the majority of shorebirds, which winter at tropical and south temperate latitudes, are limited. Baker and Baker (1973) compared niche relationships among six species of shorebirds on their breeding range in the eastern Canadian Arctic and wintering range in southern Florida. In general, feeding rates increased and niche breadth decreased in the winter versus summer season, presumably in response to diminished food availability. In eastern Buenos Aires Province, Argentina, the area of suitable wintering habitat is small compared with that available on the breeding grounds, so shorebird densities are high (Myers, 1980). Studies in both South America and Africa suggest that shorebirds wintering in the tropics have little difficulty maintaining a positive energy budget (Puttick, 1980) and experience minimal competitive stress (Duffy *et al.*, 1981). Clearly, however, much more study at many different sites is needed.

A potential source of energy stress for wintering, as well as migrating, shorebirds is the molt process. There is great variability in molt schedules both within and between species. Dunlins from Alaska, for example, complete the prebasic molt near the breeding grounds before migrating (Holmes, 1966b). Curlew Sandpipers (*Calidris ferruginea*) in Morocco (Pienkowski *et al.*, 1976) and Least Sandpipers in the central United States (T. Maxwell, personal communication) may complete the molt at migratory staging areas. The Waddenzee in Holland is well known as a stopover where many species molt before moving on to wintering estuaries in the British Isles (Boere, 1976). The most common pattern among long-distance migrants is for molt to be deferred until after migration or to be arrested while migration is in progress. This pattern has led some to conclude that the energy demands of molt and migration often make the two processes incompatible (Holmes, 1966b; McNeil and Cadieux, 1972). Payne (1972), in an extensive review of the literature pertaining to molt, concluded that molt does not normally cause a severe energy drain, but may be significant when it coincides with other energy-demanding processes. Weight loss by molting shorebirds in winter has also been interpreted as a case of incompatible energy demands. At least one observer, however, views such weight loss not as an energy drain but rather as an adaptation to reduced flight capability (Hulscher, in Boere, 1976). The energetic significance of molt in the annual cycle of shorebirds needs further investigation.

III. THREATS TO SHOREBIRD POPULATIONS

In historic times, shorebirds have been subjected to a variety of human-related activities and developments that have had adverse effects on individuals and, in some cases, populations. Such effects may operate directly on the individual in the sense of outright mortality or indirectly by interfering with reproductive success. Before the turn of the century, hunting for food or sport posed the primary threat to shorebirds. Since

1900, however, environmental contamination and habitat modification, the impacts of which are largely indirect, have increasingly predominated. There is no question that habitat modification, in particular, is the most pervasive and serious problem confronting shorebirds in the remainder of this century. The same is true with other groups of birds (e.g., Newton, 1979) and wildlife generally.

The purpose of this section is to give an overview of those activities and developments that represent threats to shorebird populations, focusing particularly on those studies that have presented quantitative results.

A. Hunting

Through the first quarter of the 20th century, the hunting of shorebirds for food or sport was common practice in coastal and interior United States. Such hunting apparently was a significant factor in the decline of several shorebird species, including the Eskimo Curlew (Banks, 1977), Hudsonian Godwit (Forbush, 1912), and Golden Plover (*Pluvialis dominica*) (Wetmore, 1926). Passage of the Migratory Bird Treaty Act in 1918, implementing the 1916 Convention between the United States and Great Britain concerning U.S.–Canada migratory birds, gave all shorebird species the status of "migratory game birds." After 1927, all species expect the American Woodcock (*Scolopax minor*) and the Common Snipe (*Gallinago gallinago*) were fully protected in the United States and Canada, respectively. Both woodcock and snipe are hunted today, with an annual take in the United States on the order of 500,000 of each species (Banks, 1979).

At present, hunting of shorebirds for food or sport (other than the snipe and woodcock) is not a serious threat to populations in the United States and Canada. Some taking of shorebirds undoubtedly occurs incidental to the hunting of waterfowl, but there is no evidence that the take is significant. [Prater (1981, p. 111) concludes that in Britain the major problems "with shooting on estuaries relate to disturbance rather than mortality."] South of the United States, hunting may be a more serious, but largely undocumented, problem. For example, there is a report from Barbados that 15,000 Lesser Yellowlegs (*Tringa flavipes*) are taken annually by sport hunters (M. Shaffer, personal communication). It is also known that the bands of many shorebirds returned from South America to the U.S. Fish and Wildlife Service Bird Banding Laboratory came from birds taken by local residents for food (G. Jonkel, personal communication).

B. Disturbance

The potential effects of disturbance on nesting, migrating, or wintering shorebirds are of concern because many coastal wetlands supporting significant numbers of shorebirds are close to urban population centers or receive heavy human visitation. We define disturbance simply as the disruption of "normal" activity patterns (e.g., flushing a feeding bird), including the direct destruction of nests, eggs, or young. In a typical coastal setting, shorebirds may be disturbed by people walking, jogging, fishing, hunting, or digging bait, off-road vehicles, aircraft, and by feral or domestic dogs, cats, and livestock.

While the effects of disturbance (especially by biologists) on colonially nesting waterbirds are well documented (e.g., Manuwal, 1978), there have been very few quantitative studies concerning effects on either breeding or nonbreeding shorebirds. Primary concerns with nesting shorebirds are that adults may be driven from their nests, thus exposing eggs or young to predators and temperature extremes, and that people, dogs, or vehicles may directly destroy nests, eggs, or young.

Disturbance is apparently one of the factors contributing to the decline of Piping Plovers breeding in North America. Cairns and McLaren (1980) reported that on a large, remote beach in Nova Scotia, 27–29 pairs of Piping Plovers produced an estimated 1.3–2.1 fledged young per pair, while at eight small, accessible beaches 15 pairs produced only 0.7–1.1 fledged young per pair. Lambert and Ratcliff (1981) also documented low productivity (0.6 fledged young per pair) among Piping Plovers at a beach heavily used by people in Michigan. Disturbance has been similarly implicated in the case of Snowy Plovers on the California coast (Page and Stenzel, 1979). Extended observation of incubating Snowy Plovers on sandy beaches in California revealed that incubating birds usually left the nest when people or dogs were active within 100 m of the nest (Halbeisen, 1977). Predators and people and/or dogs caused about the same degree of disturbance as measured by the minutes spent incubating during half-hour blocks. Disruption of incubation was greatest when both human and natural disturbance occurred in the same half-hour interval. This study did not relate these data on disturbance to reproductive success, but it does illustrate the potential for reproductive failure due to the combined influences of human and natural disturbances.

Disturbance of nesting shorebirds is not limited to narrow coastal beaches. On annually tilled cropland in North Dakota, 37% of the nest failures in 79 shorebird (mostly Killdeer, *Charadrius vociferus*) and gamebird nests were caused by farm machinery (Higgins, 1975). On rangeland in southwestern Idaho, Jenni *et al.*, (1982) rarely observed Long-billed Curlew on slopes that supported frequent off-road vehicle traffic or that were eroded as a result of past traffic. Several hundred hectares of otherwise suitable habitat were eliminated from use by off-road vehicle traffic. An even greater area of suitable nesting habitat was eliminated as a result of people littering the landscape with various types of refuse (car bodies, broken glass, dead livestock, etc.). Among other effects, litter interrupted the curlews' field of vision, provided habitat for predators, and attracted target shooters who often turned their attention to the curlews. The presence of people, generally, and off-road vehicles elicited antipredator behavior throughout the curlew breeding phenology (except the premigratory period), thus leaving the chicks vulnerable to exposure. Frequent harrassment may lead to abandoning nests or cause curlews with chicks to shift to undisturbed but less desirable habitats (Jenni *et al.*, 1982).

Concern over the disturbance of nonbreeding shorebirds is predicated on the hypothesis that undisturbed periods of resting and feeding are important to shorebirds. At Nauset Bay in the Cape Cod National Seashore, Massachusetts, Blodget (1978) reported that feeding shorebirds were so widely dispersed over the tide flats that there was little "contact" with off-road vehicles. However, during the hours before and after high tides, flocks of resting shorebirds, then concentrated in a small area, were subject to constant traffic and human activity. Comparing the presence of people and waterbirds (gulls and terns, shorebirds, waders, and waterfowl) and the responses of birds in the presence of people at the Jamaica Bay Refuge in the Gateway National Seashore, New York, Burger (1981) found that significantly fewer sites had birds at them when people were present compared to when people were absent. When both people and birds were present, people walking or jogging along the beach always disturbed birds (Table III). Burger (1981) considered shorebirds to be the group of birds most vulnerable to disturbance, as they frequently vacated the area when disturbed (unlike gulls, for example). Burger's (1981) study and those of Halbeisen (1977) and Blodget (1978) underscore the importance of distance as a factor in determining the level and frequency of disturbance.

We conclude that in some instances disturbance may have serious negative consequences for shorebird populations, most particularly in areas used heavily by people where geographically restricted breeders such as Piping Plovers nest or where large numbers of shorebirds gather to roost on high tides. Fortunately, the limited data available suggest that problems related to human disturbance are susceptible to elimination or

Table III. Effect of Human Activity on Birds at Jamaica Bay Refuge[a,b]

Human disturbance	% of time no birds in area[c]	% of time birds disturbed[d]	% of time birds undisturbed
People walking	87	13	0
Worm diggers	50	0	50
Horseback riders	50	0	50
Joggers	0	100	0
Men working	50	30	20
Supersonic transport	?	9[e]	0
When no human activity is present	28	0	72

[a] Adapted from Burger (1981).
[b] Data gathered at stations around the refuge exclusive of the ponds.
[c] Could reflect that birds have already left. Area refers to one of the sampling plots.
[d] Birds flush.
[e] Birds were usually disturbed only by the SST, and it landed once daily during the period of this study. Sometimes it took off when no birds were near, or when data were not being recorded. However, on 9% of the field days, birds were disturbed by the SST.

mitigation through management techniques that minimize disturbance during sensitive periods.

C. Environmental Contaminants

Although birds that feed high in the trophic chain (e.g., raptors, piscivorous birds) and pelagic species in general have suffered the most severe and pervasive effects of environmental contaminants (Blus *et al.*, 1977; Eastin and Hoffman, 1979), shorebirds are not totally immune to their influence. The main pollutants shown to be of concern for shorebirds are organochlorine insecticides and oil. Pesticides and herbicides, mainly organochlorines, are still applied widely in North, Central, and South America in agricultural land and coastal wetlands often used by shorebirds. Their use is poorly regulated in many Latin American countries, where DDT, banned for general use in the United States since 1972, remains one of the most commonly used insecticides. Oil spills pose local threats to shorebirds almost anywhere along coasts in the Western Hemisphere.

For raptors and herons, the greatest public attention has been focused on indirect effects of pesticides on reproductive success. The most common problem has been eggshell thinning leading to egg breakage. Most specimens of adult shorebirds collected in the wild, however, show little evidence of organochlorine accumulation sufficient to pose reproductive problems (e.g., Blus and Lamont, 1979). Eggshell thinning of 3.8 to 4.7% during the period of DDT use was found in a survey of shorebird eggs in museum collections by Morrison and Kiff (1979). Normally, 20% thinning results in reproductive failure. Among the highest levels of DDE reported was an average of 39.4 ppm wet weight in Semipalmated Plovers (*Charadrius semipalmatus*) collected near Churchill, Manitoba, in a DDT-sprayed area (Brown and Brown, 1970). Very recent studies indicate possible problems for only a few species. L. R. DeWeese (personal communication) found levels of 2.5 to 15.7 ppm DDE in Killdeers from Colorado, Montana, Oregon, and Utah. White *et al.* (1983) reported high DDE levels (40 to 64 ppm) in Long-billed Dowitchers and American Avocets (*Recurvirostra americana*) on the Texas coast in winter. They suggested that the depths from which dowitchers procured food were apt to have higher levels of DDE than mud closer to the surface. Western Sandpipers, which feed in the upper mud layers, had very low levels of DDE (0.5 ppm). Nearly all published reports of residues of other toxic organochlorines such as dieldrin, toxaphene, and polychlorinated biphenyls in shorebirds indicate little cause for concern. Similarly, heavy metal loads are typically low. However, potentially deleterious concentrations of cadmium (Evans and Moon, 1981) and mercury (Parslow, 1973) have been found in shorebirds wintering in Britain. White *et al.* (1980) reported levels of selenium in shorebirds from Texas that might be sufficient to cause reproductive failure. Five of 54 Dunlin from British Columbia contained lead shot in the gizzards, but all birds had appeared healthy (Kaiser and Fry, 1980).

Locally there are often reports of shorebird mortality at times when pesticides are being applied heavily. Flickinger *et al.* (1980) reported deaths of three Western and one Pectoral Sandpiper (*Calidris melanotos*) within 24 hr after treatment of rice fields in Texas with carbofuran, a carbamate insecticide. The same chemical has killed small numbers of shorebirds in New York (Stone, 1979). The molluscicide, sodium pentachlorophenate, has caused deaths of jacanas (*Jacana jacana*) in Surinam (Vermeer *et al.*, 1974). Several Lesser Yellowlegs, and Western and Semipalmated Sandpipers were killed by ingestion of aldrin applied to rice fields in Texas (Flickinger and King, 1972). Finally, 72 Wilson's Phalaropes died in Wyoming in a 1000-ha plot after it was sprayed with oil- and water-treated mixtures of fenthion to control mosquitoes (L. R. DeWeese, personal communication). We conclude that some pesticides can have severe local impacts on shorebirds, but reproductive failure associated with bioaccumulation has not been demonstrated.

The most readily observable direct effect of pollution on shorebirds is feather oiling resulting from contact with spilled crude oil. However, despite much concern over the potential for major catastrophes (e.g., Connors *et al.*, 1979; Gill and Handel, 1981), there have not yet been any reports of large numbers or local populations of shorebirds being seriously depleted by oil. Where major spills, such as IXTOC in the Gulf of Mexico and the Seestern in Britain, have reached shore, hundreds of severely oiled shorebirds have been found (e.g., Vermeer and Vermeer, 1975; Prater, 1981). More moderate, sublethal oiling has been reported in at least one case in the Bay of Fundy (Smith and Bleakney, 1969). Because most shorebirds are littoral feeders, oiling can usually be expected on the belly, breast, and head, unless birds happen to bathe in highly contaminated tidal pools or backwaters. P. G. Connors (personal communication) has shown that Red Phalaropes in a laboratory setting learn to avoid pans of water covered with a thin film of oil and spend more time attempting to escape from aquaria with oil on the surface than from aquaria without oil. However, even brief exposure to oil caused substantial matting of the feathers. The implications of these results for encounters with oil in the wild are not entirely clear.

Of the shorebirds, the Red and Northern (*P. lobatus*) Phalaropes, being strictly pelagic outside the breeding season, face the highest probability of suffering major losses from oil contact (see Burger, Vol. 6 of this series). This is particularly true where large numbers occur during migration. For example, in the lower Bay of Fundy many hundreds of thousands, and perhaps millions, of Northern Phalaropes congregate at the mouth of Passamaquoddy Bay every August. Deep-water oil ports being constructed in this area may result in an increase in oil tanker traffic. A major spill in August could conceivably affect a large proportion of the eastern North American population of this species.

D. Habitat Loss and Degradation

Human pressures and demands, aided by a rapidly expanding technological capacity, have greatly altered the habitat base on which shorebirds depend. The rate of loss of coastal wetlands has been related directly to human population density (Gosselink and Baumann, 1980). In view of the current estimate that 80% of the population of the United States will live within 50 miles of the coast by 1995, it appears that the trend of habitat loss and degradation will continue in the foreseeable future (NOAA, 1980).

A variety of development and other human activities result in losses or degradation of wetland habitats (Table IV). While it is beyond the scope of this chapter to describe each such activity in detail, Darnell (1976) and

Table IV. Development and Other Human Activities Causing Wetland Loss and Degradation[a,b]

Dredging and excavation (e.g., Herman and Bulger, 1981, Grays Harbor, Wash.)
Spoil disposal
Landfill (e.g., Kane, 1976, Hackensack Meadowlands, N.J.)
Waste disposal
Impounding and/or draining ("reclamation") (e.g., for industrial purposes, Evans, 1978/79, Tees estuary)
Impoundments, reservoirs (e.g., tidal power, Morrison, 1977, Bay of Fundy, Maine/Nova Scotia)
Sediment diversion due to structures and channels
Hydrologic alterations (e.g., diversions for water supply, Jehl, 1981, Mono Lake, Calif.)
Subsidence due to extraction of ground water oil, gas, sulfur, and other materials
Off-road vehicle traffic (e.g., Blodget, 1978)

[a] Adapted from Gosselink and Baumann (1980)
[b] In parentheses are shown selected examples from the literature involving direct references to shorebird populations.

Prater (1981) present reviews of those activities affecting wetland ecosystems and estuarine birds, respectively. Specific causes of wetland losses and degradation vary within regions and over time. For example, diking and filling have reduced the area of San Francisco Bay, California, marshlands by 75% since 1850 (Horwitz, 1978). Since 1940, a minimum of 16 km^2 of Grays Harbor, Washington, has been used for the disposal of dredge spoils, with at least 5.2 km^2 having been entirely removed from tidal influence (Smith *et al.*, 1976). By 1938, some 562,500 acres of tidelands from Maine to Virginia was crisscrossed by mosquito control ditches, thus reducing the extent of tidal influence. This acreage is believed to be 90% of the tidelands existing prior to 1885 (Bourn and Cottam, 1950). While man's activities are the primary cause of wetland losses, it is important to recognize that they are superimposed on natural processes and that the two often interact to accelerate wetland losses (Gosselink and Baumann, 1980).

There are no precise statistics available on wetland losses (due to man's development and natural processes) in North America, but sufficient data exist to document major losses in this century. Based on two national surveys (1922 and 1954) and a compilation of statewide assessments, Gosselink and Baumann (1980) report that coastal wetlands in 22 states (contiguous United States only) have been reduced from 4,041,000 ha in 1922 to 3,403,110 ha in the early to mid-1970s, a decline of about 16% (Fig. 2). Total national losses (coastal and inland) may be as high as 30 to 40% (Horwitz, 1978), while losses in individual states or regions may be much higher. California, for example, has lost about two-thirds

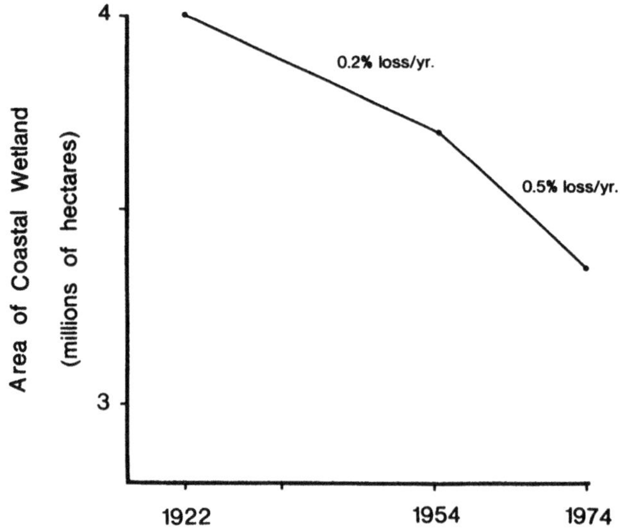

Fig. 2. Loss of coastal wetlands between 1922 and the early 1970s. [Redrawn from Gosselink and Baumann (1980).]

of the 381,000 acres (154,000 ha) of prime coastal wetlands it had at the turn of the century (Speth, 1979). By 1968, southeastern Wisconsin had lost 61% of its freshwater marshes (Marsh *et al.*, 1973).

Habitat losses affecting shorebirds are not limited to wetlands. About 50% of 770,000 km^2 of upland grassland in the prairie pothole region of the United States and Canada is now cultivated on an annual basis (Higgins, 1975). These changes are believed to be responsible for declines in populations of species like the Long-billed Curlew (Renaud, 1980). Increased use and diversion of water for irrigation purposes are in some instances lowering water tables and reducing the extent of surface water in interior river systems and wetlands. The side effects of water-intensive agriculture directly affect the extent and habitat available to nesting species such as the Snowy Plover (e.g., Boyd, 1981, Grover and Knopf, 1982).

In general, a given type of development activity results in a characteristic set of impacts in a wetland ecosystem, although the specifics will vary with individual projects and wetlands. Darnell (1976) presents a detailed review of such impacts and rates outright habitat loss as the most important impact of construction activity upon aquatic environments. He rates the addition of suspended solids to aquatic systems and modification of flow regimes and water levels as the second and third most significant impacts, respectively. Other major impacts include altered water temperature and pollution by heavy metals and pesticides.

While many investigators make reference to the impacts of development on shorebirds, the few detailed studies have focused primarily on quantifying changes in numbers of shorebirds as a result of habitat losses or degradation. For example, Joyes *et al.* (1976) documented a 90% reduction in numbers of feeding shorebirds in autumn migration after construction of a road resulted in inundation of most of the Oued Chebeika Estuary (Morocco). At a small estuary in Hawaii, Lum (1978) recorded significant declines in numbers of three of five species of shorebirds following removal of several intertidal sand and mud bars (incidentally, the bars were removed for "esthetic" reasons). In an example of habitat degradation rather than loss, at Grays Harbor, Washington, disposal of dredge spoils raised the elevation of an intertidal study plot to +2.13 m, causing a substantial decrease in numbers of benthic invertebrates such as *Corophium* amphipods. The distribution of feeding shorebirds (mostly Dunlins) following spoil disposal corresponded to that of the invertebrates. The mean density of birds feeding on an adjacent control area was more than 100 times higher than in the elevated area (Smith and Mudd, 1976).

Between the first half of the 19th century and 1974, the Tees River Estuary, England, was reduced from 2400 to 140 ha as a result of various industrial reclamation projects. Reports by Evans (1978/79) and Evans *et al.* (1979) each describe different aspects of the effects of a loss of 60% of the most important mudflats (Seal Sands) in the estuary (also see Prater, 1981). Although the effects varied among shorebird species, numbers of shorebirds generally declined relative to national population trends and the percentage decreases in numbers were greater than, or equal to, the percentage reduction in total feeding area (Evans, 1978/79). The shorebirds were also affected by a reduction of 4 hr (from 12 to 8 hr) in the time mudflats were exposed and available for feeding purposes. This reduction had little affect on Eurasian Curlews (*Numenius arquata*), for example, as their average daily feeding time was only 5–6 hr. Dunlins, on the other hand, with an average daily feeding time of 9–11 hr, were no longer able to meet daily food requirements on the Seal Sands and were forced to feed at nontidal sites (Evans, 1978/79).

Finally, in an example involving degradation of nesting habitat, Larsson (1969) compared numbers of several shorebird species nesting on grazed and ungrazed wet pastures in south Sweden. Species such as the Dunlin and Common Redshank (*Tringa totanus*) strongly favored the grazed pastures as nesting habitat and it was predicted that their breeding ranges would shrink as grazing of wet pastures was phased out.

In each of these examples it is clear that, in general, the proximate result of habitat losses or significant degradation is to displace shorebirds that had been using these habitats for feeding, roosting, or nesting. It can be assumed that such displacement will not have a beneficial effect, but the ultimate significance of such displacement is largely a matter of speculation and will vary with specific circumstances (see Section V).

In the case of a loss of feeding habitats, the displacement of shorebirds may result in higher densities of birds in alternative habitats, the choice of suboptimal habitats, or both. High bird densities can have various negative impacts on shorebirds, including reduced feeding rate (Goss-Custard, 1976, 1977), reduced prey detectability (Goss-Custard, 1970), increased aggression (Goss-Custard, 1977; Burger *et al.*, 1979), and prolonged feeding time (Goss-Custard, 1970; Goss-Custard *et al.*, 1977).

IV. MANAGEMENT OF SHOREBIRD POPULATIONS

Shorebird "management" (we use the term loosely) today consists primarily of two aspects: information gathering and the protection of shorebirds and their habitats from destruction by man. The former involves the filling of gaps in our understanding of shorebird distribution and ecological requirements and enhancing our ability to predict the effects of human activities and developments on shorebird populations. Fundamental to these tasks is the implementation of systematic programs for population estimation and monitoring. The latter aspect, the protection of shorebirds and their habitats, consists of the application of various legal-political tools that exist at national and international levels. The purpose of this section is to describe the most important legal-political tools involving the United States and current efforts to develop programs to estimate and monitor shorebird populations in the Western Hemisphere. We also briefly describe efforts to reintroduce shorebirds to historic ranges and to manipulate and enhance shorebird habitats, practices that may assume greater significance as man increasingly modifies the natural environment.

A. Legal-Political Tools

A variety of national and international laws and agreements exist that are being, or potentially may be, used to protect shorebirds and shorebird habitats. The scope, strength, and enforcement of wildlife protection laws vary widely from nation to nation. Thus, the ranges of migratory species are fragmented into political patchworks consisting of differing degrees of protection for birds and habitat. Obviously, protection of shorebirds in one part of their range may be negated by actions elsewhere, a matter that is of general concern among students of neotropical migrants (Keast and Morton, 1980). (The full names of the treaties and laws cited below are listed in Tables V and VI. Most names used in the text are abbreviated forms.)

1. Laws and Treaties

The foundations of modern efforts in the United States to protect migratory birds are four bilateral conventions implemented for the United States by the Migratory Bird Treaty Act (Table V) [see Bean (1977) for excellent discussion]. The oldest convention, concerning birds in the U. S. and Canada, was one of a number of conservation initiatives at state and national levels stimulated by the indiscriminate exploitation of various bird species (for flesh, feathers, eggs, or sport) in the 19th and early 20th centuries (e.g., Matthiessen, 1959). Each of the conventions prohibits the taking of most bird species except in certain instances. Only the most recent of them, the conventions with Japan and the Soviet Union, address both this issue and the preservation of migratory bird habitats. The Western Hemisphere Convention (Table V) encourages the preservation of Western Hemisphere flora and fauna through establishment of nature reserves and explicitly encourages actions to ensure that bird species crossing international boundaries do not become extinct. To date, relatively little has been done to implement its broad provisions (Train, 1982; but see Freese, 1982), although Congress recently reaffirmed its interest in the Western Hemisphere Convention by giving the Department of the Interior additional guidance in implementing its provisions (P.L. 97-304). The primary purpose of the International Trade Convention (Table V) is to regulate commercial traffic in endangered species. It would appear to be of relatively little importance to present-day problems in shorebird conservation because the Charadrii are not subject to the same commercial pressures as are groups such as the Psittacidae.

Interest in the Migratory Species Convention and the Ramsar Convention has been focused on Europe, the Mediterranean, and North Africa (Table V). Chile and Canada are the only Western Hemisphere nations to have ratified the Ramsar Convention. In its present form the primary value of this convention is in encouraging systematic identification of wetlands of international significance and in the attention it focuses on the international scope of problems in wetlands conservation. However, parties to the Ramsar Convention may satisfy its minimum requirements

Table V. Selected International Treaties Related to the Conservation of Birds, Including Shorebirds

Treaty	Year entered into force	Comment
Convention between the United States and Great Britain (for Canada) for the Protection of Migratory Birds (39 Stat. 1702; TS 628)	1916	
Convention between the United States and the United Mexican States for the Protection of Migratory Birds and Game Mammals (50 Stat. 1311; TS 912)	1937	
Convention on Nature Protection and Wildlife Preservation in the Western Hemisphere (56 Stat. 1354; TS 981)—Western Hemisphere Convention	1942	Action in 97th Congress provided new guidance
Convention between the Government of the United States and the Government of Japan for the Protection of Migratory Birds and Birds in Danger of Extinction, and Their Environment (25 U.S.T. 3329; T.I.A.S. 7990)	1974	First U.S. migratory bird treaty to cover habitat protection
Convention on International Trade in Endangered Species of Wild Fauna and Flora (27 U.S.T. 1087; T.I.A.S. 8249)—International Trade Convention	1975	
Convention on Wetlands of International Importance, Especially as Waterfowl Habitat (I.L.M. 11:963–976, September 1972)—Ramsar Convention	1975	The U.S. is not a party to this convention
Convention between the United States and the Union of Soviet Socialist Republics Concerning the Conservation of Migratory Birds and Their Environment (T.I.A.S. 9073)	1978	Only partially implemented by Migratory Bird Treaty Act
Convention on the Conservation of Migratory Species of Wild Animals (I.L.M. 19:11 et seg., Jan. 1980)—Migratory Species Convention	Not in force; adopted in 1979	The U.S. is not a party to this convention

by designation of a single wetland that may already be protected under preexisting domestic law, thus leaving other equally important wetlands undesignated and unprotected (Burhenne-Guilmin and de Klemm, 1980; Saeijs and Baptist, 1980).

While most conservation efforts at the turn of the century were focused on the birds themselves, there has been increasing recognition of

Table VI. Selected Domestic (U.S.) Laws Related to the Conservation of Birds and Their Habitats

Public law	Year of enactment	Comment
Fish and Wildlife Coordination Act, as amended (48 Stat. 401)	1934	Requires consultation with U.S. Fish and Wildlife Service regarding Federally sponsored or permitted water resource projects and wildlife
National Environmental Policy Act of 1969, as amended (83 Stat. 852)—NEPA	1970	
Coastal Zone Management Act of 1972, as amended (86 Stat. 1280)	1972	
Federal Water Pollution Control Act Amendments of 1972, as amended (86 Stat. 816)—Clean Water Act	1972	Section 404 particularly crucial for wetland protection
Endangered Species Act of 1973, as amended (84 Stat. 1312)	1973	
Fish and Wildlife Conservation Act of 1980 (94 Stat. 1322)	1980	Concerns nongame species. No funds have ever been appropriated by Congress pursuant to this law

the importance of protecting bird habitats as well. The great "dust bowl" drought of the 1930s aroused much concern among waterfowl hunters over loss of nesting habitats and stimulated various Federal and state programs to acquire and preserve interior wetlands (e.g., Linduska, 1964). However, it was not until the "environmental movement" of the 1960s and 1970s thrust ecological problems into the forefront of public debate that far-reaching Federal laws were enacted to conserve wildlife habitats and other environmental resources on both public and private lands.

A landmark among such legislation is the National Environmental Policy Act of 1969 (Table VI), the most significant result of which has been to integrate consideration of environmental impacts into Federal decision-making processes. This requirement has brought to public attention the impacts of large-scale public works projects (often involving modification of wetlands) and other Federal actions. From the standpoint of protecting the wetland habitats on which shorebirds depend, the most important law is the Clean Water Act (Table VI), particularly Section 404. Section 404 gives the Army Corps of Engineers, in conjunction with the

Environmental Protection Agency, the authority to issue—and deny—permits for activities that modify aquatic and wetland habitats. The Corps' statement of general policy for evaluating permit applications states that "[s]ome wetlands are vital areas that constitute a productive and valuable public resource, the unnecessary alteration or destruction of which should be discouraged as contrary to the public interest" (47 Fed. Reg. 31804). The Coastal Zone Management Act (Table VI) established a program of Federal assistance to coastal states in setting up coastal resource protection programs. As a result, some states (e.g., Massachusetts) have enacted and implemented tough laws protecting coastal wetlands and related areas of ecological concern (e.g., barrier islands).

The Endangered Species Act (Table VI) is another law oriented toward species protection. However, recognizing that protecting endangered species means protecting their habitats, the law requires that Federal agencies must ensure that their actions (e.g., construction of a public works project or approval of a Section 404 dredging permit) do not jeopardize the continued existence of an endangered species. Two shorebirds are presently listed as endangered under the Act, the Hawaiian Stilt (*Himantopus himantopus knudseni*) and the Eskimo Curlew. Anderson and Gress (1981) note that efforts to protect an endangered species such as the Brown Pelican (*Pelecanus occidentalis californicus*) often serve to benefit other nonendangered wildlife as well.

2. Wildlife Refuges and Other Reserves

While a great deal has been done in the United States to regulate the uses of both public and private lands to conserve wildlife resources, the most enduring form of protection appears to be the dedication of lands (and waters) specifically for ecological conservation purposes. As a result of acquisitions under authority of the Migratory Bird Conservation Act, the Endangered Species Act, and other laws, as well as the designation by the President or the Congress of lands already in public ownership, the United States has a National Wildlife Refuge System of 404 units and approximately 86.6 million acres (35.1 million ha) (D. Hoobler, personal communication). Many units in the system, such as the Chincoteague National Wildlife Refuge, Virginia, provide vital habitat to nesting, migrating, or wintering shorebirds. Other types of government conservation system units such as national parks (e.g., the Everglades National Park, Florida), national forests (e.g., the Copper River Delta in the Chugach National Forest, Alaska), and state parks and reserves (e.g., Cape Henlopen State Park, Delaware) may also include important shorebird habitats. (Similar types of conservation units of comparable importance for shorebirds exist in Canada as well, e.g., Fundy National Park, New Brunswick; Polar Bear Provincial Park, Ontario; and Shepody Bay National Wildlife Area, New Brunswick.) Of special note are the more than 100 million acres (42 million ha) of lands added to the Federal conservation systems by the Alaska National Interest Lands Conservation Act of 1980 (P.L. 96-487). As a result of this action, extensive and important shorebird habitats were placed in a protected status (although the degree of protection varies with different classifications and individual areas), ranging from interior mountains, where Surfbirds (*Aphriza virgata*) nest, to coastal wetlands, which support breeding Dunlins and Western Sandpipers. It is important to note with respect to these areas in Alaska, or any other lands protected by legislative or administrative action, that such action will not necessarily ensure protection of the area in perpetuity, nor are such areas free of the influence of development activities which may occur in adjacent, or even distant, areas (e.g., oil drilling offshore of a coastal wildlife refuge).

The protection of shorebird and other wildlife habitats through designation or acquisition need not be limited to agencies of the government. Many private organizations in North America, Europe, and elsewhere have vigorous land preservation programs. In the United States, for example, The Nature Conservancy presently owns more than 1.8 million acres (0.7 million ha) of land including the 35,000-acre (14,170 ha) Virginia Coast Reserve (C. Combs, personal communication), an area used heavily by migrating shorebirds. The role of private organizations in acquiring and protecting wildlife habitats is a significant one, especially during times when governmental resources for, or commitments to, such purposes are limited.

B. Habitat Manipulation

Not all modifications of "natural" habitats are necessarily detrimental to shorebirds, and, in fact, some changes may have beneficial effects (e.g., Glue, 1971; Burger *et al.*, 1982). In contrast to waterfowl and some colonially nesting waterbirds (e.g., Soots and Landin, 1978), however, shorebirds have received almost no attention from the standpoint of actively manipulating habitats for their benefit. Funds are typically targeted for endangered species, species highly susceptible to environmental contamination, and species of economic interest. However, a few local efforts have been made to improve habitats for shorebirds. For example, the Missouri Department of Conservation maintains two small impoundments specifically for shorebirds (J. Wilson, personal communication). More such efforts could be implemented by adapting management techniques for other species or by slight modification of habitat restoration programs.

Water levels in freshwater or brackish impoundments as exist on state or Federal wildlife refuges can be manipulated to produce highly productive shallow water attractive to migrating and wintering shorebirds. Such opportunities may also exist incidental to agricultural practices (Sykes and Hunter, 1978) and the operation of municipal sewage facilities (Fuller and Glue, 1980). Because the peaks of spring and fall migration of shorebirds do not coincide with migration schedules of most waterfowl, programs for maintaining optimal water depths for shorebirds do not necessarily conflict with provision of optimal water depths for waterfowl. Rundle (1980) showed that the best procedure for making freshwater impoundments attractive to shorebirds is by a regime of water drawdown in late summer, followed by disking and shallow flooding. Where a system of impoundments exists, maintaining slightly different water levels assures that new habitats will become available as others dry out. Most impoundments have water control structures that permit drawdown but not refilling. Therefore, a risk exists that rainfall will not fill impoundments prior to the onset of waterfowl migration. Some national wildlife refuges (e.g., Chincoteague National Wildlife Refuge, Virginia) have taken such risks and created habitats attracting tens of thousands of shorebirds. Because most refuges along the Atlantic and Gulf coasts have impoundments, the addition of simple water-pumping systems could eliminate the risk and result in a network of impoundments attracting both shorebirds and waterfowl at different times.

Although habitat manipulation can potentially improve the suitability of certain shorebird nesting habitats, this aspect of management has not been explored for most species. Islands created by disposal of dredged materials may create nesting and roosting habitat for some species (Soots and Landin, 1978) that is more free of disturbance than beaches on adjacent mainlands. Kantrud and Kologiski (1982) found that heavy grazing by cattle generally favors nesting Killdeers and Mountain Plovers. Lightly grazed arid grasslands and more heavily grazed mesic grasslands appeared to benefit Long-billed Curlews, while Upland Sandpiper densities were not strongly influenced by grazing. Kirsch and Higgins (1976), however, found the highest densities of the latter species in ungrazed grassland. Attempts to restore native grassland after it has been destroyed by surface mining or other activities often result in grasslands unsuitable for typical native bird species. In the case of the Long-billed Curlew, this is true because of a tendency to replant with wheatgrasses (*Agropyron* spp.) that grow taller than the native species. Jenni *et al.* (1982) have suggested reseeding reclaimed land with other grass species that will result in a grassland system that is both grazable and attractive to the curlews.

There are, no doubt, many land management practices and forms of land manipulation that can be carried out for the benefit of nesting, roosting, or feeding shorebirds. The major difficulty to date is simply a failure to consider the requirements of this group of birds. Shorebird habitat management is in its infancy and will evolve only to the extent that land managers are informed of its possibilities and its long-term significance.

C. Species Reintroduction

Reintroduction of species to historic breeding ranges is a popular and, in general, successful management tool. Efforts to reintroduce raptors such as the Peregrine Falcon (*Falco peregrinus*) and Bald Eagle (*Haliaeetus leucocephalus*) to formerly-occupied breeding ranges are examples of current, large-scale restoration projects (e.g., Temple 1983). For shorebirds as a group, reintroduction has not been extensively tested as a management tool. In the early 1950's and again in the late 1970's there were modest efforts to restore Black-necked Stilts (*Himantopus mexicanus*) as a breeding species in New Jersey. The latter effort also included American Avocets. Both projects involved use of Willets (*Catoptrophorus semipalmatus*) as foster parents by replacing Willet eggs with stilt or avocet eggs. In both cases fledging success was uncertain, but in any event neither project resulted in restoration of a breeding population.

More recently, the Kansas and Colorado Fish and Game Commissions are cooperating on a project to reintroduce the Mountain Plover as a breeding species in western Kansas. In the summer of 1982, 50 fledging-aged plovers were released on a Wallace County ranch and their dispersal into the surrounding shortgrass prairie was monitored. Success of the project cannot yet be evaluated, but it was to continue for two years after the initial season (Ptacek and Schwilling 1983).

D. Population Estimation and Monitoring

Effective population monitoring techniques are necessary if a decline in the population of a species is to be diagnosed before the population reaches a dangerously low level. For large populations, a sampling scheme that provides an index to the total population may be necessary. This is most feasible when the animals remain in one place for an extended period of time. Perhaps the best example of this type of monitoring program is the North American Breeding Bird Survey, conducted by the U.S. Fish and Wildlife Service (USFWS) and the Canadian Wildlife Service (CWS). In this survey, volunteer observers census breeding birds along randomly selected highway routes (Bystrak, 1981). Most of the species censused exhibit territorial behavior and individuals are distributed relatively evenly throughout suitable habitat.

During the breeding season, most species of North American shorebirds occupy remote habitats unsuitable for this type of a survey. If financial resources were available, their populations would probably be best monitored on Latin American wintering grounds. Wintering birds tend to return to traditional sites (Page *et al.*, 1979; Prater, 1981) and remain for the duration of the winter. They are more concentrated than in the breeding season, often in productive coastal habitats, and it may be possible to count the total number of individuals present. This type of survey has been most fully developed in the British Isles, where volunteers supervised by regional coordinators conduct high-tide censuses at all major wintering estuaries at regular intervals. Known as the Birds of Estuaries Enquiry (described in detail in Prater, 1981), this survey is believed to provide reliable estimates of the total number of individuals wintering in the entire geographic region.

The winter counts of shorebirds in South America now being conducted by CWS are not expected to continue on a regular basis. So, although they will enable identification of important wintering sites, the data cannot be applied to determination of shorebird population trends. A year-round survey of numerous sites in California in the late 1960s and early 1970s was not continued long enough to permit trend analysis (Jurek, 1974). Another survey, developed independently by the Manomet Bird Observatory (Massachusetts) and CWS in the early 1970s, provides annual data that are potentially useable toward this end. The International Shorebird Survey (ISS) (Morrison and Harrington, 1979) is patterned after the Birds of Estuaries Enquiry except that shorebirds are censused at migratory stopovers instead of wintering sites. Volunteer observers census southbound migrants at 10-day to 2-week intervals between July and November at selected sites, primarily in eastern North America. Most of the sites are located on the Atlantic coast between the Canadian Maritime Provinces and South Carolina. This is the zone through which most trans-atlantic migrant shorebirds from breeding grounds in eastern Canada pass. To date, the most important results of the ISS have been a better understanding of migration routes, identification or confirmation of major staging areas, and the production of density-distribution maps of most of the species. The utility of the data set for determination of population trends is currently being studied by USFWS.

The mobility of migrating populations introduces confounding variables into population trend analysis. Because there are probably varying rates of turnover among species and some variation in migration routes between years, surveys of migrants must be carefully structured, both spatially and temporally, to minimize these additional sources of variance (Howe, 1980). One must first assume that the migrant population censused in a geographic region is consistently representative of the total population size of the species or the subpopulation of concern. Census sites must also be distributed spatially in such a way that important segments of the population are not overlooked in any given year. Further, censuses must be conducted often enough so that portions of the migrating population do not move in and out between consecutive censuses. One must make the additional assumption that turnover rates do not vary significantly within species between years. Recent comparisons of migrant passerine populations with breeding surveys from the same years show encouraging correlations in some cases (Svensson, 1978; Hussell, 1981).

Our present knowledge of the migration biology of North American shorebirds justifies accepting the necessary assumptions for those species that require coastal habitats during migration. Species that regularly use inland habitats during migration will probably show too much geographic instability among years to permit population trend analysis. Therefore, there is the potential to monitor species such as Red Knot and Short-billed Dowitcher, but probably not species such as Pectoral and Least Sandpipers. The analyses of ISS data being conducted by USFWS are focusing on obligate coastal species, and the nature of within-site and between-site variation is being studied. Modifications of the trend analysis programs for the Breeding Bird Survey are being tested on the ISS data. If results are promising, modifications of survey design that may increase the precision of population indices will be recommended. Although a high level of precision is not anticipated, even gross population trend estimates will facilitate implementation of species-specific conservation measures.

V. DISCUSSION

A. Critical Area Concept

As population and development pressures threaten remaining coastal and interior wetlands, it becomes increasingly important that we identify those wetland areas that can be considered "critical" in the annual cycles of shorebird species. (We use the term "critical" here in the sense that a loss or degradation of a critical area would ultimately result in reduced populations of one or more species.) Such information is necessary for predicting the effects of proposed developments on shorebird populations and in allocating limited financial and manpower resources for habitat acquisition or other conservation measures.

At present our information base is not sufficient to identify systematically those areas that can be considered critical. Thus, we attach high priority to instituting a regular program for estimating and monitoring shorebird populations. Further, we propose consideration of three guidelines, each of which represents an incremental addition to the sophistication and utility of our information base. The guidelines are individually listed and discussed below.

1. Numeric Importance

1. The number of individuals estimated to use an area traditionally over the course of a year.
2. The importance of the number of a given species relative to the species' total population or subpopulation.

Application of predetermined quantitative standards to basic data on the number of shorebirds using an area is a direct, positive, and objective (albeit arbitrary) means of identifying wetlands and other areas deserving of further attention and concern. The criteria adopted at the 1980 conference of the parties to the Ramsar Convention, meeting in Cagliari, Italy, attach international importance to wetlands that "regularly" support 20,000 shorebirds (all species over entire year), or 1% of the individuals (or pairs in the case of a nesting area) in a biogeographical population of one species or subspecies (Anonymous, 1980; Atkinson-Willes *et al.*, 1980). We lack sufficient data on population sizes in the Western Hemisphere to apply either of these standards at this time. However, projects such as the California Shorebird Survey and the ISS provide sufficient data from which to compile a partial, preliminary listing of those wetlands regularly supporting 20,000 shorebirds on any given day during migration (Fig. 1, Table I). In addition to quantitative standards, recognition should also be given to wetlands supporting an identified element or subgroup—but not necessarily a subspecies, e.g., a population on the periphery of, or isolated from, a main breeding population [e.g., Snowy Plovers breeding at Cheyenne Bottoms Wildlife Area, Barton County, Kansas (Boyd, 1981)].

2. Functional Significance

The functional significance of an area in the annual cycles of shorebird species and the site-specific relationship between a species and a particular habitat.

Knowing the number of shorebirds using an area is not, in itself, a sufficient basis for predicting the effects of development schemes or for prescribing conservation measures. One needs to understand the functional significance of each area in the annual cycle of a species. Some areas are important to migrants for the purpose of replenishing fat reserves, while others serve as postbreeding aggregation areas, molting grounds, etc. In this regard it is important to recognize that an area may have a significance to a species that is out of proportion to that fraction of the annual cycle actually spent at a particular wetland area, much as a parasite may have a fleeting—but absolutely indispensable—association with a specific host at a critical stage in its life history (Senner, 1977).

Shorebirds do not use wetland or other areas uniformly in space and time in that their distribution varies with the availability of prey, location of roost sites, and environmental factors such as tides and weather. Thus, it is also important to understand the activity patterns of shorebirds within each wetland. From such information one may discover that a particular human activity or development scheme will conflict greatly or very little with the shorebirds' use of the same site [e.g., traffic by off-road vehicles across the upper intertidal zone of a mudflat may conflict little with shorebirds when they are feeding at lower tidal levels (Blodget, 1978)].

3. Effects on Populations

The likely impacts on a shorebird population if a particular habitat is lost or degraded.

In considering the impacts on a shorebird population of habitat modification, it is useful to distinguish between immediate and long-term effects. It is the latter with which we are concerned. As noted in Section III, few studies have made rigorous attempts to determine the effects of developments or human activities on shorebird populations. None have shown a clear relationship between a long-term decline of a population and the loss or degradation of a particular wetland. This is not to say that relationships do not exist. Rather, it is extremely difficult to obtain the kinds and scope of data to demonstrate such a relationship statistically. One must first demonstrate that the birds are indeed being affected, and to what degree. This can be especially difficult when the effects are sublethal, such as a deteriorating physiological state leading to lowered reproductive fitness. Further, one must be able to show that a species' population is actually declining, something we are only now beginning to have the capability of doing for species with large populations.

A key consideration in predicting effects of habitat change on shorebird populations is the availability of alternative habitats that may serve the same purposes for the birds. In some parts of the Western Hemisphere like the coast of Peru (Hughes, 1979), where suitable habitats for shorebirds are isolated by up to 1000 km from other ones, major habitat change could result in mortality of thousands of shorebirds. In other areas, although there may appear to be suitable alternative sites, they may not offer comparable feeding opportunities or safety from predators. The biological effects of feeding in suboptimal or overcrowded habitats have been mentioned (Section III) and are discussed in more detail elsewhere in this volume. Thus, we are not only concerned with the availability of alternative habitats, but also their ability to sustain fitness of the birds at the same level. In cases where the alternatives are in fact comparable, we note parenthetically that wetland losses should be measured not only against that which still remains but also that which already has been lost to previous development activities (Herman and Bulger, 1981).

Finally, it is important that in some cases alternative habitats may be created as a deliberate by-product of development projects. For example, artificial islands made of dredge spoils may create shorebird nesting and roosting sites that are free from disturbance from dogs and people (see Section IV).

The guidelines suggested here should be applied incrementally in identifying those areas critical for shorebirds. The first establishes an objective criterion for narrowing the range of habitats to those warranting special attention. Having identified areas with large numbers of shorebirds, the second guideline focuses our attention on those areas used during sensitive stages of the annual cycle. Finally, the third guideline uses the information from the first two to predict whether population declines are apt to result from loss of the habitat. Prognosis of a population decline would suggest that the area is critical in the annual cycle of the shorebird.

The Copper–Bering River Delta System: A Case Study

The combined Copper and Bering river delta system (C-BRD system) is a habitat island on the northern Gulf of Alaska coast with tide flats of more than 1000 km^2 and associated supratidal wetlands (Isleib and Kessel, 1973; Senner *et al.*, 1981). This area is potentially one that will be affected by oil spilled from passing supertankers or offshore oil drilling and other developments such as coal mining and the associated construction of port and transportation infrastructure.

During late April and the first half of May, the C-BRD system supports more than 20 million waterfowl and shorebirds (Isleib and Kessel, 1973), thus surpassing by many times the standard of 20,000. Further, about half of those spring migrants are Western Sandpipers and Dunlins. While we lack total population estimates for either species, these birds represent nearly the entire world population of Western Sandpipers and nearly the entire population of the Dunlin subspecies (*Calidris alpina pacifica*) nesting in western Alaska (Senner *et al.*, 1981). Thus, for at least these two species the C-BRD system unquestionably supports more than 1% of their population during the spring migration alone.

Although Western Sandpipers and Dunlins only use the C-BRD system intensively for a few days a year, it appears to be highly significant in their annual cycles. The C-BRD system is one of the few wetland areas of any size available for feeding and resting purposes between the Fraser River Delta, British Columbia, and breeding grounds on the Yukon–Kuskokwim Delta in western Alaska, a distance (following the coast) of more than 3000 km (Senner *et al.*, 1981). Within the C-BRD system the sandpipers are largely confined to intertidal wetlands for feeding purposes, because the supratidal wetlands are usually snow- and ice-covered during spring migration (Senner *et al.*, 1981). Feeding Dunlins concentrate at water's edge on falling and rising tides and prey heavily on a single species of bivalve mollusc, *Macoma balthica*. Western Sandpipers tend to feed at all stages of the tide, both at and away from the water's edge, taking a variety of prey including several species of molluscs, amphipods, and insect larvae (Senner, 1977). The functional significance of the C-BRD system appears to be that it enables both species of sandpipers, but most clearly Dunlins, to replenish fat reserves needed for long-distance migrations to and from the C-BRD system en route to western Alaska breeding grounds. These fat reserves are also important in enabling the sandpipers to initiate reproductive activities at a time when weather conditions are harsh and prey populations highly variable (Senner, 1979).

If the western end of the C-BRD system were to be degraded by catastrophic or chronic oil pollution, for example, it could have serious consequences for both Dunlins and Western Sandpipers. Given the isolated position of the area within their migratory pathways and that the sandpipers are essentially limited to intertidal mudflats for feeding purposes, loss of this habitat could have the effect of reducing reproductive fitness among individuals arriving at their breeding grounds. Of the two species, the Dunlin would appear to be the more vulnerable in view of its stenophagous food habits, its selection of the water's edge as its feeding microhabitat, and the long-distance migrations it undertakes to and from the C-BRD system (Senner, 1977, 1979; Senner *et al.*, 1981).

Thus, in all respects the C-BRD system would appear to be a critical habitat in the annual cycles of Dunlins and Western Sandpipers. Other species for which this area may also be critical include Surfbirds and Red Knots migrating on the Pacific coast.

B. Conclusion

Although most shorebird species are still common and even abundant, there are sensitive periods in their annual cycles during which shorebirds are particularly vulnerable to the effects of environmental modification, pollution, and disturbance. Of the threats to shorebird populations, habitat loss and degradation loom as the most significant and pervasive. However, disturbance, environmental contaminants, and hunting may also be significant in some instances.

Our ability to predict—and therefore prevent or diminish—the effects of human activities and developments on shorebird populations is hindered by our lack of knowledge about their distribution and ecological requirements throughout the annual cycle. Thus, a primary task of any effort to conserve or manage shorebird populations is to fill the gaps in our information base and to identify areas critical to shorebirds. We suggest that establishment of systematic programs to monitor population change is essential, and we propose three criteria to aid the process of delineating those areas that are critical for different species.

While information gathering is important, with each passing day new decisions are made having long-term implications for environmental quality. Thus, there is a need for substantive action at the same time as we refine our information base. There exist a variety of national and international laws and agreements that are or may be brought to bear to protect shorebirds and their habitats. In reality, however, translating the potential of these legal-political tools into constructive action is dependent on governmental commitment and, ultimately, the degree of public interest and involvement.

The scientific community has a key role in this process. Because most shorebird species are still common, we have the opportunity to bring the requirements for their protection to the attention of the public and decision-makers before it becomes a matter of attempting to salvage token remnants of once abundant species populations.

ACKNOWLEDGMENTS

We thank Brian A. Harrington and Warren B. King for reviewing earlier drafts of the manuscript and William Brown and Michael Bean for reviewing the portion of the paper dealing with legal-political tools. Lawrence R. DeWeese and Donald A. White provided unpublished information on environmental contaminants and reviewed the relevant sections. Information on Canadian parks and reserves was supplied by Barbara Campbell. Warren B. King provided us with information on the Ramsar Convention. Many persons contributed information on shorebird stopover sites in the United States and Canada. In addition to those acknowledged in Table I, we are grateful to David Blankinship, Roger Boyd, Chuck Carlson, Bernie Gollop, Bruce Harris, Ruth Jacobs, Dennis Paulson, Richard Rosche, Palmer Skaar, Peter Vickery, Jim Wilson, and Al Wisely. Permission to reproduce published material was kindly granted by Joanna Burger, John Goss-Custard, James Gosselink, and the editors of *Biological Conservation, Journal of Applied Ecology*, and *Zeitschrift fuer Geomorphologie*. Barbara Dowell and Danny Bystrak prepared the map. Finally, we thank Michael Harwood for providing the quotation from the late Rosalie Edge.

REFERENCES

Anderson, D. W., and Gress, F., 1981, The politics of pelicans, in: *Coast Alert: Scientists Speak Out* (T. C. Jackson and D. Reische, eds.), pp. 117–143, Coast Alliance/Friends of the Earth, San Francisco.

Anonymous, 1980, Conference on the conservation of wetlands of international importance, especially as waterfowl habitat, Cagliari, Italy, International Waterfowl Research Bureau, Slimbridge, England.

Ashmole, M. J., 1970, Feeding of Western and Semipalmated Sandpipers in Peruvian winter quarters, *Auk* **87**:131–135.

Atkinson-Willes, G. L., Scott, D. A., and Prater, A. J., 1980, Criteria for selecting wetlands of international importance: Proposed amendments and guidelines for use, Conference on the conservation of wetlands of international importance, especially as waterfowl habitat, Cagliari, Italy, International Waterfowl Research Bureau, Slimbridge, England.

Baker, M. C., and Baker, A. E. M., 1973, Niche relationships among six species of shorebirds on their wintering and breeding ranges, *Ecol. Monogr.* **43**:193–212.

Banks, R. C., 1977, The decline and fall of the Eskimo Curlew, or why did the curlew go extaille? *Am. Birds* **31**:127–134.

Banks, R. C., 1979, Human related mortality of birds in the United States, *U. S. Fish Wild. Serv. Spec. Sci. Rep. Wildl.* **215**:1–16.

Bean, M. J., 1977, The evolution of national wildlife law, U. S. Council on Environmental Quality, Washington, D.C.

Bird, C. G., and Bird, E. G., 1940, Some remarks on non-breeding in the Arctic, especially in north-east Greenland, *Ibis* **14**:671–678.

Blodget, B. G., 1978, The effect of off-road vehicles on Least Terns and other shorebirds, *Univ. Mass. Natl. Park Serv. Coop. Res. Unit Rep.* **26**.

Blus, L. J., and Lamont, T. G., 1979, Organochlorine residues in six species of estuarine birds, South Carolina, 1971–75, *Pestic. Monit. J.* **13**:56–60.

Blus, L. J., Wiemeyer, S. N., Kerwin, J. A., Stendell, R. C., Ohlendorf, H. M., and Stickel. L. F., 1977, Impact of estuarine pollution on birds, in: *Estuarine Pollution Control and Assessment*, Vol. I, pp. 57–71, U. S. Environmental Protection Agency, Washington, D.C.

Boere, G. C., 1976, The significance of the Dutch Waddenzee in the annual life cycle of Arctic, subarctic and boreal waders. Part I. The function as a moulting area, *Ardea* **64**:210–291.

Bourn, W. S., and Cottam, C., 1950, Some biological effects of ditching tidewater marshes, *U.S. Fish Wildl. Serv. Res. Rep.* **19**.

Boyd, R. L., 1981, Distribution and abundance of Snowy Plovers in Kansas and northern Oklahoma, *Bull. Kans. Ornithol. Soc.* **32**:25–28.

Brown, J. J., and Brown, A. W. A., 1970, Biological fate of DDT in a sub-arctic environment, *J. Wildl. Manage.* **34**:929–940.

Burger, J., 1981, The effect of human activity on birds at a coastal bay, *Biol. Conserv.* **21**:231–241.

Burger, J., Howe, M. A., Hahn, D. C., and Chase, J., 1977, Effects of tide cycles on habitat selection and habitat partitioning by migrating shorebirds, *Auk* **94**:743–758.

Burger, J., Hahn, D. C., and Chase, J., 1979, Aggressive interactions in mixed-species flocks of migrating shorebirds, *Anim. Behav.* **27**:459–469.

Burger, J., Shisler, J., and Lesser, F. H., 1982, Avian utilization on six salt marshes in New Jersey, *Biol. Conserv.* **23**:187–212.

Burhenne-Guilmin, F., and de Klemm, C., 1980, The Ramsar Convention: A legal review, Conference on the conservation on wetlands of international importance, especially as waterfowl habitat, Cagliari, Italy, International Waterfowl Research Bureau, Slimbridge, England.

Bystrak, D., 1981, The North American Breeding Bird Survey, in: *Studies in Avian Biology No. 6*, pp. 34–41.

Cairns, W. E., and McLaren, I. A., 1980, Status of the Piping Plover on the east coast of North America, *Am. Birds* **34**:206–208.

Connors, P. G., Myers, J. P., and Pitelka, F. A., 1979, Seasonal habitat use by Arctic Alaskan shorebirds, in: *Studies in Avian Biology No. 2*, pp. 101–111.

Darnell, R. M., 1976, Impacts of construction activities in wetlands of the United States, *U.S. Environmental Protection Agency Ecological Research Series*, EPA-600/3-76-045, Tereco Corporation, College Station, Tex.

Davidson, N. C., 1981, Survival of shorebirds (Charadrii) during severe weather: The role of nutritional reserves, in: *Feeding and Survival Strategies of Estuarine Organisms* (N. V. Jones and W. J. Wolff, eds.), pp. 231–250, Plenum Press, New York.

Duffy, D. C., Atkins, N., and Schneider, D. C., 1981, Do shorebirds compete on their wintering grounds? *Auk* **98**:215–229.

Dugan, P. J., Evans, P. R., Goodyer, L. R., and Davidson, N. C., 1981, Winter fat reserves in shorebirds: Disturbance of regulated levels by severe weather conditions, *Ibis* **123**:359–363.

Eastin, W. C., and Hoffman, D. J., 1979, Biological effects of petroleum on aquatic birds, in: *Proceedings of the Conference on Assessment of Ecological Impacts of Oil Spills*, pp. 561–582, American Institute of Biological Sciences.

Evans, P. R., 1978/79, Reclamation of intertidal land: Some effects on Shelduck and wader populations in the Tees Estuary, *Verh. Ornithol. Ges. Bayern* **23:**147–168.

Evans, P. R., and Moon, S. J., 1981, Heavy metals in shorebirds and their prey in northeast England, in *Heavy Metals in Northern England: Environmental and Biological Aspects* (P. J. Say and B. A. Whitton, eds.), pp. 181–190, University of Durham, U.K.

Evans, P. R., and Smith, P. C., 1975, Studies of shorebirds at Lindisfarne, Northumberland. 2. Fat and pectoral muscle as indicators of body condition in the Bar-tailed Godwit, *Wildfowl* **26:**64–75.

Evans, P. R., Herdson, D. M., Knight, P. J., and Pienkowski, M. W., 1979, Short-term effects of reclamation of part of Seal Sands, Teesmouth, on wintering waders and Shelducks. 1. Shorebird diets, invertebrate densities, and the impact of predation on the invertebrates, *Oecologia (Berlin)*, **41:**183–206.

Flickinger, E. L., and King, K. A., 1972, Some effects of aldrin-treated rice on Gulf coast wildlife, *J. Wildl. Manage.* **36:**706–726.

Flickinger, E. L., King, K. A., Stout, W. F., and Mohn, M. M., 1980, Wildlife hazards from Furadan 3G applications to rice in Texas, *J. Wildl. Manage.* **44:**190–197.

Forbush, E. H., 1912, A history of the game birds, wild-fowl, and shore birds of Massachusetts and adjacent states, Massachusetts Board of Agriculture.

Freese, C., 1982, Western Hemisphere Convention: International framework for wildlife conservation, *U.S. Fish Wildl. Serv. Endangered Species Tech. Bull.* **7:**4–7.

Fuller, R. J., and Glue, D. E., 1980, Sewage works as bird habitats in Britain, *Biol. Conserv.* **17:**165–181.

Gill, R. E., Jr., and Handel, C. M., 1981, Shorebirds of the eastern Bering Sea, in: *Eastern Bering Sea Shelf: Oceanography and Resources*, Vol. 2 (D. W. Hood and J. A. Calder, eds.), pp. 719–738, Office of Marine Pollution Assessment, National Oceanic and Atmospheric Administration.

Gill, R. E., Jr., and Jorgensen, P. D., 1979, A preliminary assessment of timing and migration of shorebirds along the northcentral Alaska peninsula, in: *Studies in Avian Biology No. 2*, pp. 113–123.

Gill, R. E. Jr., Petersen, M. R., and Jorgensen, P. D., 1981, Birds of the northcentral Alaska peninsula, 1976–1980, *Arctic* **34:**286–306.

Glue, D. E., 1971, Saltmarsh reclamation stages and their associated bird-life, *Bird Study* **18:**187–198.

Goss-Custard, J. D., 1969, The winter feeding ecology of the Redshank *Tringa totanus*, *Ibis* **111:**338–356.

Goss-Custard, J. D., 1970, The responses of Redshank (*Tringa totanus* (L.)) to spatial variations in the density of their prey, *J. Anim. Ecol.* **39:**91–113.

Goss-Custard, J. D., 1976, Variation in the dispersion of Redshank, *Tringa totanus*, on their winter feeding grounds, *Ibis* **118:**257–263.

Goss-Custard, J. D., 1977, The ecology of the Wash. III. Density-related behaviour and the possible effects of a loss of feeding grounds on wading birds (Charadrii), *J. Appl. Ecol.* **14:**721–739.

Goss-Custard, J. D., Jenyon, R. A., Jones, R. E., Newbery, P. E., and Williams, R. L., 1977, The ecology of the Wash. II. Seasonal variation in the feeding conditions of wading birds (Charadrii), *J. Appl. Ecol.* **14:**701–719.

Gosselink, J. G., and Baumann, R. H., 1980, Wetland inventories: Wetland loss along the United States coast, *Z. Geomorphol. N. F. Suppl.* **34:**173–187.

Graul, W. D., 1974, Adaptive aspects of the Mountain Plover social system, *Living Bird* **12**:69–94.
Graul, W. D., and Webster, L. E., 1976, Breeding status of the Mountain Plover, *Condor* **78**:265–267.
Green, G. H., Greenwood, J. J. D., and Lloyd, C. S., 1977, The influence of snow conditions on the date of breeding of wading birds in north-east Greenland, *J. Zool.* **183**:311–328.
Grover, P. B., and Knopf, F. L., 1982, Habitat requirements and breeding success of charadriiform birds nesting at Salt Plains National Wildlife Refuge, Oklahoma. *J. Field Ornithol.* **53**:139–148.
Halbeisen, R., 1977, Disturbance of incubating Snowy Plovers on Point Reyes, *Point Reyes Bird Observatory* **42**:2–3.
Herman, S. G., and Bulger, J. B., 1981, The distribution and abundance of shorebirds during the 1981 spring migration at Grays Harbor, Washington, U.S. Army Corps of Engineers, Seattle District, Contract No. DACW6781-M-0936.
Higgins, K. F., 1975, Shorebird and game bird nests in North Dakota croplands, *Wildl. Soc. Bull.* **3**:176–179.
Holmes, R. T., 1966a, Feeding ecology of the Red-backed Sandpiper (*Calidris alpina*) in Arctic Alaska, *Ecology* **47**:32–45.
Holmes, R. T., 1966b, Molt cycle of the Red-backed Sandpiper (*Calidris alpina*) in western North America, *Auk* **83**:517–533.
Holmes , R. T., 1970, Differences in population density, territoriality, and food supply of Dunlin on Arctic and subarctic tundra, *Symp. Br. Ecol. Soc.* **10**:303–319.
Horwitz, E. L., 1978, Our nation's wetlands, Interagency Task Force Report, U.S. Council on Environmental Quality, Washington, D.C.
Howe, M. A., 1980, Monitoring migrant shorebird populations in the eastern United States, *Atl. Nat.* **33**:23–24.
Hughes, R. A., 1979, Notes on Charadriiformes of the south coast of Peru, in: *Studies in Avian Biology No. 2*, pp. 49–54.
Hussell, D. J. T., 1981, The use of migration counts for monitoring bird population levels, in: *Studies in Avian Biology No. 6*, 92–102.
Isleib, M. E., and Kessel, B., 1973, Birds of the north Gulf Coast–Prince William Sound region, Alaska, *Biol. Pap. Univ. Alaska* **14**.
Jehl, J. R., Jr., 1981, Mono Lake: A vital way station for the Wilson's Phalarope, *Natl. Geogr. Mag.* **160**:520–525.
Jenni, D. A., Redmond, R. L., and Bicak, T. K., 1982, Behavioral ecology and habitat relationships of Long-billed Curlews in western Idaho, U.S. Bureau of Land Management Research Contract YA-512-CT7-54, University of Montana, Missoula.
Joyes, A., Knight, P. J., Leah, R. T., and Pienkowski, M. W., 1976, The blockage of the Oued Chebeika Estuary and its effects on the avifauna, *Bull. Inst. Sci.* **1**:39–47.
Jurek, R. M., 1974, California Shorebird Study, 1969–1974, Federal Aid in Wildlife Restoration Project W-54-R, Special Wildlife Investigations Project Final Report, Job III-1, California Fish and Game Department, Sacramento.
Kaiser, G. W., and Fry, K., 1980, Ingestion of lead shot by Dunlin, *Murrelet* **61**:37.
Kaiser, G. W., and Gillingham, M., 1981, Some notes on seasonal fluctuations in the weight of Dunlin *Calidris alpina* on the Fraser River Delta, British Columbia, *Wader Study Group Bull.* **31**:46–48.
Kane, R., 1976, The 1975 fall shorebird migration in the Hackensack Meadowlands, Occasional Paper 123, New Jersey Audubon Society.

Kantrud, H. A., and Kologiski, R. L., 1982, Effects of soils and grazing on breeding birds of uncultivated upland grasslands of the northern Great Plains, *U. S. Fish Wildl. Serv. Wildl. Res. Rep.* **15**:1–33.

Keast, A., and Morton, E. S. (eds.), 1980, Migrant birds in the Neotropics: Ecology, behavior, distribution, and conservation, Symp. Natl. Zool. Park, Smithsonian Institution, Washington, D.C.

Kirsch, L. M., and Higgins, K. F., 1976, Upland Sandpiper nesting and management in North Dakota, *Wildl. Soc. Bull.* **4**:16–22.

Lambert, A., and Nol, E., 1978, Status of the Piping Plover at Long Point, Unpublished report, Long Point Bird Observatory, Ontario.

Lambert, A., and Ratcliff, B., 1981, Present status of the Piping Plover in Michigan, *Jack-Pine Warbler* **59**:44–52.

Larsson, T., 1969, Land use and bird fauna on shores in S. Sweden, *Oikos* **20**:136–155.

Linduska, J. P. (ed.), 1964, *Waterfowl Tomorrow*, Fish and Wildlife Service, U. S. Department of the Interior, Washington, D.C.

Lum, A. L., 1978, Shorebird fauna changes of a small tropical estuary following habitat alteration: Biological and political impacts of environmental restoration, *Environ. Manage.* **2**:423–430.

McCaskie, G., 1970, Shorebird and waterbird use of the Salton Sea, *Calif. Fish Game* **56**:87–95.

McCollough, M., and May, T., 1980, Habitat utilization by southward migrating shorebirds in Cobscook Bay, Maine during 1979, Unpublished report, School of Forest Resources, University of Maine.

McCracken, J. D., Bradstreet, M. S. W., and Holroyd, G. L., 1981, Breeding birds of Long Point, Lake Erie: A study in community succession, *Can. Wildl. Serv. Rep. Ser.* **44**:1–72.

McNeil, R., and Burton, J., 1977, Southbound migration of shorebirds from the Gulf of St. Lawrence, *Wilson Bull.* **89**:167–171.

McNeil, R., and Cadieux, F., 1972, Fat content and flight-range capabilities of some adult spring and fall migrant North American shorebirds in relation to migration routes on the Atlantic coast, *Nat. Can.* **99**:589–605.

Manuwal, D. A., 1978, Effect of man on marine birds: A review, *Proceedings of the 4th J. S. Wright Forestry Conference*, pp. 140–160.

Marcström, V., and Mascher, J. W., 1979, Weights and fat in lapwings *Vanellus vanellus* and oystercatchers *Haematopus ostralegus* starved to death during a cold spell in spring, *Ornis Scand.* **10**:235–240.

Marsh, J. H., Marte, G. F., and Hunt, R. A., 1973, Breeding duck populations and habitat in Wisconsin, *Wis. Dep. Nat. Resour. Tech. Bull.* **68**.

Matthiessen, P., 1959, *Wildlife in America*, Viking Press, New York.

Morrison, M. L., and Kiff, L. F., 1979, Eggshell thickness in American shorebirds before and since DDT, *Can. Field Nat.* **93**:187–190.

Morrison, R. I. G., 1976, Maritimes Shorebird Survey 1974, Canadian Wildlife Service, Ottawa.

Morrison, R. I. G., 1977, Use of the Bay of Fundy by shorebirds, in: *Fundy Tidal Power and the Environment* (G. A. Daborn, ed.), pp. 187–199, The Acadia University Institute, Wolfville, Nova Scotia.

Morrison, R. I. G., and Harrington, B. A., 1979, Critical shorebird resources in James Bay and eastern North America, *Trans. North Am. Wildl. Nat. Resour. Conf.* **44**:498–507.

Myers, J. P., 1980, The pampas shorebird community: Interactions between breeding and nonbreeding members, in: *Migrant Birds in the Neotropics* (A. Keast and E. S. Morton, eds.), pp. 37–49, Smithsonian Institution, Washington, D.C.

Myers, J. P., Connors, P. G., and Pitelka, F. A., 1979, Territory size in wintering Sanderlings: The effects of prey abundance and intruder density, *Auk* **96**:551–561.

Newton, I., 1979, *Population Ecology of Raptors*, Buteo Books, Vermillion, S. D.

NOAA (National Oceanic and Atmospheric Administration), 1980, The federal coastal programs review: Report to the President, U.S. Department of Commerce, Washington, D.C.

Page, G. W., and Stenzel, L. E., 1979, Status and breeding biology of the Snowy Plover *Charadrius alexandrinus* in California, *Wader Study Group Bull.* **27**:38–39.

Page, G. W., Stenzel, L. E., and Wolfe, C. M., 1979, Aspects of the occurrence of shorebirds on a central California estuary, in: *Studies in Avian Biology No. 2*, pp. 15–32.

Pampush, G. J., 1980, Status report on the Long-billed Curlew in the Columbia and northern Great Basins, U.S. Fish and Wildlife Service Report, Portland, Ore.

Parslow, J. L. F., 1973, Mercury in waders from the Wash, *Environ. Pollut.* **5**:295–304.

Payne, R. B., 1972, Mechanisms and control of molt, in: *Avian Biology*, Vol. II (D. S. Farner and J. R. King, eds.), pp. 103–155, Academic Press, New York.

Pienkowski, M. W., 1979, Waders feeding on grain in cold weather, *Scott. Birds J. Scott. Ornithol. Club* **10**:234–236.

Pienkowski, M. W., Knight, P. J., Stanyard, D. J., and Argyle, F. B., 1976, The primary moult of waders on the Atlantic coast of Morocco, *Ibis* **118**:347–365.

Pilcher, R. E. M., Beer, J. V., and Cook, W. A., 1974, Ten years of intensive late winter surveys for waterfowl corpses on the north-west shore of the Wash, England, *Wildfowl* **25**:149–154.

Pitelka, F. A. (ed.), 1979, Shorebirds in marine environments, in: *Studies in Avian Biology No. 2*, Cooper Ornithological Society, Allen Press, Lawrence, Kans.

Pitelka, F. A., Holmes, R. T., and MacLean, S. F., Jr., 1974, Ecology and evolution of social organization in Arctic sandpipers, *Am. Zool.* **14**:185–204.

Prater, A. J., 1981, Estuary birds of Britain and Ireland, Poyser, Calton, England.

Ptacek, J., and Schwilling, M., 1983, Mountain plover reintroduction in Kansas. *Bull. Kans. Ornithol. Soc.* **34**:21–22.

Puttick, G. M., 1980, Energy budgets of Curlew Sandpipers at Langebaan Lagoon, South Africa, *Estuarine Coastal Mar. Sci.* **11**:207–216.

Renaud, W. E., 1980, The Long-billed Curlew in Saskatchewan: Status and distribution, *Blue Jay* **38**:221–237.

Rundle, W. D., 1980, Management, habitat selection and feeding ecology of migrant rails and shorebirds, Unpublished M. S. dissertation, University of Missouri.

Saeijs, H. L. F., and Baptist, H. J. M., 1980, Coastal engineering and European wintering wetland birds, *Biol. Conserv.* **17**:63–83.

Schneider, D. C., and Harrington, B. A., 1981, Timing of shorebird migration in relation to prey depletion, *Auk* **98**:801–811.

Senner, S. E., 1977, The ecology of Western Sandpipers and Dunlins during spring migration through the Copper–Bering river delta system, Alaska, Unpublished M.S. dissertation, University of Alaska.

Senner, S. E., 1979, An evaluation of the Copper River Delta as a critical habitat for migrating shorebirds, in: *Studies in Avian Biology No. 2*, pp. 131–145.

Senner, S. E., West, G. C., and Norton, D. W., 1981, The spring migration of Western Sandpipers and Dunlins in southcentral Alaska: Numbers, timing, and sex ratios, *J. Field Ornithol.* **52**:271–284.

Smith, J. L., and Mudd, D. R., 1976, Impact of dredging on the avian fauna in Grays Harbor, Appendix H, in: Maintenance Dredging and the Environment of Grays Harbor Washington, Summary Report, U.S. Army Corps of Engineers Contract no. DACW67-74C-0086, Washington State Department of Ecology, Olympia.

Smith, J. L., Mudd, D. R., and Messmer, L. W., 1976, Impact of dredging on the vegetation in Grays Harbor, Appendix F, in: Maintenance Dredging and the Environment of Grays Harbor Washington, Summary Report, U. S. Army Corps of Engineers Contract No. DACW67-74C-0086, Washington State Department of Ecology, Olympia.

Smith, P. C., and Bleakney, J. S., 1969, Observations on oil pollution and wintering Purple Sandpipers, *Erolia maritima* (Brunnich), in Nova Scotia, *Can. Field Nat.* **83**:19–22.

Soots, R. F., Jr., and Landin, M. C., 1978, Development and management of avian habitat on dredged material islands, *U.S. Army Corps of Engineers Technical Report.* DS-78-18:1–96.

Spaans, A. L., 1978, Status and numerical fluctuations of some North American waders along the Surinam coast, *Wilson Bull.* **90**:60–83.

Speth, J., 1979, Conservation and management of coastal wetlands in California, in: *Studies in Avian Biology No. 2*, pp. 151–155.

Stone, W. B., 1979, Poisoning of wild birds by organophosphate and carbamate pesticides, *N.Y. Fish Game J.* **26**:37–47.

Sugden, J. W., 1933, Range restriction of the Long-billed Curlew, *Condor* **35**:3–9.

Summers, R. W., Cooper, J., and Pringle, J. S., 1977, Distribution and numbers of coastal waders (Charadrii) in the south-western Cape, South Africa, summer 1975–76, *Ostrich* **48**:85–97.

Svensson, S. E., 1978, Efficiency of two methods for monitoring bird population levels: Breeding bird censuses contra counts of migrating birds, *Oikos* **30**:373–386.

Sykes, P. W., Jr., and Hunter, G. S., 1978, Bird use of flooded agricultural fields during summer and early fall and some recommendations for management, *Fla. Field Nat.* **6**:36–43.

Temple, S. A., 1983, Bird conservation, No. 1. Univ. Wisconsin Press, Madison.

Thompson, M. C., 1974, Migratory patterns of Ruddy Turnstones in the central Pacific region, *Living Bird* **12**:5–24.

Train, R. E., 1982, Statement of Russell E. Train, President, World Wildlife Fund-U.S., Hearings before the Subcommittee on Fisheries and Wildlife Conservation and the Environment, U.S. House of Representatives, on Endangered Species Act Reauthorization and Oversight, Serial No. 97-32:91–94, U.S. Government Printing Office, Washington, D.C.

Vermeer, K., and Vermeer, R., 1975, Oil threat to birds on the Canadian west coast, *Can. Field Nat.* **89**:278–298.

Vermeer, K., Risebrough, R. W., and Spaans, A. L., 1974, Pesticide effects on fishes and birds in rice fields of Surinam, South America, *Environ. Pollut.* **7**:217–236.

Wander, W., and Dunne, P. J., 1981, A preliminary assessment of the spring shorebird migration along the Delaware Bayshore of New Jersey, Unpublished report, U.S. Fish and Wildlife Service, Office of Migratory Bird Management.

Wetmore, A., 1926, Observations on the birds of Argentina, Paraguay, Uruguay, and Chile, *U.S. Natl. Mus. Bull.* **133**:1–448.

White, D. H., King, K. A., and Prouty, R. M., 1980, Significance of organochlorine and heavy metal residues in wintering shorebirds at Corpus Christi, Texas, 1976–77, *Pestic. Monit. J.* **14**:58–63.

White, D. H., Mitchell, C. A., and Prouty, R. M., 1983, Temporal accumulation of organochlorine pesticides in shorebirds wintering on the south Texas coast, *Arch. Environ. Contam. Toxicol.* **12**:241–245.

SPECIES INDEX

SUBJECT INDEX

GPSR Compliance

The European Union's (EU) General Product Safety Regulation (GPSR) is a set of rules that requires consumer products to be safe and our obligations to ensure this.

If you have any concerns about our products, you can contact us on ProductSafety@springernature.com

In case Publisher is established outside the EU, the EU authorized representative is:

Springer Nature Customer Service Center GmbH
Europaplatz 3
69115 Heidelberg, Germany

Batch number: 10837688

Printed by Printforce, the Netherlands